St. Louis Community College

Forest Park
Florissant Valley
Meramec

Insructional Resources
St. Louis, Missouri

DIGITAL SYSTEMS
PRINCIPLES and APPLICATIONS

THIRD EDITION

DIGITAL SYSTEMS
PRINCIPLES and APPLICATIONS

Ronald J. Tocci

PRENTICE-HALL, INC. ENGLEWOOD CLIFFS, NEW JERSEY 07632

Library of Congress Cataloging in Publication Data

Tocci, Ronald J.
  Digital systems.

  Includes index.
  1. Digital electronics.   I. Title.
TK7868.D5T62   1985   621.3819'58   84-4788
ISBN 0-13-212374-6

Editorial/production supervision
and interior design: *Shari Ingerman*
Display and frontmatter design: *Celine A. Brandes*
Cover design: *Diane Saxe*
Manufacturing buyer: *Gordon Osbourne*

Cover photo courtesy of Western Electric

Printed in the United States of America

10 9 8 7 6 5 4 3 2 1

ISBN 0-13-212374-6 01

Prentice-Hall International, Inc., *London*
Prentice-Hall of Australia Pty. Limited, *Sydney*
Editora Prentice-Hall do Brasil, Ltda., *Rio de Janeiro*
Prentice-Hall Canada Inc., *Toronto*
Prentice-Hall of India Private Limited, *New Delhi*
Prentice-Hall of Japan, Inc., *Tokyo*
Prentice-Hall of Southeast Asia Pte. Ltd., *Singapore*
Whitehall Books Limited, *Wellington, New Zealand*

C O N T E N T S

6 DIGITAL ARITHMETIC: OPERATIONS and CIRCUITS 177

7 COUNTERS and REGISTERS 219

8 INTEGRATED-CIRCUIT LOGIC FAMILIES 287

Contents

Contents

P R E F A C E

This text presents a comprehensive up-to-date study of the principles and techniques used in modern digital systems and is intended for use in 2-year and 4-year technology programs.

Although a basic electronics background is helpful, a major portion of the material requires no prior electronics training. Hence, the text is well-suited for students of computer science or other nonelectronic technologies concerned with digital systems.

GENERAL IMPROVEMENTS

This third edition contains several general improvements over the previous edition, including that all material has been updated to reflect the latest devices and practices. Actual ICs are used throughout the text; more illustrative examples have been added to each chapter, many using actual ICs; timing diagrams and waveforms have been used more extensively; review questions have been included at the end of most sections to provide the reader with more immediate reinforcement; and new end-of-chapter problems have been added, many using actual ICs.

SPECIFIC CHANGES

In addition to these general improvements, the specific chapter-by-chapter changes are that number systems and codes have been moved from Chapter 5 to Chapter 2; Chapter 3 now includes the use of alternate logic gate symbols and "bubble" notation for constructing and interpreting logic diagrams; Chapter 4 now includes *optional* Karnaugh mapping technique for circuit simplification; Chapter 5 on flip-flops now includes more detail on the internal structure of edge-triggered and master/slave flip-flops, more on latches, more on timing considerations, and more applications. The retriggerable one-shot has also been added; Chapter 6 now covers digital arithmetic operations and circuits in the same chapter for a more cohesive treatment; Chapter 7, on counters and registers, includes actual TTL and CMOS devices; Chapter 8 includes more on open-collector and tri-state outputs, more on the CMOS family including the bilateral switch, and IC interfacing; Chapter 9 now includes priority encoders, LCD displays, and multiplexer logic function generation; Chapter 10 now includes multiplying DACs, current mode DACs, flash ADCs, and sample and hold circuits; Chapter 11 on memories has been extensively expanded and updated to include ROM and RAM internal architecture and timing, EPROM programming, the EEPROM, more on dynamic memory architecture and timing, and more on magnetic bubble and charge-coupled memories; and the material from the old Chapters 12 and 13 has been combined and streamlined, with the emphasis on microcomputers.

RETAINED FEATURES

This edition, of course, retains all of the features that made the previous editions so widely accepted. It is still up-to-date in its coverage of digital IC families (TTL, MOS, CMOS, ECL, TRI-STATE LOGIC) and in its coverage of such topics as decoding, encoding, multiplexing, demultiplexing, strobing, data busing, and microprocessors as well as the standard topics. It utilizes a block diagram approach to teach the basic digital and logic operations without confusing the reader with detailed circuit diagrams. In Chapter 8, the reader is introduced to the internal IC circuitry after he/she has a firm understanding of logic principles. At that point, he/she can interpret a logic block's input and output characteristics and "fit" it properly into a complete system. Emphasis on the teaching of principles has been retained, followed by *thoroughly explained* applications of the same principles so that they can be understood and applied to new applications. Among these applications are the frequency counter, the digital clock, the keyboard entry encoder, the control sequencer, the digital voltmeter, the digital computer, and many others. As in the previous edition, this edition continues to make use of extensive use of illustrations and practical illustrative examples in order to provide a vehicle to help the reader to understand the concepts involved. Each chapter is still followed by numerous questions and problems of various levels of complexity that give students the opportunity to test and examine his understanding of the material. Many of the problems also present modifications to, extension of, or new applications for the material covered in the chapter, thereby offering the student a chance to expand his/her knowledge of the material.

SEQUENCING

It is a rare instructor who uses the chapters of a textbook in the exact sequence in which they are presented. In fact, I must admit that, for many different reasons, I do not even use my own textbooks in that way. This textbook was written so that, for the most part, each chapter builds on previous material, but it is possible to alter the sequence without causing problems.

For example, the first part of Chapter 6 (arithmetic operations) can be covered right after Chapter 2 (number systems), although this would then produce quite a delay before the arithmetic circuitry is covered. Likewise, much of the material in Chapter 8 (IC families) can be covered earlier (e.g., after Chapter 4 or 5), without any loss of effectiveness. Other topical rearrangements can be made at the discretion of the instructor.

ACKNOWLEDGEMENTS

I am indebted to the many users of the previous editions whose comments and suggestions have been incorporated into this edition. A special note of thanks is due the following group of reviewers for their valuable critiques and attention to detail: Dee F. Robertson, John L. Morgan, Vincent F. Alisouskas, and A. Lehtila of DeVry Institute of Technology; Robert A. Cuiffetti of GTE Sylvania Technical School; and Gene Adair of Indiana IVY Tech.

Ronald J. Tocci

DIGITAL SYSTEMS
PRINCIPLES and APPLICATIONS

1
INTRODUCTORY CONCEPTS

OUTLINE

This book is concerned with the study of digital technology. In this field, as in others, most of the principles and applications are based on certain underlying concepts. We will introduce some of these concepts in this chapter, expanding upon them as needed later in the text. We will also introduce some of the terminology that is such an important part of digital technology. The reader will gain a broader understanding of the terminology after seeing it used throughout the text.

1.1 NUMERICAL REPRESENTATIONS

In science, technology, business, and, in fact, in most other fields of endeavor, we are constantly dealing with *quantities*. These quantities are measured, monitored, recorded, manipulated arithmetically, observed, or in some other way utilized in most physical systems. It is important when dealing with various quantities that we be able to represent their values efficiently and accurately. There are basically two ways of representing the numerical value of quantities: *analog* and *digital*.

Analog Representations In *analog representation* one quantity is represented by another which is directly proportional to the first. An example is an automobile speedometer, in which the deflection of the needle is proportional to the speed of the auto. The angular position of the needle represents the value of the auto's speed, and the needle follows any changes that occur as the auto speeds up or slows down.

Another example is the common room thermostat, in which the bending of the bimetallic strip is proportional to the room temperature. As the temperature changes gradually, the curvature of the strip changes proportionally.

Still another example of an analog quantity is found in the familiar audio microphone. In this device an output voltage is generated in proportion to the amplitude of the sound waves that impinge on the microphone. The variations in the output voltage follow the same variations as the input sound.

Analog quantities such as those cited above have an important characteristic: *they can gradually vary over a continuous range of values*. The automobile speed can have *any* value between zero and, say, 100 mph. Similarly, the microphone output might be anywhere within a range of zero to 10 mV (for example, 1 mV, 2.3724 mV, 9.9999 mV).

Digital Representations In *digital representation* the quantities are represented not by proportional quantities but by symbols called *digits*. As an example, consider the digital clock, which provides the time of day in the form of decimal digits which represent hours and minutes (and sometimes seconds). As we know, the time of day continuously changes, but the digital clock reading does not change continuously; rather, it changes in steps of one per minute (or per second). In other words, this digital representation of the time of day changes in *discrete* steps, as compared to the analog representation of time provided by a wristwatch, where the dial reading changes continuously.

The major difference between analog and digital quantities, then, can be simply stated as follows:

$$\text{analog} \equiv \text{continuous}$$

$$\text{digital} \equiv \text{discrete (step by step)}$$

Because of the discrete nature of digital representations, there is no ambiguity when reading the value of a digital quantity, whereas the value of an analog quantity is often open to interpretation.

EXAMPLE 1.1

Which of the following involve analog quantities and which involve digital quantities?
(a) Resistor substitution box.
(b) Current meter.
(c) Temperature.
(d) Sand grains on a beach.
(e) Radio volume control.

Solution:
(a) Digital.
(b) Analog.
(c) Analog.
(d) Digital, since the number of grains can be only certain discrete (integer) values and not any value over a continuous range.
(e) Analog.

REVIEW QUESTION

1. Concisely describe the major difference between analog and digital quantities.

1.2 DIGITAL AND ANALOG SYSTEMS

A *digital system* is a combination of devices (electrical, mechanical, photoelectric, etc.) arranged to perform certain functions in which quantities are represented digitally. In an analog system the physical quantities are principally analog in nature. Many practical systems are *hybrid*, in that both digital and analog quantities are present and continual conversion between the two types of representation takes place.

Some of the more common digital systems are digital computers and calculators, digital voltmeters, and numerically controlled machinery. In these systems the electrical and mechanical quantities change only in discrete steps. Examples of *analog systems* are analog computers, radio broadcast systems, and audio tape recording. In these systems the quantities change gradually over a continuous range. For example, in AM radio broadcasting you can tune your radio to any frequency over a continuous band from 535 to 1605 kHz.

Generally speaking, digital systems offer the advantages of programmability, greater speed and accuracy, and the capability of memory. In addition, digital systems

FIGURE 1.1 Block diagram of typical hybrid process control.

are less susceptible than analog systems to fluctuations in the characteristics of the system components and are generally more versatile in a wider range of applications.

In the real world most quantities are analog in nature, and it is these quantities which are often being measured or monitored or controlled. Thus, if the advantages of digital techniques are to be utilized, it is obvious that many hybrid systems must exist. Chief among these are industrial process control systems in which analog quantities such as temperature, pressure, liquid level, and flow rate are measured and controlled. Figure 1.1 shows the symbolic block diagram of such a system. As the diagram shows, the analog quantity is measured and the measured value is then converted to a digital quantity by an *analog-to-digital (A/D) converter*. The digital quantity is then operated on by the central processor, which is completely digital and whose output is then recoverted to an analog quantity in a *digital-to-analog (D/A) converter*. This analog output is fed to a controller, which then exerts some type of action on the process to adjust the value of the original measured analog quantity.

1.3 DIGITAL NUMBER SYSTEMS

Many number systems are in use in digital technology. The most common are the decimal, binary, octal, and hexadecimal systems. The decimal system is clearly the most familiar to us because it is a tool that we use every day. Examining some of its characteristics will help us to better understand the other systems.

Decimal System   The *decimal system* is composed of *10* numerals or symbols. These 10 symbols are 0, 1, 2, 3, 4, 5, 6, 7, 8, 9; using these symbols as *digits* of a number, we can express any quantity. The decimal system, also called the *base-10* system, because it has 10 digits, has evolved naturally as a result of the fact that man has 10 fingers. In fact, the word "digit" is the Latin word for "finger."

The decimal system is a *positional-value system* in which the value of a digit depends on its position. For example, consider the decimal number 453. We know that the digit 4 actually represents 4 *hundreds,* the 5 represents 5 *tens,* and the 3 represents 3 *units*. In essence, the 4 carries the most weight of the three digits; it is referred to as the *most significant digit (MSD)*. The 3 carries the least weight and is called the *least significant digit (LSD)*.

Consider another example, 27.35. This number is actually equal to two tens plus

Positional values (weights)

MSD

Decimal point

LSD

FIGURE 1.2 Decimal position values as powers of 10.

seven units plus three tenths plus five hundredths, or $2 \times 10 + 7 \times 1 + 3 \times 0.1 + 5 \times 0.01$. The decimal point is used to separate the integer and fractional parts of the number.

More rigorously, the various positions relative to the decimal point carry weights that can be expressed as powers of 10. This is illustrated in Figure 1.2, where the number 2745.214 is represented. The decimal point separates the positive powers of 10 from the negative powers. The number 2745.214 is thus equal to

$$(2 \times 10^{+3}) + (7 \times 10^{+2}) + (4 \times 10^{1}) + (5 \times 10^{0})$$
$$+ (2 \times 10^{-1}) + (1 \times 10^{-2}) + (4 \times 10^{-3})$$

In general, any number is simply the sum of the products of each digit value times its positional value.

Decimal Counting When counting in the decimal system, we start with 0 in the units position and take each symbol (digit) in progression until we reach 9. Then we add a 1 to the next higher position and start over with zero in the first position (see Figure 1.3). This process continues until the count of 99 is reached. Then we add a 1 to the third position and start over with zeros in the first two positions. The same pattern is followed continuously as high as we wish to count.

It is important to note that in decimal counting the units position (LSD) changes upward with each step in the count, the tens position changes upward every 10 steps in the count, the hundreds position changes upward every 100 steps in the count, and so on.

Another characteristic of the decimal system is that using only two decimal places we can count through $10^2 = 100$ different numbers (0 to 99).* With three places we can count through 1000 numbers (0 to 999); and so on. In general, with N places or digits we can count through 10^N different numbers, starting with and including zero. The largest number will always be $10^N - 1$.

Binary System Unfortunately, the decimal number system does not lend itself to convenient implementation in digital systems. For example, it is very difficult to design electronic equipment so that it can work with 10 different voltage levels (each

*Zero is counted as a number.

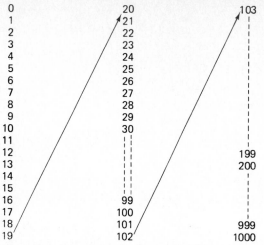

0	20	103
1	21	
2	22	
3	23	
4	24	
5	25	
6	26	
7	27	
8	28	
9	29	
10	30	
11		
12		199
13		200
14		
15		
16	99	
17	100	
18	101	999
19	102	1000

FIGURE 1.3   Decimal counting.

one representing one decimal character, 0–9). On the other hand, it is very easy to design simple, accurate electronic circuits that operate with only two voltage levels. For this reason, almost all digital systems use the binary number system (base 2) as the basic number system of its operations, although other systems are often used in conjunction with binary.

In the *binary system* there are only two symbols or possible digit values, 0 and 1. Even so, this base-2 system can be used of represent any quantity that can be represented in decimal or other number systems. In general though, it will take a greater number of binary digits to express a given quantity.

All the statements made earlier concerning the decimal system are equally applicable to the binary system. The binary system is also a positional-value system, wherein each binary digit has its own value or weight expressed as powers of 2. This is illustrated in Figure 1.4. Here, places to the left of the *binary point* (counterpart of the decimal point) are positive powers of 2 and places to the right are negative powers of 2. The number 1011.101 is shown represented in the figure. To find its equivalent in the decimal system we simply take the sum of the products of each digit value (0 or 1) times its positional value.

$$1011.101_2 = (1 \times 2^3) + (0 \times 2^2) + (1 \times 2^1) + (1 \times 2^0)$$

$$+ (1 \times 2^{-1}) + (0 \times 2^{-2}) + (1 \times 2^{-3})$$

$$= 8 + 0 + 2 + 1 + 0.5 + 0 + 0.125$$

$$= 11.625_{10}$$

Notice in the preceding operation that subscripts (2 and 10) were used to indicate the base in which the particular number is expressed. This convention is used to avoid confusion whenever more than one number system is being employed.

In the binary system, the term *binary digit* is often abbreviated to the term *bit* which we will use henceforth. Thus, in the number expressed in Figure 1.4 there are 4 bits to the left of the binary point, representing the integer part of the number, and 3 bits to the right of the binary point, representing the fractional part. The most

Positional values

2^3 2^2 2^1 2^0 2^{-1} 2^{-2} 2^{-3}

1	0	1	1	1	0	1

MSB

Binary point

LSB

FIGURE 1.4 Binary position values as powers of 2.

significant bit (MSB) is the leftmost bit (largest weight). The least significant bit (LSB) is the rightmost bit (smallest weight). These are indicated in Figure 1.4.

Binary Counting When we deal with binary numbers, we will usually be restricted to a specific number of bits. This restriction is based on the circuitry that is used to represent these binary numbers. Let's use 4-bit binary numbers to illustrate the method for counting in binary.

The sequence (shown in Figure 1.5) begins with all bits at 0; this is called the *zero count*. For each successive count, the units (2^0) position *toggles;* that is, it changes from one binary value to the other. Each time it changes from a 1 to a 0, the twos (2^1) position will toggle. Each time the twos position changes from 1 to 0, the fours (2^2) position will toggle. Likewise, each time the fours position goes from 1 to 0, the eights (2^3) position toggles. This same process would be continued for the higher-order bit positions if the binary number had more than 4 bits.

The binary counting sequence has an important characteristic, as shown in Figure 1.5. The units bit (LSB) changes either from 0 to 1 or 1 to 0 with *each* count.

FIGURE 1.5 Binary counting sequence.

The second bit (twos position) stays at 0 for two counts, then 1 for two counts, then 0 for two counts, and so on. The third bit (fours position) stays at 0 for four counts, then 1 for four counts; and so on. The fourth bit (eights position) stays at 0 for eight counts, then at 1 for eight counts. If we wanted to count further we would add more places, and this pattern would continue with 0s and 1s alternating in groups of 2^{N-1}. For example, using a fifth binary place, the fifth bit would alternate sixteen 0s, then sixteen 1s, and so on.

As we saw for the decimal system, it is also true of the binary system that by using N bits or places we can go through 2^N counts. For example, with 2 bits we can go through $2^2 = 4$ counts (00_2 through 11_2); with 4 bits we can go through $2^4 = 16$ counts (0000_2 through 1111_2); and so on. The last count will always be all 1s and is equal to $2^N - 1$ in the decimal system. For example, using 4 bits, the last count is $1111_2 = 2^4 - 1 = 15_{10}$.

EXAMPLE 1.2

What is the largest number that can be represented using 8 bits?

Solution: $2^N - 1 = 2^8 - 1 = 255_{10} = 11111111_2$.

REVIEW QUESTIONS

1. What is the decimal equivalent of 1101011_2? (*Ans.* 107)
2. What is the next binary number following 10111_2 in the counting sequence? (*Ans.* 11000)

1.4 REPRESENTING BINARY QUANTITIES

In digital systems the information that is being processed is usually present in binary form. Binary quantities can be represented by any device that has only two operating states or possible conditions. For example, a switch has only two states: open or closed. We can arbitrarily let an open switch represent binary 0 and a closed switch represent binary 1. With this assignment we can now represent any binary number as illustrated in Figure 1.6(a), where the states of the various switches represent 10010_2.

Another example is shown in Figure 1.6(b), where holes punched in paper are

(a)   (b)

FIGURE 1.6 Using (a) switches and (b) punched paper tape to represent binary numbers.

used to represent binary numbers. A punched hole is a binary 1 and absence of a hole is a binary 0.

There are numerous other devices which have only two operating states or can be operated in two extreme conditions. Among these are: light bulb (bright or dark), diode (conducting or nonconducting), relay (energized or deenergized), transistor (cut off or saturated), photocell (illuminated or dark), thermostat (open or closed), mechanical clutch (engaged or disengaged), and magnetic tape (magnetized or demagnetized).

In electronic digital systems, binary information is represented by voltages (or currents) that are present at the inputs and outputs of the various circuits. Typically, the binary 0 and 1 are represented by two nominal voltage levels. For example, zero volts (0 V) might represent binary 0, and +5 V might represent binary 1. In actuality, because of circuit variations, the 0 and 1 would be represented by voltage ranges. This is illustrated in Figure 1.7(a), where any voltage between 0 and 0.8 V represents a 0 and any voltage between 2 and 5 V represents a 1. All input and output signals will fall in either of these ranges. Figure 1.7(b) shows a typical digital waveform as it sequences through the binary value 01010.

(a)   (b)

FIGURE 1.7   (a) Typical voltage assignments in digital system; (b) typical digital signal.

We can now see another significant difference between digital and analog systems. In digital systems, the exact value of a voltage *is not* important; for example, a voltage of 3.6 V means the same as a voltage of 4.3 V. In analog systems, the exact value of a voltage *is* important. For instance, if the analog voltage is proportional to the temperature measured by a transducer, 3.6 V would represent a different temperature than would 4.3 V. In other words, the voltage value carries significant information. This characteristic means that the design of accurate analog circuitry is generally more difficult than the design of digital circuits.

1.5   DIGITAL CIRCUITS

As explained in the last section, digital circuits are designed to produce output voltages that fall within the prescribed 0 and 1 voltage ranges such as those defined in Figure

1.7. Likewise, digital circuits are designed to respond predictably to input voltages that are within the defined 0 and 1 ranges. What this means is that a digital circuit will respond in the same way to all input voltages that fall within the allowed 0 range; similarly, it will not distinguish between input voltages that lie within the allowed 1 range.

To illustrate, Figure 1.8 represents a typical digital circuit with input v_i and output v_o. The output is shown for two different input signal waveforms. Note that v_o is the same for both cases because the two input waveforms, while they differ in their exact voltage levels, are at the same binary levels.

LOGIC CIRCUITS. The manner in which a digital circuit responds to an input is referred to as the circuit's *logic*. Each type of digital circuit obeys a certain set of logic rules. For this reason, digital circuits are also called *logic circuits*. We will use both terms interchangeably throughout the text. In Chapter 3 we will see more clearly what is meant by a circuit's "logic."

We will be studying all the types of logic circuits that are currently used in digital systems. Initially, our attention will be focused only on the logical operation which these circuits perform—that is, the relationship between the circuit inputs and outputs. We will defer any discussions of the actual circuitry used to construct logic circuits until Chapter 8. In this way we can develop a thorough understanding of logic-circuit operation without being burdened by circuit details.

DIGITAL INTEGRATED CIRCUITS. Almost all the digital circuits used in modern digital systems are integrated circuits (ICs). The wide variety of available logic ICs

FIGURE 1.8 A digital circuit responds to an input's binary level (0 or 1) and not to its actual voltage.

has made it possible to construct complex digital systems that are smaller and more reliable than the discrete-component counterparts.

Several integrated-circuit fabrication technologies are used to produce digital ICs, the most common being TTL, CMOS, NMOS, and ECL. Each differs in the type of circuitry used to provide the desired logic operation. For example, TTL (Transistor-Transistor-Logic) uses the bipolar transistor as its main circuit element, while CMOS (Complementary Metal Oxide Semiconductor) uses the enhancement MOSFET as its principal circuit element.

Detailed descriptions of each of these IC "logic families" will not be presented until Chapter 8. However, many illustrative examples in the earlier chapters will employ specific devices from the TTL and CMOS families. This will give a practical flavor to our discussions and will also provide actual device numbers that can be used for laboratory verification of the theory.

REVIEW QUESTIONS

1. True or false: The exact value of an input voltage is critical for a digital circuit.
2. Can a digital circuit produce the same output voltage for different input voltage values?
3. A digital circuit is also referred to as a _____ circuit.

1.6 PARALLEL VERSUS SERIAL REPRESENTATION

In a digital system, binary numbers are represented by the outputs of digital circuits. The binary numbers are transmitted from one part of the system to another by connecting these circuit outputs to other digital circuit inputs. There are basically two ways for representing and transmitting binary information: *parallel* and *serial*.

When parallel representation is used, each bit of the binary number is represented by the output of a separate circuit, as illustrated in Figure 1.9(a), where *five* circuit outputs represent a 5-bit binary number. The output of circuit A is the MSB, while that of circuit E is the LSB. At any time, the outputs of each circuit can represent either a 0 or a 1 (0 V or +5 V). In the example shown, the binary number is 10110. This is called parallel representation because all the bits of the 5-bit number are present simultaneously (in parallel).

When serial representation is used, only *one* circuit output is needed to represent all the bits of the binary number, as shown in Figure 1.9(b). Here the output of the circuit produces a digital signal whose voltage levels change (during specific time intervals) in accordance with the binary number being represented. During the first time interval, T_0, the signal is at 0 V (binary 0); during interval T_1 the signal is at +5 V (binary 1), and so on.

The principal trade-off between parallel and serial representations is one of speed versus circuit simplicity. The transmission of binary data from one part of a digital system to another can be done more quickly using parallel representation because all the bits are transmitted simultaneously, while serial representation transmits one bit at a time. On the other hand, parallel requires more signal lines connected between the sender and receiver of the binary data than does serial. In other words, parallel is faster and serial requires fewer signal lines. This comparison between parallel and

FIGURE 1.9 (a) Parallel representation requires one circuit output for each bit; (b) serial requires only one circuit for any number of bits.

serial methods for representing binary information will be encountered many times in discussions throughout the text.

_____ REVIEW QUESTION

1. Describe the relative advantages of parallel and serial transmission of binary data.

1.7 MEMORY

When an input signal is applied to most devices or circuits, the output somehow changes in response to the input, and when the input signal is removed, the output returns to its original state. These circuits do not exhibit the property of *memory,* since their outputs revert back to normal. In digital circuitry certain types of devices and circuits do have memory. When an input is applied to such a circuit, the output will change its state, but it will remain in the new state even after the input is removed. This property of retaining its response to a momentary input is called memory. Figure 1.10 illustrates nonmemory and memory operations.

Memory devices and circuits play an important role in digital systems because they provide means for storing binary numbers either temporarily or permanently, with the ability to change the stored information at any time. As we shall see, the

FIGURE 1.10 Comparison of nonmemory and memory operation.

various memory elements include magnetic types and those which utilize electronic latching circuits (called *flip-flops*).

1.8 DIGITAL COMPUTERS

Digital techniques have found their way into innumerable areas of technology, but the area of automatic *digital computers* is by far the most notable and most extensive. Although digital computers affect some part of all of our lives, it is doubtful that many of us know exactly what a computer does. In simplest terms, *a computer is a system of hardware that performs arithmetic operations, manipulates data (usually in binary form), and makes decisions.*

For the most part, human beings can do whatever computers can do, but computers can do it with much greater speed and accuracy. This is in spite of the fact that computers perform all their calculations and operations one step at a time. For example, a human being can take a list of 10 numbers and find their sum all in one operation by listing the numbers one over the other and adding them column by column. A computer, on the other hand, can add numbers only two at a time, so that adding this same list of numbers will take nine actual addition steps. Of course, the fact that the computer requires only a microsecond or less per step makes up for this apparent inefficiency.

A computer is faster and more accurate than people are, but unlike most people it has to be given a complete set of instructions that tell it *exactly* what to do at each step of its operation. This set of instructions, called a *program,* is prepared by one or more persons for each job the computer is to do. Programs are placed in the computer's memory unit in binary-coded form, with each instruction having a unique code. The computer takes these instruction codes from memory *one at a time* and performs the operation called for by the code. Much more will be said about this later.

Major Parts of a Computer   There are several types of computer systems, but each can be broken down into the same functional units. Each unit performs specific functions, and all units function together to carry out the instructions given in the program. Figure 1.11 shows the five major functional parts of a digital computer and their interaction. The solid lines with arrows represent the flow of information. The dashed lines with arrows represent the flow of timing and control signals.

The major functions of each unit are:

  1. *Input unit:* Through this unit a complete set of instructions and data are fed into the computer system and into the memory unit, to be stored until needed. The informa-

FIGURE 1.11   Functional diagram of a digital computer.

tion typically enters the input unit by means of punched cards, magnetic tape, magnetic disk, or a keyboard.

2. *Memory unit:* The memory stores the instructions and data received from the input unit. It stores the results of arithmetic operations received from the arithmetic unit. It also supplies information to the output unit.

3. *Control unit:* This unit takes instructions from the memory unit one at a time and interprets them. It then sends appropriate signals to all the other units to cause the specific instruction to be executed.

4. *Arithmetic unit:* All arithmetic calculations and logical decisions are performed in this unit, which can then send results to the memory unit to be stored.

5. *Output unit:* This unit takes data from the memory unit and prints out, displays, or otherwise presents the information to the operator (or process, in the case of a process control computer).

How Many Types of Computers Are There?   The answer depends on the criteria used to classify them. Computers are often classified according to physical size, which usually, but not always, is also indicative of their relative capabilities. The *micro-computer* is the smallest and newest member of the computer family. It generally consists of several IC chips, including a *microprocessor* chip, *memory* chips, and *input/output interface* chips. These chips are a result of tremendous advances in large-scale integration (LSI), which have also served to bring prices down so that it is possible to buy a fully assembled microcomputer now for less than $100.

Minicomputers are larger than microcomputers and have prices that go into the tens of thousands of dollars (including input/output peripheral equipment). Minis are widely used in industrial control systems, scientific applications for schools and research laboratories, and in business applications for small businesses. Although more expensive than microcomputers, minicomputers continue to be widely used because they are generally faster and possess more capabilities. These differences in speed and capability, however, are rapidly shrinking.

The largest computers are those found in large corporations, banks, universities, and scientific laboratories. These "maxicomputers" can cost as much as several million dollars and include complete systems of peripheral equipment such as magnetic tape units, magnetic disc units, punched card punchers and readers, keyboards, printers, and many more. Applications of maxicomputers range from computationally oriented science and engineering problem solving to the data-oriented business applications, where emphasis is on maintaining and updating large quantities of data and information.

REVIEW QUESTIONS

1. Explain how a digital circuit that has memory differs from one that does not.
2. Name the five major functional units of a computer.
3. Name the three types of computers and briefly describe how they differ.

PROBLEMS

1.1 Which of the following involve analog quantities and which are digital?
  (a) Pressure gauge.
  (b) Atoms in a sample of material.
  (c) Radio tuning dial.
  (d) Aircraft altitude.
  (e) TV VHF channel selector.

1.2 Convert the following binary numbers to their equivalent decimal values:
  (a) 11001
  (b) $1001.1001_2 = $ _____ $_{10}$
  (c) $10011011001.10110 = $ _____ $_{10}$

1.3 Using 6 bits, show the binary counting sequence from 000000 to 111111.

1.4 What is the maximum number that we can count up to using 10 bits?

1.5 How many bits are needed to count up to a maximum of 511?

1.6 Suppose that the decimal integer values from 0 to 15 are to be represented by digital circuits.
  (a) How many circuits will be needed if parallel representation is used?
  (b) How many will be needed if serial representation is used?

2
NUMBER SYSTEMS and CODES

OUTLINE

The binary number system is the most important one in digital systems, but several others also are important. The decimal system is important because it is universally used to represent quantities outside a digital system. This means that there will be situations where decimal values have to be converted to binary values before they are entered into the digital system. For example, when you punch a decimal number into your hand calculator (or computer), the circuitry inside the device converts the decimal number to a binary value.

Likewise, there will be situations where the binary values at the outputs of a digital circuit have to be converted to decimal values for presentation to the outside world. For example, your calculator (or computer) uses binary numbers to calculate answers to a problem, then converts the answers to a decimal value before displaying them.

In addition to binary and decimal, two other number systems find widespread applications in digital systems. The *octal* (*base-8*) and *hexadecimal* (*base-16*) number systems are both used for the same purpose—to provide an efficient means for representing large binary numbers. As we shall see, both these number systems have the advantage that they can be easily converted to and from binary.

In a digital system, three or four of these number systems may be in use at the same time, so that an understanding of the system operation requires the ability to convert from one number system to another. This chapter will show you how to perform these conversions. Although some of them will not be of immediate use in our study of digital systems, you will need them when you begin to study microprocessors.

This chapter will also introduce some of the *binary codes* that are used to represent various kinds of information. These binary codes will use 1s and 0s, but in a way that differs somewhat from the binary number system.

2.1 BINARY-TO-DECIMAL CONVERSIONS

As explained in Chapter 1, the binary number system is a positional system where each binary digit (bit) carries a certain weight based on its position relative to the LSB. Any binary number can be converted to its decimal equivalent simply by summing together the weights of the various positions in the binary number which contain a 1. To illustrate:

$$1 \quad 1 \quad 0 \quad 1 \quad 1 \quad (binary)$$
$$2^4 + 2^3 + 0 + 2^1 + 2^0 = 16 + 8 + 2 + 1$$
$$= 27_{10} \quad (decimal)$$

Let's try another example with a greater number of bits.

$$1 \quad 0 \quad 1 \quad 1 \quad 0 \quad 1 \quad 0 \quad 1_2 =$$
$$2^7 \quad + 2^5 + 2^4 \quad + 2^2 \quad + 2^0 = 181_{10}$$

Note that the procedure is to find the weights (i.e., powers of 2) for each bit position that contains a 1, then add them up. Also note that the MSB has a weight of 2^7 even though it is the 8th bit; this is because the LSB is the first bit and it always has a weight of 2^0.

―― **REVIEW QUESTIONS**

1. Convert 100011011011_2 to its decimal equivalent. (*Ans.* 2267)
2. What is the weight of the MSB of a 16-bit number? (*Ans.* 32768)

2.2 DECIMAL-TO-BINARY CONVERSIONS

There are two ways to convert a decimal number to its equivalent binary-system representation. A method that is convenient for small numbers is the reverse of the process described in Section 2.1. The decimal number is simply expressed as a sum of powers of 2 and then 1s and 0s are written in the appropriate bit positions. To illustrate:

$$45_{10} = 32 + 8 + 4 + 1 = 2^5 + 0 + 2^3 + 2^2 + 0 + 2^0$$
$$= 1 \quad 0 \quad 1 \quad 1 \quad 0 \quad 1_2$$

Note that a 0 is placed in the 2^1 and 2^4 positions, since all positions must be accounted for. Another example is shown below:

$$76_{10} = 64 + 8 + 4 = 2^6 + 0 + 0 + 2^3 + 2^2 + 0 + 0$$
$$= 1 \quad 0 \quad 0 \quad 1 \quad 1 \quad 0 \quad 0_2$$

Repeated Division   Another method that is better for larger decimal numbers uses repeated division by 2. The conversion, illustrated below for 25_{10}, requires repeatedly dividing the decimal number by 2 and writing down the remainders after each division until a quotient of 0 is obtained. Note that the binary result is obtained by writing the first remainder as the LSB and the last remainder as the MSB.

If a calculator is being used to perform the divisions by 2, the remainders can be determined by noting whether the quotient has a fractional part or not. For example, the calculator would produce $25/2 = 12.5$. The .5 indicates that there is a remainder of 1. The calculator would also produce $12/2 = 6.0$, which indicates a remainder of 0.

Another example is shown below as it would occur using a calculator.

$$\frac{37}{2} = 18.5 \longrightarrow \text{remainder of 1 (LSB)}$$

$$\frac{18}{2} = 9.0 \longrightarrow \qquad\qquad 0$$

$$\frac{9}{2} = 4.5 \longrightarrow \qquad\qquad 1$$

$$\frac{4}{2} = 2.0 \longrightarrow \qquad\qquad 0$$

$$\frac{2}{2} = 1.0 \longrightarrow \qquad\qquad 0$$

$$\frac{1}{2} = 0.5 \longrightarrow \qquad\qquad 1 \text{ (MSB)}$$

Thus, $37_{10} = 100101_2$.

REVIEW QUESTION

1. Convert 729_{10} to binary using both methods. (*Ans.* 1011011001_2)

2.3 THE OCTAL NUMBER SYSTEM

The octal number system is very important in digital computer work. The octal number system has a base of *eight,* meaning that it has eight possible digits: 0, 1, 2, 3, 4, 5, 6, and 7. Thus, each digit of an octal number can have any value from 0 to 7. The digit positions in an octal number have weights as follows:

8^4	8^3	8^2	8^1	8^0	8^{-1}	8^{-2}	8^{-3}	8^{-4}	8^{-5}

octal point

Octal-to-Decimal Conversion   An octal number, then, can be easily converted to its decimal equivalent by multiplying each octal digit by its positional weight. For example,

$$372_8 = 3 \times (8^2) + 7 \times (8^1) + 2 \times (8^0)$$
$$= 3 \times 64 + 7 \times 8 + 2 \times 1$$
$$= 250_{10}$$

Another example:

$$24.6_8 = 2 \times (8^1) + 4 \times (8^0) + 6 \times (8^{-1})$$
$$= 20.75_{10}$$

Decimal-to-Octal Conversion   A decimal integer can be converted to octal by using the same repeated-division method that we used in the decimal-to-binary conversion, but with a division factor of 8 instead of 2. An example is shown below.

Note that the first remainder becomes the least significant digit (LSD) of the octal number, and the last remainder becomes the most significant digit (MSD).

    If a calculator is used to perform the divisions in the above process, the result will include a decimal fraction instead of a remainder. The remainder can be obtained, however, by multiplying the decimal fraction by 8. For example, 266/8 produces 33.25. The remainder becomes $.25 \times 8 = 2$. Similarly, 33/8 will be 4.125 and the remainder becomes $0.125 \times 8 = 1$.

Conversion from Octal to Binary   The primary advantage of the octal number system is the ease with which conversion can be made between binary and octal numbers. The conversion from octal to binary is performed by converting *each* octal digit to its 3-bit binary equivalent. The eight possible digits are converted as indicated in Table 2.1.

TABLE 2.1

Octal digit	0	1	2	3	4	5	6	7
Binary equivalent	000	001	010	011	100	101	110	111

Using these conversions, any octal number is converted to binary by individually converting each digit. For example, we can convert 472_8 to binary as follows:

$$
\begin{array}{ccc}
4 & 7 & 2 \\
\downarrow & \downarrow & \downarrow \\
100 & 111 & 010
\end{array}
$$

Hence, octal 472 is equivalent to binary 100111010. As another example, consider converting 5431 to binary:

$$
\begin{array}{cccc}
5 & 4 & 3 & 1 \\
\downarrow & \downarrow & \downarrow & \downarrow \\
101 & 100 & 011 & 001
\end{array}
$$

Thus, $5431_8 = 101100011001_2$.

Binary-to-Octal Conversion   Converting from binary integers to octal integers is simply the reverse of the foregoing process. The bits of the binary number are grouped into groups of *three* bits starting at the LSB. Then each group is converted to its octal equivalent (Table 2.1). To illustrate, consider the conversion of 100111010_2 to octal.

$$
\begin{array}{ccc}
\underbrace{1\ 0\ 0} & \underbrace{1\ 1\ 1} & \underbrace{0\ 1\ 0} \\
\downarrow & \downarrow & \downarrow \\
4 & 7 & 2_8
\end{array}
$$

Sometimes the binary number will not have even groups of three bits. For those cases, we can add one or two 0s to the left of the MSB of the binary number to fill out the last group. This is illustrated below for the binary number 11010110.

$$
\begin{array}{ccc}
\underbrace{0\ 1\ 1} & \underbrace{0\ 1\ 0} & \underbrace{1\ 1\ 0} \\
\downarrow & \downarrow & \downarrow \\
3 & 2 & 6_8
\end{array}
$$

Note that a 0 was placed to the left of the MSB in order to produce even groups of three bits.

Counting in Octal   The largest octal digit is 7, so that when counting in octal, a digit is incremented upward from 0 to 7. Once it reaches 7, it recycles to 0 on the next count and causes the next-higher digit to be incremented. This is illustrated in the following sequences of octal counting: (a) 65, 66, 67, 70, 71; (b) 275, 276, 277, 300.

With N octal digits, we can count from zero up to $8^N - 1$, for a total of 8^N different counts. For example, with three octal digits we can count from 000_8 to 777_8, which is a total of $8^3 = 512_{10}$ different octal numbers.

Usefulness of Octal System   The ease with which conversions can be made between octal and binary make the octal system attractive as a "shorthand" means of expressing large binary numbers. In computer work, binary numbers with up to 64 bits are not uncommon. These binary numbers, as we shall see, do not always represent a numerical quantity but are often some type of code which conveys nonnumerical information. In computers, binary numbers might represent (1) actual numerical data; (2) numbers corresponding to a location (address) in memory; (3) an instruction code; (4) a code representing alphabetic and other nonnumerical characters; or (5) a group of bits representing the status of devices internal or external to the computer.

When dealing with a large quantity of binary numbers of many bits, it is convenient and more efficient for us to write the numbers in octal rather than binary. Keep in mind, however, that the digital circuits and systems work strictly in binary; we are using octal only as a convenience for the operators of the system.

————————————————————————— REVIEW QUESTIONS

1. Convert 614_8 to decimal. (*Ans.* 396)
2. Convert 146_{10} to octal, then from octal to binary. (*Ans.* 222 and 010010010)
3. Convert 10011101_2 to octal. (*Ans.* 235)
4. Write the next three numbers in this octal counting sequence: 624, 625, 626, ___, ___, ___. (*Ans.* Last number is 631)

2.4 THE HEXADECIMAL NUMBER SYSTEM

The hexadecimal system uses base 16. Thus, it has 16 possible digit symbols. It uses the digits 0 through 9 plus the letters A, B, C, D, E, and F as the 16 digit symbols. Table 2.2 shows the relationships among hexadecimal, decimal, and binary. Note that each hexadecimal digit represents a group of four binary digits. It is important to remember that hex (abbreviation for hexadecimal) digits A through F are equivalent to the decimal values 10 through 15.

TABLE 2.2

Hexadecimal	Decimal	Binary
0	0	0000
1	1	0001
2	2	0010
3	3	0011
4	4	0100
5	5	0101
6	6	0110
7	7	0111
8	8	1000
9	9	1001
A	10	1010
B	11	1011
C	12	1100
D	13	1101
E	14	1110
F	15	1111

Hex-to-Decimal Conversion   A hex number can be converted to its decimal equivalent by using the fact that each hex digit position has a weight that is a power of 16. The LSD has a weight of $16^0 = 1$, the next higher digit has a weight of $16^1 = 16$, the next higher digit has a weight of $16^2 = 256$, and so on. The conversion process is demonstrated in the examples below.

$$356_{16} = 3 \times 16^2 + 5 \times 16^1 + 6 \times 16^0$$

$$= 768 + 80 + 6$$

$$= 854_{10}$$

$$2AF_{16} = 2 \times 16^2 + 10 \times 16^1 + 15 \times 16^0$$
$$= 512 + 160 + 15$$
$$= 687_{10}$$

Note that in the second example the value 10 was substituted for A and the value 15 for F in the conversion to decimal.

Decimal-to-Hex Conversion   Recall that we did decimal-to-binary conversion using repeated division by 2, and decimal-to-octal using repeated division by 8. Likewise, decimal-to-hex conversion can be done using repeated division by 16. The examples below will illustrate. Again note how the remainders of the division processes form the digits of the hex number. Also note that any remainders that are greater than 9 are represented by the letters A–F.

EXAMPLE 2.1

Convert 423_{10} to hex.

Solution:

$$\frac{423}{16} = 26 + \text{remainder of } 7$$

$$\frac{26}{16} = 1 + \text{remainder of } 10$$

$$\frac{1}{16} = 0 + \text{remainder of } 1$$

$$423_{10} = \boxed{1A7_{16}}$$

EXAMPLE 2.2

Convert 214_{10} to hex.

Solution:

$$\frac{214}{16} = 13 + \text{remainder of } 6$$

$$\frac{13}{16} = 0 + \text{remainder of } 13$$

$$214_{10} = \boxed{D6_{16}}$$

If a calculator is being used to perform the divisions in the conversion process, the results will include a decimal fraction instead of a remainder. The remainder can be obtained by multiplying the fraction by 16. To illustrate, in Example 2.2 the calculator would have produced

$$\frac{214}{16} = 13.375$$

The remainder becomes $(.375) \times 16 = 6$.

Conversion from Hex to Binary   Like the octal number system, the hexadecimal number system is used primarily as a "shorthand" method for representing binary numbers. It is a relatively simple matter to convert a hex number to binary. *Each* hex digit is converted to its 4-bit binary equivalent (Table 2.2). This is illustrated below for $9F2_{16}$.

$$9\ F\ 2_{16} = \qquad 9 \qquad\qquad F \qquad\qquad 2$$
$$\downarrow \qquad\qquad \downarrow \qquad\qquad \downarrow$$
$$1\ 0\ 0\ 1 \quad 1\ 1\ 1\ 1 \quad 0\ 0\ 1\ 0$$
$$= 100111110010_2$$

Conversion from Binary to Hex   This conversion is just the reverse of the process above. The binary number is grouped into groups of *four* bits, and each group is converted to its equivalent hex digit.

$$1\ 0\ 1\ 1\ 1\ 0\ 1\ 0\ 0\ 1\ 1\ 0_2 = \underbrace{1\ 0\ 1\ 1}_{B}\ \underbrace{1\ 0\ 1\ 0}_{A}\ \underbrace{0\ 1\ 1\ 0}_{6}$$
$$= BA6_{16}$$

In order to perform these conversions between hex and binary, it is necessary to know the 4-bit binary numbers (0000–1111) and their equivalent hex digits. Once these are mastered, the conversions can be performed quickly without the need for any calculations. This is why hex (and octal) are so useful in representing large binary numbers.

Counting in Hexadecimal   When counting in hex, each digit position can be incremented (increased by 1) from 0 to F. Once a digit position reaches the value F, it is reset to 0 and the next digit position is incremented. This is illustrated in the hex counting sequences below:

(a) 38, 39, 3A, 3B, 3C, 3D, 3E, 3F, 40, 41, 42
(b) 6F8, 6F9, 6FA, 6FB, 6FC, 6FD, 6FE, 6FF, 700

Note that when there is a 9 in a digit position, it becomes an A when it is incremented.

REVIEW QUESTIONS

1. Convert $24CE_{16}$ to decimal. (*Ans.* 9422)
2. Convert 3117_{10} to hex, then from hex to binary. (*Ans.* C2D and 110000101101)

Sec. 2.4   The Hexadecimal Number System

3. Convert 10010111110110101_2 to hex. (*Ans.* 97B5)
4. Write the next four numbers in this hex counting sequence: E9A, E9B, E9C, E9D, ___, ___, ___, ___. (*Ans.* Last number is EA1)

2.5 THE BCD CODE

When numbers, letters, or words are represented by a special group of symbols, this is called *encoding*, and the group of symbols is called a *code*. Probably one of the most familiar codes is the Morse Code, where series of dots and dashes represent letters of the alphabet.

We have seen that any decimal number can be represented by an equivalent binary number. The group of 0s and 1s in the binary number can be thought of as a code representing the decimal number. When a decimal number is represented by its equivalent binary number, we call it *straight binary coding*.

Digital systems all use some form of binary numbers for their internal operation, but the external world is decimal in nature. This means that conversions between the decimal and binary systems are being performed often. We have seen that the conversions between decimal and binary can become long and complicated for large numbers. For this reason, a means of encoding decimal numbers that combines some features of both the decimal and binary systems is used in certain situations.

Binary-Coded-Decimal Code   If *each* digit of a decimal number is represented by its binary equivalent, this produces a code called *binary-coded-decimal* (hereafter abbreviated BCD). Since a decimal digit can be as large as 9, 4 bits are required to code each digit (the binary code for 9 is 1001).

To illustrate the BCD code, take a decimal number such as 874. Each digit is changed to its binary equivalent as follows:

$$
\begin{array}{cccl}
8 & 7 & 4 & \text{(decimal)}\\
\downarrow & \downarrow & \downarrow & \\
1000 & 0111 & 0100 & \text{(BCD)}
\end{array}
$$

As another example, let us change 943 to its BCD-code representation:

$$
\begin{array}{cccl}
9 & 4 & 3 & \text{(decimal)}\\
\downarrow & \downarrow & \downarrow & \\
1001 & 0100 & 0011 & \text{(BCD)}
\end{array}
$$

Once again, each decimal digit is changed to its straight binary equivalent. Note that 4 bits are *always* used for each digit.

The BCD code, then, represents each digit of the decimal number by a 4-bit binary number. Clearly, only the 4-bit binary numbers from 0000 through 1001 are used. The BCD code does not use the numbers 1010, 1011, 1100, 1101, 1110, and 1111. In other words, only 10 of the 16 possible 4-bit binary code groups are used. If any of these "forbidden" 4-bit numbers ever occurs in a machine using the BCD code, it is usually an indication that an error has occurred.

EXAMPLE 2.3

Convert 0110100000111001 (BCD) to its decimal equivalent.

Solution: Divide the BCD number into 4-bit groups and convert each to decimal.

$$\underbrace{0110}_{6} \quad \underbrace{1000}_{8} \quad \underbrace{0011}_{3} \quad \underbrace{1001}_{9}$$

EXAMPLE 2.4

Convert the BCD number 011111000001 to its decimal equivalent.

Solution:

$$\underbrace{0111}_{7} \quad \underbrace{1100}_{\downarrow} \quad \underbrace{0001}_{1}$$

forbidden code group indicates error in BCD number

Comparison of BCD and Binary It is important to realize that BCD is not another number system like binary, octal, decimal, and hexadecimal. It is, in fact, the decimal system with each digit encoded in its binary equivalent. It is also important to understand that a BCD number is *not* the same as a straight binary number. A straight binary code takes the *complete* decimal number and represents it in binary; the BCD code converts *each* decimal *digit* to binary individually. To illustrate, take the number 137 and compare its straight binary and BCD codes:

$$137_{10} = 10001001_2 \qquad \text{(binary)}$$

$$137_{10} = 0001 \quad 0011 \quad 0111 \qquad \text{(BCD)}$$

The BCD code requires 12 bits while the straight binary code requires only 8 bits to represent 137. BCD requires more bits than straight binary to represent decimal numbers of more than one digit. This is because BCD does not use all possible 4-bit groups, as pointed out earlier and is therefore somewhat inefficient.

The main advantage of the BCD code is the relative ease of converting to and from decimal. Only the 4-bit code groups for the decimal digits 0–9 need be remembered. This ease of conversion is especially important from a hardware standpoint because in a digital system it is the logic circuits that perform the conversions to and from decimal.

BCD is used in digital machines whenever decimal information is either applied as inputs or displayed as outputs. Digital voltmeters, frequency counters, and digital clocks all use BCD because they display output information in decimal. Electronic calculators use BCD because the input numbers are entered in decimal via the keyboard and the output numbers are displayed in decimal.

BCD is not often used in modern high-speed digital computers for two good reasons. First, as was already pointed out, the BCD code for a given decimal number requires more bits than the straight binary code and is therefore less efficient. This is important in digital computers because the number of places in memory where these bits can be stored is limited. Second, the arithmetic processes for numbers represented in BCD code are more complicated than straight binary and thus require more complex circuitry. The more complex circuitry contributes to a decrease in the speed at which arithmetic operations take place. Calculators that use BCD are therefore considerably slower in their operation than computers.

REVIEW QUESTIONS

1. Represent the decimal value 178 by its straight binary equivalent. Then encode the same decimal number using BCD. (*Ans.* 10110010_2 and 000101111000 [BCD])
2. How many bits are required to represent an 8-digit decimal number in BCD?
3. What is an advantage of encoding a decimal number in BCD as compared to straight binary? What is a disadvantage?

2.6 EXCESS-3 CODE

The *excess-3 code* is related to the BCD code and is sometimes used instead because it possesses advantages in certain arithmetic operations. The excess-3 code for a decimal number is performed in the same manner as BCD except that *3* is added to each decimal digit before encoding it in binary. For example, to encode the decimal number 4 into excess-3 code, we must first add 3 to obtain 7. Then the 7 is encoded in its equivalent 4-bit binary code 0111.

As another example, let us convert 48 into its excess-3-code representation.

$$
\begin{array}{cc}
4 & 8 \\
+\ 3 & +\ 3 \qquad \text{add 3 to each digit} \\
\hline
7 & 11 \\
\downarrow & \downarrow \\
0111 & 1011 \qquad \text{convert to 4-bit binary code}
\end{array}
$$

Table 2.3 lists the BCD and excess-3-code representations for the decimal digits. Note that both codes use only 10 of the 16 possible 4-bit code groups. The excess-3 code, however, does not use the same code groups. For excess-3, the invalid code groups are 0000, 0001, 0010, 1101, 1110, and 1111.

TABLE 2.3

Decimal	BCD	Excess-3
0	0000	0011
1	0001	0100
2	0010	0101
3	0011	0110
4	0100	0111
5	0101	1000
6	0110	1001
7	0111	1010
8	1000	1011
9	1001	1100

2.7 THE GRAY CODE

The *Gray code* belongs to a class of codes called *minimum-change codes,* in which only *one* bit in the code group changes when going from one step to the next. The Gray code is an *unweighted* code, meaning that the bit positions in the code groups do not have any specific weight assigned to them. Because of this, the Gray code is not suited for arithmetic operations but finds application in input/output devices and some types of analog-to-digital converters.

Table 2.4 shows the Gray-code representation for the decimal numbers 0 through 15, together with the straight binary code. If we examine the Gray-code groups for each decimal number, it can be seen that in going from *any* one decimal number to the next, only *one* bit of the Gray code changes. For example, in going from 3 to 4, the Gray code changes from 0010 to 0110, with only the second bit from the left changing. Going from 14 to 15 the Gray-code bits change from 1001 to 1000, with only the last bit changing. This is the principal characteristic of the Gray code. Compare this with the binary code, where anywhere from one to all of the bits change in going from one step to the next.

TABLE 2.4

Decimal	Binary Code	Gray Code	Decimal	Binary Code	Gray Code
0	0000	0000	8	1000	1100
1	0001	0001	9	1001	1101
2	0010	0011	10	1010	1111
3	0011	0010	11	1011	1110
4	0100	0110	12	1100	1010
5	0101	0111	13	1101	1011
6	0110	0101	14	1110	1001
7	0111	0100	15	1111	1000

The Gray code is often used in situations where other codes, such as binary, might produce erroneous or ambiguous results during those transitions in which more than one bit of the code is changing. For instance, using binary code and going from 0111 to 1000 requires that all four bits change simultaneously. Depending on the device or circuit that is generating the bits, there may be a significant difference in the transition times of the different bits. If so, the transition from 0111 to 1000 could produce one or more intermediate states. For example, if the most significant bit changes faster than the rest, the following transitions will occur:

```
0111  —  decimal 7
 ┌- - - - - - - - - - - - - - - - ┐
 │ 1111  —  erroneous code        │
 └─ - - - - - - - - - - - - - - - ┘
1000  —  decimal 8
```

TABLE 2-5 PARTIAL LISTING OF ASCII CODE

Character	7-Bit ASCII	Octal	Hex
A	100 0001	101	41
B	100 0010	102	42
C	100 0011	103	43
D	100 0100	104	44
E	100 0101	105	45
F	100 0110	106	46
G	100 0111	107	47
H	100 1000	110	48
I	100 1001	111	49
J	100 1010	112	4A
K	100 1011	113	4B
L	100 1100	114	4C
M	100 1101	115	4D
N	100 1110	116	4E
O	100 1111	117	4F
P	101 0000	120	50
Q	101 0001	121	51
R	101 0010	122	52
S	101 0011	123	53
T	101 0100	124	54
U	101 0101	125	55
V	101 0110	126	56
W	101 0111	127	57
X	101 1000	130	58
Y	101 1001	131	59
Z	101 1010	132	5A
0	011 0000	060	30
1	011 0001	061	31
2	011 0010	062	32
3	011 0011	063	33
4	011 0100	064	34
5	011 0101	065	35
6	011 0110	066	36
7	011 0111	067	37
8	011 1000	070	38
9	011 1001	071	39
blank	010 0000	040	20
.	010 1110	056	2E
(	010 1000	050	28
+	010 1011	053	2B
$	010 0100	044	24
*	010 1010	052	2A
)	010 1001	051	29
—	010 1101	055	2D
/	010 1111	057	2F
,	010 1100	054	2C
=	011 1101	075	3D

The occurrence of 1111 is only momentary but it could conceivably produce erroneous operation of the elements that are being controlled by the bits. Obviously, using the Gray code would eliminate this problem, since only one bit change occurs per transition and no "race" between bits can occur.

REVIEW QUESTION

1. What is the principal advantage of the Gray code over straight binary?

2.8 ALPHANUMERIC CODES

In order to be very useful, a computer must be capable of handling nonnumeric information. In other words, a computer must be able to recognize codes that represent numbers, letters, and special characters. These codes are classified as *alphanumeric codes*. A complete and adequate set of necessary characters includes: (1) 26 lowercase letters; (2) 26 uppercase letters; (3) 10 numeric digits; and (4) about 25 special characters, such as $+$, $/$, $\#$, $\%$, and so on.

This totals up to 87 characters. To represent 87 characters with some type of binary code would require at least 7 bits. With 7 bits, there are $2^7 = 128$ possible binary numbers; 87 of these arrangements of 0 and 1 bits serve as the code groups representing the 87 different characters. For example, the code group 1010101 might represent the letter U.

The most common alphanumeric code, known as the American Standard Code for Information Interchange (ASCII), is used by most minicomputer and microcomputer manufacturers. Table 2.5 shows a *partial* listing of the 7-bit ASCII code. For each character, the octal and hex equivalents are also shown. The ASCII code is used in the transmission of alphanumeric information between a computer and external input/output devices like a teletypewriter (TTY) or video terminal.

The ASCII code is widely used in microcomputers to represent the various keys that an operator can actuate on the microcomputer keyboard. For example, when the operator types the letter "G" (e.g., as part of a GOTO statement in a BASIC program), the ASCII code for G will be entered into the computer's memory. Of course, this code will be stored as a group of 0s and 1s.

EXAMPLE 2.5

A computer operator enters the following ASCII-coded statement from the keyboard. Determine what this statement is telling the computer.

1010011,1010100,1001111,1010000

Solution: Convert each 7-bit code to its hex equivalent. The results are 53,54,4F,50. Now check Table 2-5 for the characters represented by each of these codes. The results are S,T,O,P. Thus, the operator is telling the computer to STOP execution of the program.

1. Encode the following message in ASCII code using the hex representation: COST=$72. (*Ans.* 43,4F,53,54,3D,24,37,32)

2.9 PARITY METHOD FOR ERROR DETECTION

The transmission of binary data from one location to another is commonplace in all digital systems. Here are several examples:

1. Binary-data output from a computer is recorded on magnetic tape.
2. Binary data are transmitted over telephone lines, such as between a computer and a remote computer.
3. A number is taken from the computer memory and placed in the arithmetic unit, where it is added to another number. The sum is then transferred back into memory.
4. Punched IBM cards are fed to a card reader, which transfers the information from the cards into the computer.

The process of transferring data is subject to error, although modern equipment has been designed to reduce the probability of error. However, even relatively infrequent errors can cause useless results, so it is desirable to detect them whenever possible. One of the most widely used schemes for error detection is the *parity method*.

The Parity Bit   A *parity bit* is an extra bit that is attached to a code group that is being transferred from one location to another. The parity bit is made either 0 or 1, depending on the number of 1s that are contained in the code group. Two different methods are used.

In the *even-parity* method, the value of the parity bit is chosen so that the total number of 1s in the code group (including the parity bit) is an *even* number. For example, suppose that the code group is 1000011. This is the ASCII character C. The code group has *three* 1s. Therefore, we will add a parity bit of 1 to make the total number of 1s an even number. The *new* code group, including the parity bit, thus becomes

If the code group contains an even number of 1s to begin with, the parity bit is given a value of 0. For example, if the code group were 1000001 (the ASCII code for

---

*The parity bit can be placed at either end of the code word but is usually placed to the left of the MSB.

"A"), the assigned parity bit would be 0, so the new code, including the parity bit, would be 01000001.

The *odd-parity* method is used in exactly the same way except that the parity bit is chosen so the total number of 1s (including the parity bit) is an *odd* number. For example, for the code group 1000001, the assigned parity bit would be a 1. For the code group 1000011, the parity bit would be a 0.

Regardless of whether even parity or odd parity is used, the parity bit becomes an actual part of the code word. For example, adding a parity bit to the 7-bit ASCII code produces an 8-bit code. Thus, the parity bit is treated just like any other bit in the code.

The parity bit is used to detect any *single-bit* errors that occur during the transmission of a code from one location to another (e.g., from a computer to a video terminal). For example, suppose the character "A" is being transmitted and *odd* parity is being used. The transmitted code would be:

$$\boxed{1}\ 1\quad 0\quad 0\quad 0\quad 0\quad 0\quad 1$$

When the receiver circuit receives this code, it will check to see that the code contains an odd number of 1s (including the parity bit). If so, the receiver will assume that the code has been correctly received. Now, suppose that because of some noise or malfunction, the receiver actually receives the following code:

$$\boxed{1}\ 1\quad 0\quad 0\quad 0\quad 0\quad 0\quad 0$$

The receiver will find that this code has an *even* number of 1s. This tells the receiver that there must be an error in the code, since presumably the transmitter and receiver have agreed to use odd parity. There is no way, however, that the receiver can tell which bit is in error, since it does not know what the code is supposed to be.

It should be apparent that this parity method would not work if *two* bits were in error, because two errors would not change the "oddness" or "evenness" of the number of 1s in the code. In practice, the parity method is used only in situations where the probability of a single error is very low, and the probability of double errors is essentially zero.

REVIEW QUESTIONS

1. Attach an even parity bit to the ASCII code for the symbol "$" and express the result in hex. (*Ans.* 24)
2. Attach an odd parity bit to the BCD code for decimal 69 and show the result. (*Ans.* 101101001)

PROBLEMS

2.1 Convert these binary numbers to decimal:
(a) 10110   (b) 10001101   (c) 100100001001   (d) 1111010111

2.2 Convert the following decimal values to binary:
(a) 37   (b) 14   (c) 189   (d) 205   (e) 2313

2.3 What is the largest decimal value that can be represented by an 8-bit binary number? A 16-bit number?

Chap. 2   Problems

2.4 Convert each octal number to its decimal equivalent:
(a) 743 (b) 36 (c) 3777 (d) 257

2.5 Convert each of the following decimal numbers to octal:
(a) 59 (b) 372 (c) 919 (d) 65536

2.6 Convert each of the octal values from Problem 2.4 to binary.

2.7 Convert the binary numbers in Problem 2.1 to octal.

2.8 List the octal numbers in sequence from 165_8 to 200_8.

2.9 When a large decimal number is to be converted to binary, it is sometimes easier to convert it first to octal, and then from octal to binary. Try this procedure for 2313_{10} and compare it to the procedure used in Problem 2.2(e).

2.10 Convert these hex values to decimal:
(a) 92 (b) 1A6 (c) 37FD (d) 2C0

2.11 Convert these decimal values to hex:
(a) 75 (b) 314 (c) 2048 (d) 25619

2.12 Convert the binary numbers in Problem 2.1 to hexadecimal.

2.13 Convert the hex values in Problem 2.10 to binary.

2.14 In most microcomputers the *addresses* of memory locations are specified in hexadecimal. These addresses are sequential numbers that identify each memory circuit
(a) A particular microcomputer can store an 8-bit number in each memory location. If the memory addresses range from 0000_{16} to $FFFF_{16}$, how many memory locations are there?
(b) Another microcomputer is specified to have 4096 memory locations. What range of hex addresses does this computer use?

2.15 List the hex numbers in sequence from 280 to 2A0.

2.16 Encode these decimal numbers in BCD:
(a) 47 (b) 962 (c) 187 (d) 42,689,627

2.17 How many bits are required to represent the decimal numbers in the range from 0 to 999 using straight binary code? Using BCD code?

2.18 The following numbers are in BCD. Convert them to decimal.
(a) 1001011101010010
(b) 000110000100

2.19 Represent the statement "X=25/Y" in ASCII code.

2.20 Attach an *even* parity bit to each of the ASCII codes for Problem 2.19 and give the results in hex.

2.21 The following code groups are being transmitted. Attach an *even*-parity bit to each group.
(a) 10110110 (b) 00101000 (c) 11110111

2.22 Convert the following decimal numbers to BCD code and then attach an *odd*-parity bit.
(a) 74 (b) 38 (c) 165 (d) 9201

2.23 In a certain digital system, the decimal numbers from 000 through 999 are represented in BCD code. An *odd*-parity bit is also included at the end of each code group. Examine each of the code groups below and assume that each one has just been transferred from one location to another. Some of the groups contain errors. Assume that *no more than* two errors have occurred for each group. Determine which of the code groups has a single error and which of them *definitely* has a double error. (*Hint:* Remember that this is a BCD code.)

(a) 1001010110000

       ↑— parity bit

(b) 0100011101100

(c) 0111110000011

(d) 1000011000101

2.24 Perform each of the following conversions. For some of them, you may want to try several methods to see which one works best for you. For example, a binary-to-decimal conversion may be done directly, or it may be done as a binary-to-octal conversion followed by an octal-to-decimal conversion.

(a) $1417_{10} = \underline{\hspace{1cm}}_2$

(b) $255_{10} = \underline{\hspace{1cm}}_2$

(c) $11010001_2 = \underline{\hspace{1cm}}_{10}$

(d) $1110101000100111_2 = \underline{\hspace{1cm}}_{10}$

(e) $2497_{10} = \underline{\hspace{1cm}}_8$

(f) $511_{10} = \underline{\hspace{1cm}}_8$

(g) $235_8 = \underline{\hspace{1cm}}_{10}$

(h) $4316_8 = \underline{\hspace{1cm}}_{10}$

(i) $7A9_{16} = \underline{\hspace{1cm}}_{10}$

(j) $3E1C_{16} = \underline{\hspace{1cm}}_{10}$

(k) $1600_{10} = \underline{\hspace{1cm}}_{16}$

(l) $38,187_{10} = \underline{\hspace{1cm}}_{16}$

(m) $865_{10} = \underline{\hspace{1cm}}(BCD)$

(n) $100101000111(BCD) = \underline{\hspace{1cm}}_{10}$

(o) $465_8 = \underline{\hspace{1cm}}_{16}$

(p) $B34_{16} = \underline{\hspace{1cm}}_8$

(q) $01110100(BCD) = \underline{\hspace{1cm}}_2$

(r) $111010_2 = \underline{\hspace{1cm}}(BCD)$

3

LOGIC GATES and BOOLEAN ALGEBRA

O U T L I N E

As pointed out in Chapter 1, digital (logic) circuits operate in the binary mode where each input and output voltage is either a 0 or 1; the 0 and 1 designations represent predefined voltage ranges. This characteristic of logic circuits allows us to use *Boolean algebra* as a tool for the analysis and design of digital systems. In this chapter we will study *logic gates,* which are the most fundamental logic circuits, and we will see how their operation can be described using Boolean algebra. We will also see how logic gates can be combined to produce logic circuits, and how these circuits can be described and analyzed using Boolean algebra.

Throughout the chapter our attention will be focused on the logical operation which each gate can perform; we will not be concerned with the internal circuitry that allows the gate to operate the way it does. We saw earlier that most logic circuits are ICs, so it is not necessary to connect discrete components together to make a logic circuit. Nevertheless, it is important to understand the internal circuitry of these ICs if they are to be used properly, and we will study this topic in Chapter 8.

3.1 BOOLEAN CONSTANTS AND VARIABLES

Boolean algebra differs in a major way from ordinary algebra in that Boolean constants and variables are allowed to have only two possible values, 0 or 1. A Boolean variable is a quantity that may, at different times, be equal to either 0 or 1. The Boolean variables are often used to represent the voltage level present on a wire or at the input/output terminals of a circuit. For example, in a certain digital system the Boolean value of 0 might be assigned to any voltage in the range from 0 to 0.8 V while the Boolean value of 1 might be assigned to any voltage in the range 2 to 5 V.*

Thus, Boolean 0 and 1 do not represent actual numbers but instead represent the state of a voltage variable or what is called its *logic level.* A voltage in a digital circuit is said to be at the logic level 0 or the logic level 1, depending on its actual numerical value. In the digital logic field several other terms are used synonomously with 0 and 1. Some of the more common ones are shown in Table 3.1. We will use the 0/1 and LOW/HIGH designations most of the time.

TABLE 3.1

Logic 0	Logic 1
False	True
Off	On
Low	High
No	Yes
Open switch	Closed switch

Boolean algebra is used to express the effects that various digital circuits have on logic inputs and to manipulate logic variables for the purpose of determining the best method for performing a given circuit function. In all our work to follow we shall

---

*Voltages between 0.8 and 2 V are undefined (neither 0 nor 1) and under normal circumstances should not occur.

use letter symbols to represent logic variables. For example, A might represent a certain digital circuit input or output, and at any time we must have either $A = 0$ or $A = 1$: if not one, then the other.

Because only two values are possible, Boolean algebra is relatively easy to work with as compared to ordinary algebra. In Boolean algebra there are no fractions, decimals, negative numbers, square roots, cube roots, logarithms, imaginary numbers, and so on. In fact, in Boolean algebra there are only *three* basic operations:

1. *Logical addition,* also called *OR addition* or simply the *OR operation.* The common symbol for this operation is the plus sign ($+$).
2. *Logical multiplication,* also called *AND multiplication* or simply the *AND operation.* The common symbol for this operation is the multiplication sign ($\cdot$).
3. *Logical complementation or inversion,* also called the *NOT operation.* The common symbol for this operation is the overbar ($^-$).

3.2 TRUTH TABLES

Many logic circuits have more than one input and only one output. A *truth table* shows how the logic circuit's output responds to the various combinations of logic levels at the inputs. The format for two-, three-, and four-input truth tables is shown in Figure 3.1.

In each truth table the possible combinations of 0 and 1 logic levels for the inputs (A, B, C, D) are listed on the left, and the resultant logic level for output x is listed on the right. The entries for x are shown as "?" for now, because these values will be different for each type of logic circuit.

(a)

A	B	x
0	0	?
0	1	?
1	0	?
1	1	?

(b)

A	B	C	x
0	0	0	?
0	0	1	?
0	1	0	?
0	1	1	?
1	0	0	?
1	0	1	?
1	1	0	?
1	1	1	?

(c)

A	B	C	D	x
0	0	0	0	?
0	0	0	1	?
0	0	1	0	?
0	0	1	1	?
0	1	0	0	?
0	1	0	1	?
0	1	1	0	?
0	1	1	1	?
1	0	0	0	?
1	0	0	1	?
1	0	1	0	?
1	0	1	1	?
1	1	0	0	?
1	1	0	1	?
1	1	1	0	?
1	1	1	1	?

FIGURE 3.1 Truth tables for (a) two-input, (b) three-input, and (c) four-input cases.

Note that there are 4 table entries for the two-input truth table, 8 entries for a three-input truth table, and 16 entries for the four-input truth table. The number of input combinations will equal 2^N for an N-input truth table. Also note that the list of all possible input combinations follows the binary counting sequence, so it is an easy matter to write down all the combinations without missing any.

3.3 THE OR OPERATION

Let A and B represent two independent logic variables. When A and B are combined using OR addition, the result, x, can be expressed as

$$x = A + B$$

In this expression the $+$ sign does not stand for ordinary addition; it stands for OR addition, whose rules are given in the truth table shown in Figure 3.2(a).

It should be apparent from the truth table that except for the case where $A = B = 1$, the OR operation is the same as ordinary addition. However, for $A = B = 1$ the OR sum is 1 (not 2 as in ordinary addition). This is easy to remember if we recall that only 0 and 1 are possible values in Boolean algebra, so that the largest value we can get is 1. This same result is true if we have $x = A + B + C$, for the case where $A = B = C = 1$. That is,

$$x = 1 + 1 + 1 = 1$$

We can therefore say that in OR addition the result will be 1 if any one *or* more variables is a 1. This is also clear in the table in Figure 3.2(a).

The expression $x = A + B$ can either be read as "x equals A plus B" or as "x equals A OR B." Both expressions are in common use. The key thing to remember is that the $+$ sign stands for OR addition, as defined by the truth table in Figure 3.2(a), and not for ordinary addition.

The OR Gate In digital circuitry an *OR gate* is a circuit that has two or more inputs and whose output is equal to the OR sum of the inputs. Figure 3.2(b) shows the symbol for a two-input OR gate. The inputs A and B are logic voltage levels and the output x is a logic voltage level whose value is the result of the OR addition of A and B; that is, $x = A + B$. In other words, the OR gate operates such that its output is high (logic 1) if either input A or B or *both* are at a logic 1 level. The OR gate output will be low (logic 0) only if all its inputs are at logic 0.

A	B	$x = A + B$
0	0	0
0	1	1
1	0	1
1	1	1

(a)

OR Gate

(b)

FIGURE 3.2 (a) Truth table defining the OR operation; (b) circuit symbol for a two-input OR gate.

A	B	C	$x = A + B + C$
0	0	0	0
0	0	1	1
0	1	0	1
0	1	1	1
1	0	0	1
1	0	1	1
1	1	0	1
1	1	1	1

FIGURE 3.3 Symbol and truth table for a three-input OR gate.

This same idea can be extended to more than two inputs. Figure 3.3 shows a three-input OR gate and its truth table. Examination of this truth table shows again that the output will be 1 for every case where one or more inputs is 1. This general principle is the same for OR gates with any number of inputs.

Using the language of Boolean algebra, the output x can be expressed as $x = A + B + C$, where again it must be emphasized that the + means OR addition. The output of any OR gate, then, can be expressed as the OR sum of its various inputs. We will put this to use when we subsequently analyze logic circuits.

Summary of the OR Operation   The important points to remember concerning the OR operation and OR gates are:

1. The OR operation produces a result of 1 when *any* of the input variables is 1.
2. The OR operation produces a result of 0 *only* when all the input variables are 0.
3. In OR addition, $1 + 1 = 1$, $1 + 1 + 1 = 1$, and so on.

EXAMPLE 3.1

In many industrial control systems it is required to activate an output function whenever any one of several inputs is activated. For example, in a chemical process it may be desired that an alarm be activated whenever the process temperature exceeds a maximum value *or* whenever the pressure goes above a certain limit. Figure 3.4 is a block diagram of this situation. The temperature

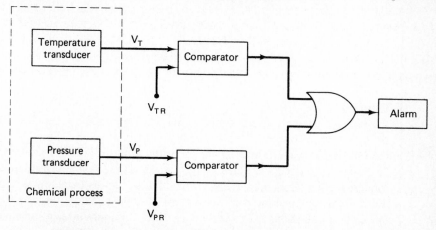

FIGURE 3.4 Example of the use of OR gate in alarm system.

Sec. 3.3   The OR Operation

transducer circuit produces an output voltage proportional to the process temperature. This voltage, V_T, is compared to a temperature reference voltage, V_{TR}, in a voltage comparator circuit. The comparator output is normally a low voltage (logical 0), but it switches to a high voltage (logical 1) whenever V_T exceeds V_{TR}, indicating that process temperature is excessive. A similar arrangement is used for the pressure measurement, so its associated comparator output goes from low to high when the pressure is excessive.

Since we want the alarm to be activated when either temperature *or* pressure is too high, it should be apparent that the two comparator outputs can be fed to a two-input OR gate. The OR gate output thus goes high (1) for either alarm condition and will activate the alarm. This same idea can be obviously extended to situations with more than two process variables.

EXAMPLE 3.2

For the situation depicted in Figure 3.5, determine the waveform at the OR-gate output.

FIGURE 3.5

Solution: The three OR-gate inputs A, B, and C are varying, as shown by their waveform diagrams. The OR-gate output is determined by realizing that it will be high whenever *any* of the three inputs is at a high level. Using this reasoning, the OR-output waveform is as shown in the figure. Particular attention should be paid to what occurs at time t_1. The diagram shows that at that instant of time, input A is going from high to low while input B is going from low to high. Since these inputs are making their transitions at approximately the same time and since these transitions take a certain amount of time, there is a short interval of time when these OR-gate inputs are both in the undefined range between 0 and 1. When this occurs, the OR-gate output also becomes a value in this range, as evidenced by the glitch or spike on the output waveform at t_1. The occurrence of this glitch and its size (amplitude and width) depends on the speed with which the input transitions occur. It should be noted that if the C input had been sitting high while A and B were changing, the glitch would not occur because the high at C would keep the OR output high.

REVIEW QUESTIONS

1. What is the only set of input conditions that will produce a LOW output for any OR gate?
2. Write the Boolean expression for a six-input OR gate.
3. If the A input in Figure 3.5 is permanently kept at the 1 level, sketch the resultant output waveform. (*Ans.* Output will be a constant HIGH)

3.4   THE AND OPERATION

If two logic variables A and B are combined using AND multiplication, the result, x, can be expressed as

$$x = A \cdot B$$

In this expression the $\cdot$ sign stands for the Boolean operation of AND multiplication, whose rules are given in the truth table shown in Figure 3.6(a).

It should be apparent from the table that AND multiplication is *exactly* the same as ordinary multiplication. Whenever A or B are 0, their product is zero; when both A *and* B are 1, their product is 1. We can therefore say that in the AND operation the result will be 1 *only if* all the inputs are 1; for all other cases the result is 0.

The expression $x = A \cdot B$ is read "x equals A *and* B." The multiplication sign is generally omitted as in ordinary algebra, so the expression becomes $x = AB$. The key thing to remember is that the AND operation is the same as ordinary multiplication, where the variables can be either 0 or 1.

The AND Gate   A two-input AND gate is shown symbolically in Figure 3.6(b). The AND-gate output is equal to the AND product of the logic inputs; that is, $x = AB$. In other words, the AND gate is a circuit that operates such that its output is high only when all its inputs are high. For all other cases the AND gate output is low.

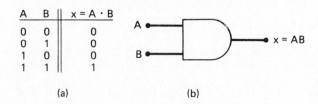

A	B	x = A · B
0	0	0
0	1	0
1	0	0
1	1	1

(a)

(b)

FIGURE 3.6   (a) Truth table for the AND operation; (b) AND-gate symbol.

A	B	C	x = ABC
0	0	0	0
0	0	1	0
0	1	0	0
0	1	1	0
1	0	0	0
1	0	1	0
1	1	0	0
1	1	1	1

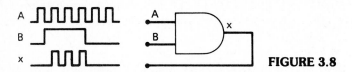

FIGURE 3.7 Truth table and symbol for a three-input AND gate.

This same operation is characteristic of AND gates with more than two inputs. For example, a three-input AND gate and its accompanying truth table are shown in Figure 3.7. Once again, note that the gate output is 1 only for the case where $A = B = C = 1$. The expression for the output is $x = ABC$. For a four-input AND gate, the output is $x = ABCD$, and so on.

Note the difference between the symbols for the AND gate and the OR gate. Whenever you see the AND symbol on a logic circuit diagram, it tells you that the output will go HIGH *only* when *all* inputs are HIGH. Whenever you see the OR symbol, it means that the output will go HIGH when *any* input is HIGH.

Summary of the AND Operation

1. The AND operation is performed exactly like ordinary multiplication of 1s and 0s.
2. An output equal to 1 occurs only for the single case where all inputs are 1.
3. The output is 0 for any case where one or more inputs are 0.

EXAMPLE 3.3

Determine the output waveform for the AND gate shown in Figure 3.8.

FIGURE 3.8

Solution: The output x will be at 1 only when A and B are both high at the same time. Using this fact, the x waveform can be determined as shown in the figure.

Notice that the x waveform is 0 whenever B is 0, regardless of the signal at A. Also notice that the x waveform is the same as A whenever B is 1. Thus, we can think of the B input as a *control* input whose logic level determines whether the A waveform gets through to the x output or not. In this situation, the AND gate is used as an *inhibit circuit*. We can say that $B = 0$ is the inhibit condition producing a 0 output. Conversely, $B = 1$ is the *enable* condition, which enables A to reach the output. This inhibit operation is an important application of AND gates.

EXAMPLE 3.4

What will happen to the x output waveform in Figure 3.8 if the B input is kept at the 0 level?

Solution: With B kept LOW, the x output will also stay LOW. This can be reasoned in two different ways. First, with $B = 0$ we have $x = A \cdot B = A \cdot 0 = 0$, since anything multiplied (ANDed) by 0 will be 0. Another way to look at it is that an AND gate requires that all inputs be HIGH in order for the output to be HIGH, and this cannot happen if B is kept LOW.

REVIEW QUESTIONS

1. What is the only input combination that will produce a HIGH at the output of a five-input AND gate?
2. What logic level should be applied to the second input of a two-input AND gate if the logic signal at the first input is to be inhibited (prevented) from reaching the output?
3. True or false: An AND-gate output will always differ from an OR-gate output for the same input conditions.

3.5 THE NOT OPERATION

The NOT operation is unlike the OR or AND operations in that it can be performed on a single input variable. For example, if the variable A is subjected to the NOT operation, the result x can be expressed as

$$x = \overline{A}$$

where the overbar represents the NOT operation. This expression is read "x equals NOT A" or "x equals the *inverse* of A" or "x equals the *complement* of A." Each of these are in common usage and they all indicate that the logic value of $x = \overline{A}$ is *opposite to* the logic value of A. The truth table in Figure 3.9(a) clarifies this for the two cases $A = 0$ and $A = 1$. That is,

$$\overline{1} = 0 \qquad \text{because NOT 1 is 0}$$

and

$$\overline{0} = 1 \qquad \text{because NOT 0 is 1}$$

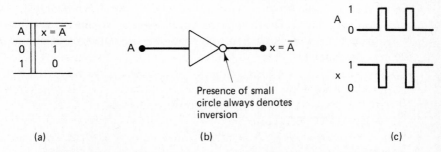

(a)

(b)

(c)

FIGURE 3.9 **(a) Truth table and (b) symbol for the NOT circuit.**

The NOT operation is also referred to as *inversion* or *complementation,* and these terms will be used interchangeably throughout the text. Although we will always use the overbar indicator to represent inversion, it is important to mention that another indicator for inversion is the prime symbol (′). That is,

$$A' = \overline{A}$$

Both should be recognized as indicating the inversion operation.

The NOT Circuit (INVERTER) Figure 3.9(b) shows the symbol for a NOT circuit, which is more commonly called an INVERTER. This circuit *always* has only a single input, and its output logic level is always opposite to the logic level of this input. Figure 3.9(c) shows how the INVERTER affects an input signal. It inverts (changes) the input signal at all points on the waveform.

Summary of Boolean Operations The rules for the OR, AND, and NOT operations may be summarized as follows:

OR	AND	NOT
$0 + 0 = 0$	$0 \cdot 0 = 0$	$\overline{0} = 1$
$0 + 1 = 1$	$0 \cdot 1 = 0$	$\overline{1} = 0$
$1 + 0 = 1$	$1 \cdot 0 = 0$	
$1 + 1 = 1$	$1 \cdot 1 = 1$	

REVIEW QUESTIONS

1. The output of the INVERTER of Figure 3.9 is connected to the input of a second INVERTER. Determine the output level of the second INVERTER for each level of input *A*. (*Ans.* In each case it is the same as *A*)
2. The output of the AND gate in Figure 3.6 is connected to the input of an INVERTER. Write the truth table showing the INVERTER output, *y*, for each combination of inputs *A* and *B*.

3.6 DESCRIBING LOGIC CIRCUITS ALGEBRAICALLY

Any logic circuit, no matter how complex, may be completely described using the Boolean operations previously defined, because the OR gate, AND gate, and NOT circuit are the basic building blocks of digital systems. For example, consider the circuit in Figure 3.10. This circuit has three inputs, *A*, *B*, and *C*, and a single output, *x*. Utilizing the Boolean expression for each gate, we can easily determine the expression for the output.

FIGURE 3.10   Logic circuit with its Boolean expression.

46   Chap. 3   Logic Gates and Boolean Algebra

FIGURE 3.11 **Logic circuit whose output expression requires parentheses.**

The expression for the AND-gate output is written $A \cdot B$. This AND output is connected as an input to the OR gate along with C, another input. The OR gate operates on its inputs such that its output is the OR sum of the inputs. Thus, we can express the OR output as $x = A \cdot B + C$. (This final expression could also be written as $x = C + A \cdot B$, since it does not matter which term of the OR sum is written first.)

Occasionally, there may be confusion as to which operation in an expression is performed first. The expression $A \cdot B + C$ can be interpreted in two different ways: (1) $A \cdot B$ is ORed with C, or (2) A is ANDed with the term $B + C$. To avoid this confusion, it will be understood that if an expression contains both AND and OR operations, the AND operations are performed first, unless there are *parentheses* in the expression, in which case the operation inside the parentheses is to be performed first. This is the same rule that is used in ordinary algebra to determine the order of operations.

To illustrate further, consider the circuit in Figure 3.11. The expression for the OR-gate output is simply $A + B$. This output serves as an input to the AND gate along with another input, C. Thus, we express the output of the AND gate as $x = (A + B) \cdot C$. Note the use of parentheses here to indicate that A and B are ORed *first,* before their OR sum is ANDed with C. Without the parentheses it would be interpreted *incorrectly,* since $A + B \cdot C$ means A is ORed with the product $B \cdot C$.

Circuits Containing INVERTERs Whenever an INVERTER is present in a logic-circuit diagram, its output expression is simply equal to the input expression with a bar over it. Figure 3.12 shows two examples using INVERTERs. In Figure 3.12(a) the input A is fed through an INVERTER, whose output is therefore $\bar{A}$. The INVERTER output is fed to an OR gate together with B, so the OR output is equal to $\bar{A} + B$. Note that the bar is only over the A, indicating that A is first inverted and then ORed with B.

In Figure 3.12(b) the output of the OR gate is equal to $A + B$ and is fed through an INVERTER. The INVERTER output is therefore equal to $\overline{(A + B)}$, since it inverts the *complete* input expression. Note that the bar covers the entire expression $(A + B)$. This is important because, as will be shown later, the expressions $\overline{(A + B)}$ and $(\bar{A} + \bar{B})$ are *not* equivalent. The expression $\overline{(A + B)}$ means that A is ORed with B and

(a)   (b)

FIGURE 3.12 **Circuits using INVERTERs.**

then their OR sum is inverted, whereas the expression $(\bar{A} + \bar{B})$ indicates that A is inverted and B is inverted and the results are then ORed together.

Figure 3.13 shows two more examples, which should be studied carefully. Note especially the use of *two* separate sets of parentheses in Figure 3.13(b). Also notice in Figure 3.13(a) that the input variable A is connected as input to two different gates.

_____ **REVIEW QUESTION**

1. In Figure 3.13(a) change each AND gate to an OR gate, and change the OR gate to an AND gate. Then write the expression for output x. (Ans. $x = \bar{A} + B + C + \bar{A \cdot D}$)
_____

3.7 EVALUATING LOGIC-CIRCUIT OUTPUTS

Once the Boolean expression for a circuit output is obtained, the logic level of the output can be determined for any values of circuit inputs. For example, suppose that we want to know the logic level of the output x for the circuit in Figure 3.13(a) for the case where $A = 0$, $B = 1$, $C = 1$, and $D = 1$. As in ordinary algebra, the value of x can be found by plugging the values of the variables into the expression and performing the indicated operations as follows:

$$x = \bar{A}BC(\overline{A + D})$$
$$= \bar{0} \cdot 1 \cdot 1 \cdot (\overline{0 + 1})$$
$$= 1 \cdot 1 \cdot 1 \cdot (\overline{0 + 1})$$
$$= 1 \cdot 1 \cdot 1 \cdot (\bar{1})$$
$$= 1 \cdot 1 \cdot 1 \cdot 0$$
$$= 0$$

(a)

(b)

FIGURE 3.13   More examples.

As another illustration, let us evaluate the output of the circuit in Figure 3.13(b) for $A = 0$, $B = 0$, $C = 1$, $D = 1$, and $E = 1$.

$$x = [D + \overline{(A + B)C}] \cdot E$$
$$= [1 + \overline{(0 + 0) \cdot 1}] \cdot 1$$
$$= [1 + \overline{0 \cdot 1}] \cdot 1$$
$$= [1 + \overline{0}] \cdot 1$$
$$= [1 + 1] \cdot 1$$
$$= 1 \cdot 1$$
$$= 1$$

As a general rule, the following rules must always be followed when evaluating a Boolean expression:

1. First, perform all inversions of single terms; that is, $\overline{0} = 1$ or $\overline{1} = 0$.
2. Then perform all operations within parentheses.
3. Perform an AND operation before an OR operation unless parentheses indicate otherwise.
4. If an expression has a bar over it, perform the operations of the expression first and then invert the result.

Determining Output Level from a Diagram   The output logic level for given input levels can also be determined directly from the circuit diagram *without* using the Boolean expression. This technique is often used by technicians during the troubleshooting or testing of a logic system since it also tells them what each gate output is supposed to be as well as the final output. To illustrate, the circuit of Figure 3.13(a) is redrawn in Figure 3.14 with the input levels $A = 0$, $B = 1$, $C = 1$, and $D = 1$. The procedure is to start from the inputs and to proceed through each INVERTER and gate, writing down each of their outputs in the process until the final output is reached.

In Figure 3.14, AND gate 1 has *all* three inputs at the 1 level because the INVERTER changes the $A = 0$ to a 1. This condition produces a 1 at the AND output since $1 \cdot 1 \cdot 1 = 1$. The OR gate has inputs of 1 and 0, which produces a 1 output since $1 + 0 = 1$. This 1 is inverted to 0 and applied to AND gate 2 along with the 1 from the first AND output. The 0 and 1 inputs to AND gate 2 produce an output of 0, because $0 \cdot 1 = 0$.

FIGURE 3.14 Determining output level from circuit diagram.

1. Use the expression for x to determine the output of the circuit in Figure 3.13(a) for the conditions $A = 0$, $B = 1$, $C = 1$, and $D = 0$. (*Ans.* $x = 1$)
2. Use the expression for x to determine the output of the circuit in Figure 3.13(b) for the conditions $A = B = E = 1$, $C = D = 0$. (*Ans.* $x = 1$)
3. Determine the answers to questions 1 and 2 by finding the logic levels present at each gate input and output as was done in Figure 3.14.

3.8 IMPLEMENTING CIRCUITS FROM BOOLEAN EXPRESSIONS

If the operation of a circuit is defined by a Boolean expression, a logic-circuit diagram can be implemented directly from that expression. For example, if we needed a circuit that was defined by $x = A \cdot B \cdot C$, we would immediately know that all that was needed was a three-input AND gate. If we needed a circuit that was defined by $x = A + \overline{B}$, we would use a two-input OR gate with an INVERTER on one of the inputs. The same reasoning used for these simple cases can be extended to more complex circuits.

Suppose that we wanted to construct a circuit whose output is $y = AC + B\overline{C} + \overline{A}BC$. This Boolean expression contains three terms (AC, $B\overline{C}$, $\overline{A}BC$), which are ORed together. This tells us that a three-input OR gate is required with inputs that are equal to AC, $B\overline{C}$, and $\overline{A}BC$, respectively. This is illustrated in Figure

(a)

(b)

FIGURE 3.15 Constructing a logic circuit from a Boolean expression.

3.15(a), where a three-input OR gate is drawn with inputs labeled as AC, $B\overline{C}$, and $\overline{A}BC$.

Each OR-gate input is an AND product term, which means that an AND gate with appropriate inputs can be used to generate each of these terms. This is shown in Figure 3.15(b), which is the final circuit diagram. Note the use of INVERTERs to produce the $\overline{A}$ and $\overline{C}$ terms required in the expression.

This same general approach can always be followed, although we shall find that there are some clever, more efficient techniques that can be employed. For now, however, this straightforward method will be used to minimize the number of new things that are to be learned.

EXAMPLE 3.5

Draw the circuit diagram that implements the expression $x = AB + \overline{B}C$.

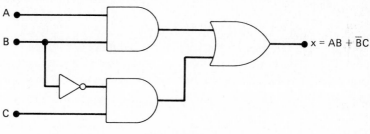

FIGURE 3.16

Solution: This expression indicates that the terms AB and $\overline{B}C$ are inputs to an OR gate, and each of these two terms is generated from a separate AND gate. The result is shown in Figure 3.16.

REVIEW QUESTION

1. Draw the circuit diagram that implements the expression $x = \overline{A}BC(\overline{A} + D)$ and compare it to Figure 3.13(a).

3.9 NOR GATES AND NAND GATES

Two other types of logic gates, NOR gates and NAND gates, are used extensively in digital circuitry. These gates actually combine the basic operations AND, OR, and NOT, which makes it relatively easy to describe them using the Boolean algebra operations learned previously.

The NOR Gate The symbol for a two-input NOR gate is shown in Figure 3.17(a). It is the same as the OR-gate symbol except that it has a small circle on the output. This small circle represents the inversion operation. Thus, the NOR gate operates like

		OR	NOR
A	B	$A + B$	$\overline{A + B}$
0	0	0	1
0	1	1	0
1	0	1	0
1	1	1	0

(c)

FIGURE 3.17 (a) NOR symbol; (b) equivalent circuit; (c) truth table.

an OR gate followed by an INVERTER, so that the circuits in Figure 3.17(a) and (b) are equivalent, and the output expression for the NOR gate is $x = \overline{A + B}$.

The truth table in Figure 3.17(c) shows that the NOR-gate output is the exact inverse of the OR-gate output for all possible input conditions. Whereas an OR-gate output goes HIGH when any input is HIGH, the NOR-gate output goes LOW when any input is HIGH. This same operation can be extended to NOR gates with more than two inputs.

EXAMPLE 3.6

Determine the waveform at the output of a NOR gate for the input waveforms shown in Figure 3.18.

FIGURE 3.18

Solution: There are several ways to determine the NOR output waveform. One way is to first find the OR output waveform and then invert it (change all 1s to 0s and vice versa). Another way utilizes the fact that a NOR-gate output will be HIGH *only* when all inputs are LOW. Thus you can examine the input waveforms, find those time intervals where they are all LOW, and make the NOR output HIGH for those intervals. The NOR output will be LOW for all other time intervals. The resultant output waveform is shown in the figure.

EXAMPLE 3.7

Determine the Boolean expression for a three-input NOR gate followed by an INVERTER.

Solution: Refer to Figure 3.19, where the circuit diagram is shown. The expression at the NOR output is $(\overline{A + B + C})$, which is then fed through an INVERTER to produce

$$x = \overline{(\overline{A + B + C})}$$

The presence of the double inversion signs indicates that the quantity $(A + B + C)$ has been inverted and then inverted again. It should be clear that this simply results in the expression $(A + B + C)$ being unchanged. That is,

$$x = \overline{(\overline{A + B + C})} = A + B + C$$

Whenever two inversion bars are over the same variable or quantity, they cancel each other out, as in the example above. However, in cases such as $\overline{A} + \overline{B}$ the inversion bars do not cancel. This is because the smaller inversion bars invert the single variables A and B, respectively, while the wider bar inverts the quantity $(\overline{A} + \overline{B})$. Thus, $\overline{\overline{A} + \overline{B}} \neq A + B$. Similarly, $\overline{\overline{A}\overline{B}} \neq AB$.

FIGURE 3.19

The NAND Gate The symbol for a two-input NAND gate is shown in Figure 3.20(a). It is the same as the AND-gate symbol except for the small circle on its output. Once again this small circle denotes the inversion operation. Thus, the NAND operates like an AND gate followed by an INVERTER, so that the circuits of Figure 3.20(a) and (b) are equivalent, and the output expression for the NAND gate is $x = \overline{AB}$.

FIGURE 3.20   (a) NAND symbol; (b) equivalent circuit; (c) truth table.

The truth table in Figure 3.20(c) shows that the NAND-gate output is the exact inverse of the AND gate for all possible input conditions. The AND output goes HIGH only when all inputs are HIGH, while the NAND output goes LOW only when all inputs are HIGH. This same characteristic is true of NAND gates having more than two inputs.

EXAMPLE 3.8

Determine the output waveform of a NAND gate having the inputs shown in Figure 3.21.

FIGURE 3.21

Solution: The output can be determined in several ways. One way is to first draw the output for an AND gate, and then invert it. Another way utilizes the fact that a NAND output will be LOW only when all inputs are HIGH. Thus, you can find those time intervals during which the inputs are all HIGH, and make the NAND output LOW for those intervals. The output will be HIGH at all other times.

EXAMPLE 3.9

Implement the logic circuit that has the expression $x = \overline{AB \cdot (\overline{C + D})}$ using only NOR and NAND gates.

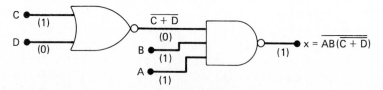

FIGURE 3.22

Solution: The $(\overline{C + D})$ term is the expression for the output of a NOR gate. This term is ANDed with A and B and the result is inverted; this, of course, is the NAND operation. Thus, the circuit is implemented as shown in Figure 3.22. Note that the NAND gate first ANDs the A, B, and $(\overline{C + D})$ terms, and then inverts the *complete* result.

EXAMPLE 3.10

Determine the output level in Figure 3.22 for $A = B = C = 1$ and $D = 0$.

Solution: In the first method we use the expression for x.

$$x = \overline{AB(\overline{C} + D)}$$
$$= \overline{1 \cdot 1 \cdot (\overline{1} + 0)}$$
$$= \overline{1 \cdot 1 \cdot (\overline{1})}$$
$$= \overline{1 \cdot 1 \cdot 0}$$
$$= \overline{0} = 1$$

In the second method we write down the input logic levels on the circuit diagram (shown in parentheses) and follow these levels through each gate to the final output. The NOR gate has inputs of 1 and 0 to produce an output of 0 (an OR would have produced an output of 1). The NAND gate thus has input levels of 0, 1, and 1 to produce an output of 1 (an AND would have produced an output of 0).

REVIEW QUESTIONS

1. What is the only set of input conditions that will produce a HIGH output from a three-input NOR gate?
2. What type of gate is equivalent to a NAND gate followed by an INVERTER? (*Ans.* An AND gate)
3. Change the NOR gate of Figure 3.22 to a NAND gate, and change the NAND to a NOR. What is the new expression for x? (*Ans.* $x = \overline{A + B + \overline{CD}}$)

3.10 BOOLEAN THEOREMS

We have seen how Boolean algebra can be used to help analyze a logic circuit and express its operation mathematically. We will continue our study of Boolean algebra by investigating the various Boolean theorems (rules) that can help us to simplify logic expressions and logic circuits. The first group of theorems is given in Figure 3.23. In each theorem, x is a logic variable that can be either a 0 or 1. Each theorem is accompanied by a logic-circuit diagram that demonstrates its validity.

Theorem (1) states that if any variable is ANDed with 0, the result has to be 0. This is easy to remember because the AND operation is just like ordinary multiplication, where we know that anything multiplied by 0 is 0. We also know that the output of an AND gate will be 0 whenever any input is 0, regardless of the level on the other input.

Theorem (2) is also obvious by comparison with ordinary multiplication.

Theorem (3) can be proved by trying each case. If $x = 0$, then $0 \cdot 0 = 0$; if $x = 1$, then $1 \cdot 1 = 1$. Thus, $x \cdot x = x$.

Theorem (4) can be proved in the same manner. However, it can also be

FIGURE 3.23 Single-variable theorems

(1) $x \cdot 0 = 0$

(2) $x \cdot 1 = x$

(3) $x \cdot x = x$

(4) $x \cdot \overline{x} = 0$

(5) $x + 0 = x$

(6) $x + 1 = 1$

(7) $x + x = x$

(8) $x + \overline{x} = 1$

reasoned that at any time either x or its inverse $\overline{x}$ has to be at the 0 level, so their AND product always has to be 0.

Theorem (5) is straightforward since 0 *added* to anything does not affect its value, either in regular addition or in OR addition.

Theorem (6) states that if any variable is ORed with 1, the result will always be 1. Checking this for both values of x: $0 + 1 = 1$ and $1 + 1 = 1$. Equivalently, we can remember that an OR-gate output will be 1 when *any* input is 1, regardless of the value of the other input.

Theorem (7) can be proved by checking for both values of x: $0 + 0 = 0$ and $1 + 1 = 1$.

Theorem (8) can be proved similarly, or we can just reason that at any time either x or $\overline{x}$ has to be at the 1 level so that we are always ORing a 0 and a 1, which always results in 1.

Before introducing any more theorems, it should be pointed out that in applying theorems (1)–(8) the variable x may actually represent an expression containing more than one variable. For example, if we have $A\overline{B}(\overline{A\overline{B}})$, we can invoke theorem (4) by letting $x = A\overline{B}$. Thus, we can say that $A\overline{B}(\overline{A\overline{B}}) = 0$. The same idea can be applied to the use of any of these theorems.

Multivariable Theorems   The theorems presented below involve more than one variable:

(9) $x + y = y + x$

(10) $x \cdot y = y \cdot x$

(11) $x + (y + z) = (x + y) + z = x + y + z$

(12) $x(yz) = (xy)z = xyz$

(13a) $x(y + z) = xy + xz$

(13b) $(w + x)(y + z) = wy + xy + wz + xz$

(14) $x + xy = x$

(15) $x + \overline{x}y = x + y$

Theorems (9) and (10) are called the *commutative laws*. These laws indicate that the order in which we OR or AND two variables is unimportant; the result is the same.

Theorems (11) and (12) are the *associative laws*, which state that we can group the variables in an AND expression or OR expression any way we want.

Theorem (13) is the *distributive law*, which states that an expression can be expanded by multiplying term by term just the same as in ordinary algebra. This theorem also indicates that we can factor an expression. That is, if we have a sum of two (or more) terms, each of which contains a common variable, the common variable can be factored out just like in ordinary algebra. For example, if we have the expression $A\overline{B}C + A\overline{B}\overline{C}$, we can factor out the $\overline{B}$ variable:

$$A\overline{B}C + A\overline{B}\overline{C} = \overline{B}(AC + A\overline{C})$$

As another example, consider the expression $ABC + ABD$. Here the two terms have the variables A and B in common, so $A \cdot B$ can be factored out of both terms. That is,

$$ABC + ABD = AB(C + D)$$

Theorems (9)–(13) are easy to remember and use since they are identical to those of ordinary algebra. Theorems (14) and (15), on the other hand, do not have any counterparts in ordinary algebra. Each can be proved by trying all possible cases for x and y. This is illustrated for theorem (14) as follows:

Case 1:   For $x = 0$, $y = 0$,

$$x + xy = x$$

$$0 + 0 \cdot 0 = 0$$

$$0 = 0$$

Case 2: For $x = 0$, $y = 1$,

$$x + xy = x$$
$$0 + 0 \cdot 1 = 0$$
$$0 + 0 = 0$$
$$0 = 0$$

Case 3: For $x = 1$, $y = 0$,

$$x + xy = x$$
$$1 + 1 \cdot 0 = 1$$
$$1 + 0 = 1$$
$$1 = 1$$

Case 4: For $x = 1$, $y = 1$,

$$x + xy = x$$
$$1 + 1 \cdot 1 = 1$$
$$1 + 1 = 1$$
$$1 = 1$$

Theorem (14) can also be proved by factoring and using theorems (6) and (2) as follows:

$$x + xy = x(1 + y)$$
$$= x \cdot 1 \quad \text{[using theorem (6)]}$$
$$= x \quad \text{[using theorem (2)]}$$

All of these Boolean theorems can be useful in simplifying a logic expression—that is, in reducing the number of terms in the expression. When this is done, the reduced expression will produce a circuit that is less complex than the one which the original expression would have produced. A good portion of the next chapter will be devoted to the process of circuit simplification. For now, the following examples will serve to illustrate how the Boolean theorems can be applied.

EXAMPLE 3.11

Simplify the expression $y = A\bar{B}D + A\bar{B}\bar{D}$.

Solution: Factor out the common variables $A\bar{B}$ using theorem (13):

$$y = A\bar{B}(D + \bar{D})$$

Using theorem (8), the term in parentheses is equivalent to 1. Thus,

$$y = A\overline{B} \cdot 1$$
$$y = A\overline{B} \quad \text{[using theorem (2)]}$$

EXAMPLE 3.12

Simplify $z = (\overline{A} + B)(A + B)$.

Solution: The expression can be expanded by multiplying out the terms [theorem (13)].

$$z = \overline{A} \cdot A + \overline{A} \cdot B + B \cdot A + B \cdot B$$

Invoking theorem (4), the term $\overline{A} \cdot A = 0$. Also, $B \cdot B = B$ [theorem (3)].

$$z = 0 + \overline{A} \cdot B + B \cdot A + B = \overline{A}B + AB + B$$

Factoring out the variable B [theorem (13)], we have

$$z = B(\overline{A} + A + 1)$$

Finally, using theorem (6),

$$z = B$$

EXAMPLE 3.13

Simplify $x = ACD + \overline{A}BCD$.

Solution: Factoring out the common variables CD, we have

$$x = CD(A + \overline{A}B)$$

Utilizing theorem (15), we can replace $A + \overline{A}B$ by $A + B$, so

$$x = CD(A + B)$$
$$= ACD + BCD$$

REVIEW QUESTIONS

1. Use theorems (13) and (14) to simplify $y = A\overline{C} + AB\overline{C}$. (*Ans.* $y = A\overline{C}$)
2. Use theorems (13) and (8) to simplify $y = \overline{A}\,\overline{B}C\overline{D} + \overline{A}\,\overline{B}\,\overline{C}\,\overline{D}$. (*Ans.* $y = \overline{A}\,\overline{B}\,\overline{D}$)

3.11 DEMORGAN'S THEOREMS

Two of the most important theorems of Boolean algebra were contributed by a great mathematician named DeMorgan. *DeMorgan's theorems* are extremely useful in simplifying expressions in which a product or sum of variables is inverted. The two theorems are:

(16) $(\overline{x + y}) = \overline{x} \cdot \overline{y}$

(17) $\quad (\overline{x \cdot y}) = \overline{x} + \overline{y}$

Theorem (16) says that when the OR sum of two variables is inverted, this is the same as inverting each variable individually and then ANDing these inverted variables. Theorem (17) says that when the AND product of two variables is inverted, this is the same as inverting each variable individually and then ORing them. Each of DeMorgan's theorems can be readily proven by checking for all possible combinations of x and y. This will be left as an end-of-chapter exercise.

Although these theorems have been stated in terms of single variables x and y, they are equally valid for situations where x and/or y are expressions that contain more than one variable. For example, let's apply them to the expression $(A\overline{B} + C)$ as shown below:

$$\overline{(A\overline{B} + C)} = \overline{(A\overline{B})} \cdot \overline{C}$$

Note that here we treated $A\overline{B}$ as x and C as y. The result can be further simplified since we have a product $A\overline{B}$ that is inverted. Using theorem (17), the expression becomes

$$\overline{A\overline{B}} \cdot \overline{C} = (\overline{A} + \overline{\overline{B}}) \cdot \overline{C}$$

Notice that we can replace $\overline{\overline{B}}$ by B, so we finally have

$$(\overline{A} + B) \cdot \overline{C} = \overline{A}\,\overline{C} + B\overline{C}$$

This final result contains only inverter signs that invert a single variable.

EXAMPLE 3.14

Simplify the expression $z = \overline{(\overline{A} + C) \cdot (B + \overline{D})}$.

Solution: Using theorem (17), we can rewrite this as

$$z = \overline{(\overline{A} + C)} + \overline{(B + \overline{D})}$$

We can think of this as breaking the large inverter sign down the middle and changing the AND sign ($\cdot$) to an OR sign ($+$). Now the term $\overline{(\overline{A} + C)}$ can be simplified by applying theorem (16). Likewise, $\overline{(B + \overline{D})}$ can be simplified.

$$z = \overline{(\overline{A} + C)} + \overline{(B + \overline{D})}$$
$$= (\overline{\overline{A}} \cdot \overline{C}) + \overline{B} \cdot \overline{\overline{D}}$$

Here we have broken the larger inverter signs down the middle and replaced the ($+$) with a ($\cdot$). Canceling out the double inversions, we have finally

$$z = A\overline{C} + \overline{B}D$$

Example 3.14 points out that when using DeMorgan's theorems to reduce an expression, an inverter sign may be broken at any point in the expression and the operator at that point in the expression is changed to its opposite ($+$ is changed to $\cdot$, and vice versa). This procedure is continued until the expression is reduced to one in which only single variables are inverted. Two more examples are given below.

1. $z = \overline{A + \overline{B} \cdot C}$

$= \overline{A} \cdot (\overline{\overline{B} \cdot C})$

$= \overline{A} \cdot (\overline{\overline{B}} + \overline{C})$

$= \overline{A} \cdot (B + \overline{C})$

2. $q = \overline{(A + BC) \cdot (D + EF)}$

$= \overline{(A + BC)} + \overline{(D + EF)}$

$= (\overline{A} \cdot \overline{BC}) + (\overline{D} \cdot \overline{EF})$

$= [\overline{A} \cdot (\overline{B} + \overline{C})] + [\overline{D} \cdot (\overline{E} + \overline{F})]$

$= \overline{A}\overline{B} + \overline{A}\overline{C} + \overline{D}\overline{E} + \overline{D}\overline{F}$

DeMorgan's theorems are easily extended to more than two variables. For example, it can be proved that

$$\overline{x + y + z} = \overline{x} \cdot \overline{y} \cdot \overline{z}$$

$$\overline{x \cdot y \cdot z} = \overline{x} + \overline{y} + \overline{z}$$

and so on for more variables. Again, realize that any one of these variables can be an expression rather than a single variable.

Implications of DeMorgan's Theorems Let us examine these theorems (16) and (17) from the standpoint of logic circuits. First, consider theorem (16),

$$\overline{x + y} = \overline{x} \cdot \overline{y}$$

The left-hand side of the equation can be viewed as the output of a NOR gate whose inputs are x and y. The right-hand side of the equation, on the other hand, is the result of first inverting both x and y and then putting them through an AND gate. These two representations are equivalent and are illustrated in Figure 3.24(a). What this means is that an AND gate with INVERTERs on each of its inputs is equivalent to a NOR gate. In fact, both representations are used to represent the NOR function. When the AND gate with inverted inputs is used to represent the NOR function, it is usually drawn as shown in Figure 3.24(b), where the small circles on the inputs represent the INVERTERs.

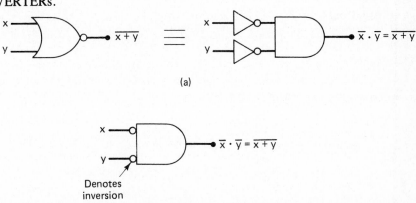

FIGURE 3.24 (a) Equivalent circuits implied by theorem (16); (b) alternative symbol for the NOR function.

(a)

$$\bar{x} + \bar{y} = \overline{xy}$$

(b)

FIGURE 3.25 (a) Equivalent circuits implied by theorem (17); (b) alternative symbol for the NAND function.

Now consider theorem (17),

$$\overline{x \cdot y} = \bar{x} + \bar{y}$$

The left side of the equation can be implemented by a NAND gate with inputs x and y. The right side can be implemented by first inverting inputs x and y and then putting them through an OR gate. These two equivalent representations are shown in Figure 3.25(a). The OR gate with INVERTERs on each of its inputs is equivalent to the NAND gate. In fact, both representations are used to represent the NAND function. When the OR gate with inverted inputs is used to represent the NAND function, it is usually drawn as shown in Figure 3.25(b), where the circles again represent inversion.

EXAMPLE 3.15

Implement a circuit having the output expression $Z = \bar{A} + \bar{B} + C$ using a NAND gate and an INVERTER.

Solution: $z = \bar{A} + \bar{B} + C$ can be written as

$$z = \overline{\overline{\bar{A} \cdot \bar{B} \cdot \bar{C}}} = \overline{A \cdot B \cdot \bar{C}}$$

by invoking theorem (17) and by canceling the double inversions over A and B. In this new form, it is easy to see how a NAND gate can be used to produce z. The circuit is drawn in Figure 3.26.

$$z = \overline{A \cdot B \cdot \bar{C}} = \bar{A} + \bar{B} + \overline{\bar{C}} = \bar{A} + \bar{B} + C$$

FIGURE 3.26 Example 3.15

1. Use DeMorgan's theorems to convert the expression $z = \overline{(A + B) \cdot \overline{C}}$ to one that has only single-variable inversions. (*Ans.* $z = \overline{A}\overline{B} + C$)
2. Repeat question 1 for the expression $y = \overline{R\overline{S}T + \overline{Q}}$. (*Ans.* $y = [\overline{R} + S + \overline{T}] \cdot Q$)
3. Implement a circuit having output expression $z = \overline{A}BC$ using only a NOR gate and an INVERTER. (*Ans.* Same as Figure 3.25 except NAND gate is replaced by NOR gate)

3.12 UNIVERSALITY OF NAND GATES AND NOR GATES

All Boolean expressions consist of various combinations of the basic operations of OR, AND, and INVERT. Therefore, any expression can be implemented using OR gates, AND gates, and INVERTERs. It is possible, however, to implement any logic expression using *only* NAND gates and no other type of gate. This is because NAND gates, in the proper combination, can be used to perform each of the Boolean operations OR, AND, and INVERT. This is demonstrated in Figure 3.27.

First, in Figure 3.27(a) we have a two-input NAND gate whose inputs are purposely connected together so that the variable A is applied to both. In this configuration, the NAND simply acts as an INVERTER, since its output is $x = \overline{A \cdot A} = \overline{A}$.

In Figure 3.27(b) we have two NAND gates connected so that the AND operation is performed. NAND gate 2 is used as an INVERTER to change $\overline{AB}$ to $\overline{\overline{AB}} = AB$, which is the desired AND function.

The OR operation can be implemented using NAND gates connected as shown in Figure 3.27(c). Here NAND gates 1 and 2 are used as INVERTERs to invert the inputs, so the final output is $x = \overline{\overline{A} \cdot \overline{B}}$, which can be simplified to $x = A + B$ using DeMorgan's theorem.

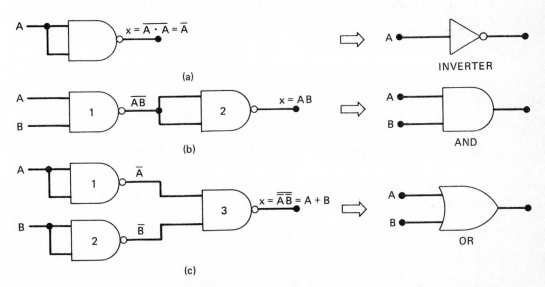

FIGURE 3.27 NAND gates can be used to implement any Boolean function.

In a similar manner, it can be shown that NOR gates can be arranged to implement any of the Boolean operations. This is illustrated in Figure 3.28. Part (a) shows that a NOR gate with its inputs connected together behaves as an INVERTER, since the output is $x = \overline{A + A} = \overline{A}$.

In Figure 3.28(b) two NOR gates are arranged so that the OR operation is performed. NOR gate 2 is used as an INVERTER to change $\overline{A + B}$ to $\overline{\overline{A + B}} = A + B$, which is the desired OR function.

The AND operation can be implemented with NOR gates as shown in Figure 3.28(c). Here NOR gates 1 and 2 are used as INVERTERs to invert the inputs, so the final output is $x = \overline{\overline{A} + \overline{B}}$, which can be simplified to $x = A \cdot B$ by use of DeMorgan's theorem.

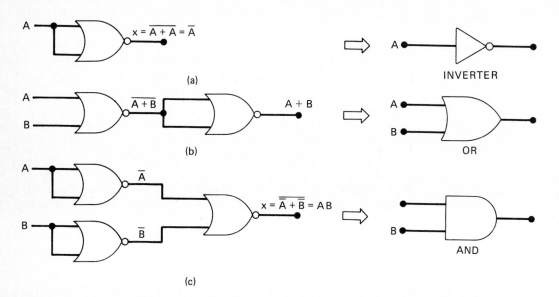

FIGURE 3.28   NOR gates used to implement Boolean operations.

Since any of the Boolean operations can be implemented using only NAND gates, then any logic circuit can be constructed using only NAND gates. The same is true for NOR gates. This characteristic of NAND and NOR gates can be very useful in logic-circuit design as the following example illustrates.

EXAMPLE 3.16

A logic-circuit designer has to implement a circuit that satisfies the expression $x = AB + CD$ using the minimum number of ICs. He has the TTL ICs shown in Figure 3.29 available for his use. Each of these ICs is a *quad,* which means that it contains *four* identical two-input gates on one chip.

The straightforward method for implementing the given expression uses two

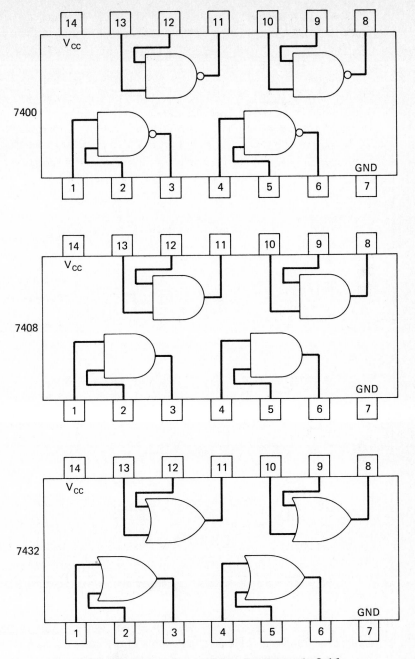

FIGURE 3.29   ICs available for Example 3.16.

FIGURE 3.30 Possible implementations for Example 3.16.

AND gates and an OR gate as shown in Figure 3.30(a). This implementation requires the 7408 and 7432 ICs, and some of the gates on each IC are wasted since they are not used.

Another implementation can be accomplished by taking the circuit of Figure 3.30(a) and replacing each AND gate and OR gate by its equivalent NAND-gate implementation from Figure 3.27. The result is shown in Figure 3.30(b).

At first glance this new circuit looks like it requires seven NAND gates. However, NAND gates 3 and 5 are connected as INVERTERs in series and can be eliminated from the circuit since they perform a double inversion of the signal out of NAND gate 1. Similarly, NAND gates 4 and 6 can be eliminated. The final circuit, after eliminating the double INVERTERs, is drawn in Figure 3.30(c).

This final circuit is more efficient than the one in Figure 3.30(a) because it uses three two-input NAND gates that can be implemented from one IC, the 7400.

REVIEW QUESTIONS

1. How many different ways do we now have to implement the inversion operation in a logic circuit? (*Ans.* Three)
2. Implement the expression $x = (A + B)(C + D)$ using OR and AND gates. Then implement the expression using only NOR gates by converting each OR and AND gate to its NOR implementation from Figure 3.28. Which circuit is more efficient? (*Ans.* The NOR circuit is more efficient because it can be implemented from a single IC, the 7402, which is a QUAD two-input NOR chip)
3. Write the output expression for the circuit of Figure 3.30(c) and use DeMorgan's theorems to show that it is equivalent to the expression for the circuit of Figure 3.30(a).

3.13 ALTERNATE LOGIC-GATE REPRESENTATIONS

We have introduced the five basic logic gates (AND, OR, INVERTER, NAND, and NOR) and the standard symbols used to represent them on logic-circuit diagrams. Although you may find that many circuit diagrams still use these standard symbols exclusively, it has become increasingly more common to find circuit diagrams that utilize an alternate set of symbols *in addition* to the standard symbols.

Before discussing the reasons for using an alternate symbol for a logic gate, we will present the alternate symbols for each gate and show that they are equivalent to the standard symbols. Refer to Figure 3.31. The left side of the illustration shows the standard symbol for each logic gate, and the right side shows the alternate symbol. The alternate symbol for each gate is obtained from the standard symbol by doing the following:

1. Invert each input and output of the standard symbol. This is done by adding bubbles (small circles) on input and output lines that do not have bubbles, and by removing bubbles that are already there.
2. Change the operation symbol from AND to OR, or from OR to AND. (In the special case of the INVERTER, the operation symbol is not changed.)

For example, the standard NAND symbol is an AND symbol with a bubble on its output. Following the steps outlined above, remove the bubble from the output, and add a bubble to each input. Then change the AND symbol to an OR symbol. The result is an OR symbol with bubbles on its inputs.

We can easily prove that this alternate symbol is equivalent to the standard symbol by using DeMorgan's theorems and recalling that the bubble represents an inversion operation. The output expression from the standard NAND symbol is

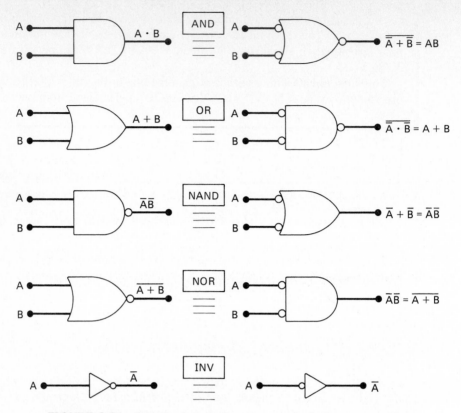

FIGURE 3.31 Standard and alternate symbols for various logic gates and inverter.

$\overline{AB} = \overline{A} + \overline{B}$, which is the same as the output expression for the alternate symbol. This same procedure can be followed for each pair of symbols in Figure 3.32.

Several points should be stressed regarding the logic symbol equivalences:

1. The equivalences are valid for gates with *any* number of inputs.
2. None of the standard symbols have bubbles on their inputs, and all the alternate symbols do.
3. The standard and alternate symbols for each gate represent the same physical circuit; there is no difference in the circuits represented by the two symbols.
4. NAND and NOR gates are inverting gates, so both the standard and alternate symbols for each will have a bubble on *either* the input or the output. AND and OR gates are *noninverting* gates, so both the standard and alternate symbols for each will have bubbles on *both* inputs and output.

Logic-Symbol Interpretation   Each of the logic-gate symbols of Figure 3.31 provides a unique interpretation of how the gate operates. Before we can demonstrate these interpretations, we must first establish the concept of *active* logic levels.

When an input or output line on a logic gate symbol has *no bubble* on it, that line is said to be active-HIGH. When an input or output line *does* have a *bubble* on

Output goes LOW only when *all* inputs are HIGH.

A
B
active-HIGH

$\overline{AB}$

LOW state is the *active* state.

(a)

Output is HIGH when *any* input is LOW.

A
B
active-LOW

$\overline{A} + \overline{B} = \overline{AB}$

HIGH state is the *active* state.

(b)

FIGURE 3.32 Interpretation of the two NAND gate symbols.

it, that line is said to be active-LOW. The presence or absence of a bubble, then, determines the active-HIGH/active-LOW status of a gate's inputs and output, and is used to interpret the gate operation.

To illustrate, Figure 3.32(a) shows the standard symbol for a NAND gate. The standard symbol has a bubble on its output and no bubbles on its inputs. Thus, it has an active-LOW output and active-HIGH inputs. The logic operation represented by this symbol can therefore be interpreted as follows:

The output goes LOW only when all the inputs are HIGH

Note that this says that the output will go to its active state only when *all* the inputs are in their active states. The word "all" is used because of the AND symbol.

The alternate symbol for a NAND gate shown in Figure 3.32(b) has an active-HIGH output and active-LOW inputs so its operation can be stated as

The output goes HIGH only when any input is LOW

Again, this says that the output will be in its active state whenever *any* of the inputs are in their active states. The word "any" is used because of the OR symbol.

With a little thought, it can be seen that the two interpretations for the NAND symbols in Figure 3.32 are different ways of saying the same thing.

EXAMPLE 3.17

Give the interpretation of the two OR-gate symbols.

Solution: The results are shown in Figure 3.33. Note again how the word *any* is used when the gate symbol includes an OR symbol and the word *all* is used when it includes an AND symbol.

FIGURE 3.33   Interpretation of the two OR-gate symbols.

SUMMARY

At this point you are probably wondering why there is a need to have two different symbols and interpretations for each logic gate. Hopefully, the reasons will become clear after reading the next section. For now, let us summarize the important points concerning the logic-gate representations.

1. To obtain the alternate symbol for a logic gate, take the standard symbol and change its operation symbol (OR to AND, or AND to OR) and change the bubbles on both inputs and output (that is, delete bubbles that are present, and add bubbles where there are none).

2. To interpret the logic-gate operation, first note which logic state, 0 or 1, is the active state for the inputs, and which is the active state for the output. Then, realize that the output's active state is produced by having *all* the inputs in their active state (if an AND symbol is used), or by having *any* of the inputs in their active state (if an OR symbol is used).

REVIEW QUESTIONS

1. Write the interpretation of the operation performed by the standard NOR-gate symbol in Figure 3.31. (*Ans.* The output goes LOW only when any input is HIGH.)
2. Repeat for the alternate NOR-gate symbol. (*Ans.* The output goes HIGH only when all inputs are LOW)
3. Repeat for the alternate AND-gate symbol. (*Ans.* The output goes LOW when any input is LOW)
4. Repeat for the standard AND-gate symbol. (*Ans.* The output goes HIGH only when all inputs are HIGH)

3.14   WHICH GATE REPRESENTATION TO USE

Some logic-circuit designers and many textbooks use only the standard logic-gate symbols in their circuit schematics. While this practice is not incorrect, it does nothing to make the circuit operation easier to follow. Proper use of the alternate gate symbols

A	B	C	D	Z
0	0	0	0	0
0	0	0	1	0
0	0	1	0	0
0	0	1	1	1
0	1	0	0	0
0	1	0	1	0
0	1	1	0	0
0	1	1	1	1
1	0	0	0	0
1	0	0	1	0
1	0	1	0	0
1	0	1	1	1
1	1	0	0	1
1	1	0	1	1
1	1	1	0	1
1	1	1	1	1

(d)

FIGURE 3.34 **(a) Original circuit using standard NAND symbols;**
(b) equivalent representation where output z is active-HIGH;
(c) equivalent representation where output z is active-LOW;
(d) truth table.

in the circuit diagram can make the circuit operation much clearer. This can be illustrated by considering the example shown in Figure 3.34.

The circuit in Figure 3.34(a) contains three NAND gates connected to produce an output Z that depends on inputs A, B, C, D. The circuit diagram uses the standard symbol for each of the NAND gates. While this diagram is logically correct, it does not facilitate any understanding of how the circuit functions. The improved circuit representations given in Figure 3.34(b) and (c), however, can be analyzed more easily to determine the circuit operation.

The representation of Figure 3.34(b) is obtained from the original circuit diagram by replacing NAND gate 3 with its alternate symbol. In this diagram output Z is taken from a NAND-gate symbol that has an active-HIGH output. Thus, we can say that Z will go HIGH when either X or Y is LOW. Now, since X and Y appear at the output of NAND symbols having active-LOW outputs, we can say that X will go LOW only if $A = B = 1$, and Y will go LOW only if $C = D = 1$. Putting this all together, we can describe the circuit operation as follows:

Output Z will go HIGH whenever either $A = B = 1$ or $C = D = 1$

This description can be translated to truth-table form by setting $Z = 1$ for those cases where $A = B = 1$, and for those cases where $C = D = 1$. For all other cases, Z is made a 0. The resultant truth table is shown in Figure 3.34(d).

The representation of Figure 3.34(c) is obtained from the original circuit diagram by replacing NAND gates 1 and 2 by their alternate symbols. In this equivalent representation the Z output is taken from a NAND gate that has an active-LOW output. Thus, we can say that Z will go LOW only when $X = Y = 1$. Since X and Y are active-HIGH outputs, we can say that X will be HIGH when either A or B is LOW, and Y will be HIGH when either C or D is LOW. Putting this all together, we can describe the circuit operation as follows:

Output Z will go LOW only when A or B is LOW and C or D is LOW

This description can be translated to truth-table form by making $Z = 0$ for all cases where at least one of the A or B inputs is LOW at the same time that at least one of the C or D inputs is LOW. For all other cases, Z is made a 1. The resultant truth table is the same as that obtained for the circuit diagram of Figure 3.34(b).

Which Circuit Diagram Should Be Used?   The answer to this question depends on the particular function being performed by the circuit output. If the circuit is being used to cause some action (e.g., turn ON a device, activate another logic circuit) when output Z goes to the 1 state, then we say that Z is active-HIGH, and the circuit diagram of Figure 3.34(b) should be used. On the other hand, if the circuit is being used to cause some action when Z goes to the 0 state, then Z is active-LOW and the diagram of Figure 3.34(c) should be used.

Of course, there will be situations where *both* output states are used to produce different actions and either one can be considered to be the active state. For these cases, either circuit representation can be used.

Bubble Placement   Refer to the circuit representation of Figure 3.34(b) and note that the symbols for NAND gates 1 and 2 were chosen to have active-LOW outputs to match the active-LOW inputs of NAND gate 3. Refer to the circuit representation of Figure 3.34(c) and note that the symbols for NAND gates 1 and 2 were chosen to have active-HIGH outputs to match the active-HIGH inputs of NAND gate 3. This leads to the following general rule for preparing logic-circuit schematics:

Whenever possible, connect bubble outputs to bubble inputs, and nonbubble outputs to nonbubble inputs

The following examples will show how this rule can be applied.

EXAMPLE 3.18

The logic circuit in Figure 3.35(a) is being used to activate an ALARM when its output Z goes HIGH. Modify the circuit diagram so that it more effectively represents the circuit operation.

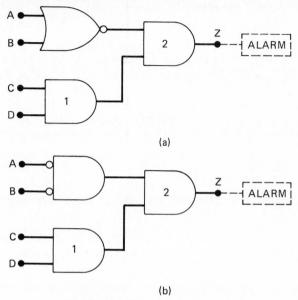

(a)

(b)

FIGURE 3.35   Example 3.18

Solution: Since $Z = 1$ will activate the ALARM, Z is active-HIGH. Thus, the AND-gate 2 symbol does not have to be changed. The NOR-gate symbol should be changed to the alternate symbol with a nonbubble (active-HIGH) output to match the nonbubble input of AND gate 2. The result is shown in Figure 3.35(b).

EXAMPLE 3.19

When the output of the logic circuit in Figure 3.36(a) goes LOW, it activates another logic circuit. Modify the circuit diagram to more effectively represent the circuit operation.

Solution: Since Z is active-LOW, the symbol for OR gate 2 has to be changed to its alternate symbol as shown in Figure 3.36(b). The new OR-gate 2 symbol has bubble inputs, so the AND-gate and OR-gate 1 symbols have to be changed to bubbled outputs as shown in Figure 3.36(b). The INVERTER already has a bubble output.

FIGURE 3.36   Example 3.19

Analyzing Circuits   When a logic-circuit schematic is drawn using the rules we followed in these examples, it is much easier for an engineer or technician (or student) to follow the signal flow through the circuit and to determine the input conditions that are needed to activate the output. This will be illustrated in the following examples—which, incidentally, use circuit diagrams taken from the logic schematics of an actual microcomputer.

EXAMPLE 3.20

The logic circuit in Figure 3.37 generates an output, *MEM,* that is used to activate the memory ICs in a particular microcomputer. Determine the input conditions necessary to activate *MEM.*

Solution: One way to do this would be to write the expression for *MEM* in terms of the inputs *RD, ROM-A, ROM-B,* and *RAM,* and to evaluate it for the 16 possible combinations of these inputs. While this method would work, it would require a lot more work than is necessary.

A more efficient method is to interpret the circuit diagram using the ideas we have been developing in the last two sections. These are the steps:

FIGURE 3.37   Example 3.20

1. *MEM* is active-LOW, and it will go LOW only when X and Y are HIGH.
2. X will be HIGH only when $RD = 0$.
3. Y will be HIGH when either W or V is HIGH.
4. V will be HIGH when $RAM = 0$.
5. W will be HIGH when either *ROM-A* or *ROM-B* = 0.
6. Putting this all together, *MEM* will go LOW only when $RD = 0$ *and* at least one of the three inputs *ROM-A*, *ROM-B*, or *RAM* is LOW.

EXAMPLE 3.21

The logic circuit in Figure 3.38 is used to control the motor on a cassette recorder when the microcomputer is sending data to or receiving data from the recorder. The circuit will turn on the motor when $CASS = 1$. Determine the input conditions necessary to turn on the motor.

FIGURE 3.38   Example 3.21

Solution: Once again we will interpret the diagram in a step-by-step fashion:

1. *CASS* is active-HIGH, and it will go HIGH only when $X = Y = 0$.
2. X will be LOW when either *IN* or *OUT* is HIGH.
3. Y will be LOW only when $W = 0$ and $A_0 = 0$.
4. W will be LOW only when A_1–A_7 are all HIGH.
5. Putting this all together, *CASS* will be HIGH when $A_1 = A_2 = A_3 = A_4 = A_5 = A_6 = A_7 = 1$ and $A_0 = 0$, and either IN or OUT or both are 1.

Note the strange symbol for the eight-input NAND gate (7430) and the fact that the signal A_7 is connected to two of the NAND inputs.

SUMMARY

The examples in this section have shown how to draw a logic-circuit diagram so that the circuit operation may be easily determined. By choosing the proper symbol for each logic gate, we may quickly analyze the circuit diagram to determine the input conditions required to activate the output.

We will use this method for many of the circuit diagrams presented throughout the text. In some cases, however, for various reasons we will use only standard gate symbols in the circuit diagram.

REVIEW QUESTIONS

1. Use the method of Examples 3.20 and 3.21 to determine the input conditions needed to activate the output of the circuit in Figure 3.35(b). (*Ans. Z* will go HIGH when $A = B = 0$ and $C = D = 1$)
2. Repeat for the circuit of Figure 3.36(b). (*Ans. Z* will go LOW when $A = B = 0$, $E = 1$, and either C or D or both are LOW)
3. How many NAND gates are in Figure 3.37? (*Ans.* 2)
4. How many NOR gates are in Figure 3.38? (*Ans.* 2)

PROBLEMS

3.1 Draw the output waveform for the circuit of Figure 3.39.

FIGURE 3.39

3.2 Suppose that the A input in Figure 3.39 is unintentionally shorted to ground (i.e., $A = 0$). Draw the resulting output waveform.

3.3 Suppose that the A input in Figure 3.39 is unintentionally shorted to the $+5$ V supply line (i.e., $A = 1$). Draw the resulting output waveform.

3.4 Read the statements below concerning an OR gate. At first they may appear to be valid, but after some thought you should realize that neither one is *always* true. Prove this by showing a specific example to refute each statement.
   (a) If the output waveform from an OR gate is the same as the waveform at one of its inputs, then the other input is being held permanently LOW.
   (b) If the output waveform from an OR gate is always HIGH, then one of its inputs is being held permanently HIGH.

3.5 How many different sets of input conditions will produce a HIGH output from a five-output OR gate?

3.6 Change the OR gate in Figure 3.39 to an AND gate.
   (a) Draw the output waveform.
   (b) Draw the output waveform if the A input is permanently shorted to ground.
   (c) Draw the output waveform if A is permanently shorted to $+5$ V.

3.7 Refer to Figure 3.4. Modify the circuit so that the alarm is to be activated only when the pressure and temperature exceed their maximum limits at the same time.

3.8 Change the OR gate in Figure 3.5 to an AND gate and draw the output waveform.

3.9 Add an INVERTER to the output of the OR gate from Figure 3.39. Draw the waveform at the INVERTER output.

3.10 Write the Boolean expression for the output x in Figure 3.40(a). Determine the value of x for all possible input conditions and list them in a truth table.

(a)

(b)

FIGURE 3.40

3.11 Repeat Problem 3.10 for the circuit in Figure 3.40(b).

3.12 Change each OR to an AND, and each AND to an OR in Figure 3.13(b). Then write the expression for the output.

3.13 Determine the complete truth table for the circuit of Figure 3.14 by finding the logic levels present at each gate output for each of the 16 possible combinations of input levels.

3.14 For each of the following expressions, construct the corresponding logic circuit, using AND and OR gates and INVERTERs.
 (a) $x = \overline{AB(C + D)}$
 (b) $z = (\overline{A + B + \overline{CD\overline{E}}}) + \overline{B}C\overline{D}$
 (c) $y = (\overline{M + N} + \overline{P}Q)$

3.15 Apply the input waveforms of Figure 3.41 to a NOR gate and draw the output waveform.

3.16 Repeat Problem 3.15 with the C input held permanently LOW. Then repeat for C held permanently HIGH.

FIGURE 3.41

3.17 Repeat Problem 3.15 for a NAND gate.

3.18 Repeat Problem 3.16 for a NAND gate.

3.19 Write the output expression for the circuit of Figure 3.42.

3.20 Determine the complete truth table for the circuit of Figure 3.42.

3.21 Modify the circuits that were constructed in Problem 3.14 so that NAND gates and NOR gates are used wherever appropriate.

3.22 Prove theorem (15) by trying all possible cases.

3.23 Simplify the following expression using theorems (13b), (3) and (4).

$$x = (M + N)(\overline{M} + P)(\overline{N} + \overline{P})$$

3.24 Simplify the following expression using theorems (13a), (8), and (6).

$$z = \overline{A}B\overline{C} + AB\overline{C} + B\overline{C}D$$

3.25 Prove DeMorgan's theorems by trying all possible cases.

3.26 Simplify each of the following expressions using DeMorgan's theorems.

(a) $\overline{\overline{AB\overline{C}}}$ (b) $\overline{\overline{A} + \overline{B}C}$ (c) $\overline{AB\overline{CD}}$ (d) $A\overline{(B + \overline{C})D}$

3.27 Use DeMorgan's theorems to simplify the expression for the output of Figure 3.42.

3.28 Convert the circuit of Figure 3.40(b) to one using only NAND gates. Then write the output expression for the new circuit, simplify it using DeMorgan's theorems, and compare it to the expression for the original circuit.

3.29 Convert the circuit of Figure 3.40(a) to one using only NOR gates. Then write the expression for the new circuit, simplify it using DeMorgan's theorems, and compare it to the expression for the original circuit.

3.30 Show how a two-input NAND gate can be constructed from a two-input NOR gate.

3.31 Show how a two-input NOR gate can be constructed from a two-input NAND gate.

3.32 Redraw the circuit diagram of Figure 3.42 using the alternate symbols for each gate and INVERTER.

3.33 For each statement below, draw the corresponding logic-gate symbol, and indicate the type of gate.
(a) A HIGH output occurs only when all three inputs are LOW.

FIGURE 3.42

78   Chap. 3   Logic Gates and Boolean Algebra

FIGURE 3.43

    (b) A LOW output occurs when any of the four inputs is LOW.

    (c) A LOW output occurs only when all eight inputs are HIGH.

3.34 The output of the circuit of Figure 3.42 is supposed to turn on an indicator lamp when it goes LOW.

    (a) Modify the circuit diagram so that it more effectively represents the circuit operation.

    (b) Use the new circuit diagram to determine the input conditions needed to activate the output. Compare the results to the truth table obtained in Problem 3.20.

3.35 Refer to Figure 3.35(b). Assume that Z is supposed to be active-LOW rather than active-HIGH, and modify the circuit diagram accordingly. Then use the new circuit diagram to determine the input conditions needed to produce $Z = 0$.

3.36 Modify the circuit of Figure 3.38 so that $A_1 = 0$ is needed to produce $CASS = 1$ instead of $A_1 = 1$.

3.37 Determine the input conditions needed to cause the output in Figure 3.43 to go to its active state.

3.38 Use the results of Problem 3.37 and write the complete truth table for the circuit of Figure 3.43.

4

COMBINATORIAL LOGIC CIRCUITS

OUTLINE

In Chapter 3 we studied the operation of all the basic logic gates, and we used Boolean algebra to describe and analyze circuits that were made up of combinations of logic gates. These circuits can be classified as *combinatorial* logic circuits because, at any time, the logic level at the output depends on the combination of logic levels present at the inputs. A combinatorial circuit has no *memory* characteristic, and so its output depends *only* on the current value of its inputs.

In this chapter we will continue our study of combinatorial logic circuits. To start, we will go further into the simplification of logic circuits. Two methods will be used: one will use Boolean algebra theorems, the other a *mapping* technique. In addition, we will study simple techniques for designing logic circuits to satisfy a given set of requirements. A complete study of logic-circuit design is not one of our objectives, but the methods we introduce are more than sufficient for the types of design situations that a technician will encounter.

4.1 SUM-OF-PRODUCTS FORM

The methods of logic-circuit simplification and design that we will study require the logic expression to be in a *sum-of-products* form. Some examples of this form are

1. $ABC + \overline{A}B\overline{C}$
2. $AB + \overline{A}B\overline{C} + \overline{C}\overline{D} + D$
3. $\overline{A}B + C\overline{D} + EF + GK + H\overline{L}$

Each of these sum-of-products expressions consists of two or more AND terms (products) that are ORed together. Each AND term consists of one or more variables appearing in either complemented or uncomplemented form. For example, in the sum-of-products expression $ABC + \overline{A}B\overline{C}$, the first AND product contains the variables A, B, and C in their uncomplemented (not inverted) form. The second AND term contains A and C in their complemented (inverted) form. Note that in a sum-of-products expression, an inversion sign *cannot* appear over more than one variable in a term (e.g., we cannot have $\overline{ABC}$ or $\overline{RST}$).

4.2 SIMPLIFYING LOGIC CIRCUITS

Once the expression for a logic circuit has been obtained, we may be able to reduce it to a simpler form containing fewer terms or fewer variables in one or more terms. The new expression can then be used to implement a circuit that is equivalent to the original circuit but that contains fewer gates and connections.

To illustrate, the circuit of Figure 4.1(a) can be simplified to produce the circuit of Figure 4.1(b). Since both circuits perform the same logic, it should be obvious that the simpler circuit is more desirable because it contains fewer gates and will therefore be smaller and cheaper than the original. Furthermore, the circuit reliability will improve because there are fewer interconnections that can be potential circuit faults.

In subsequent sections we will study two methods for simplifying logic circuits. One method will utilize the Boolean algebra theorems and, as we shall see, is greatly dependent on inspiration and experience. The other method (Karnaugh mapping) is a

(a)

(b)

FIGURE 4.1

systematic, cookbook approach. Some instructors may wish to skip over this latter method because it is somewhat mechanical and does not contribute to a better understanding of Boolean algebra. This can be done without affecting the continuity or clarity of the rest of the text.

4.3 ALGEBRAIC SIMPLIFICATION

The Boolean algebra theorems that we studied in Chapter 3 can be used to help us simplify the expression for a logic circuit. Unfortunately, it is not always obvious which theorems should be applied in order to produce the simplest result. Furthermore, there is no easy way to tell whether the simplified expression is in its simplest form or whether it could have been simplified further. Thus, algebraic simplification often becomes a process of trial and error. With experience, however, one can become adept at obtaining reasonably good results.

The examples that follow will illustrate many of the ways in which the Boolean theorems can be applied in trying to simplify an expression. You should notice that these examples contain two essential steps:

1. The expression is put into the sum-of-products form.
2. Once it is in this form, the product terms are checked for common factors, and factoring is performed wherever possible. Hopefully, the factoring results in the elimination of one or more terms.

EXAMPLE 4.1

Simplify the logic circuit shown in Figure 4.2(a).

Solution: The first step is to determine the expression for the output. The result is

$$z = ABC + A\bar{B} \cdot (\overline{\bar{A}\bar{C}})$$

Once the expression is determined, it is usually a good idea to break down all large inverter signs using DeMorgan's theorems and then multiply out all terms.

(a)

(b)

FIGURE 4.2

$$z = ABC + A\overline{B}(\overline{\overline{A}} + \overline{\overline{C}}) \qquad \text{[theorem (17)]}$$
$$= ABC + A\overline{B}(A + C) \qquad \text{[cancel double inversions]}$$
$$= ABC + A\overline{B}A + A\overline{B}C \qquad \text{[multiply out]}$$
$$= ABC + A\overline{B} + A\overline{B}C \qquad [A \cdot A = A]$$

With the expression in sum-of-products form, we should look for common variables among the various terms with the intention of factoring. The first and third terms above have AC in common, which can be factored out:

$$z = AC(B + \overline{B}) + A\overline{B}$$

Since $B + \overline{B} = 1$, then

$$z = AC(1) + A\overline{B}$$
$$= AC + A\overline{B}$$

We can now factor out A, which results in

$$z = A(C + \overline{B})$$

This result can be simplified no further. Its circuit implementation is shown in Figure 4.2(b). It is obvious that the circuit in (b) is a great deal simpler than the original circuit in (a).

EXAMPLE 4.2

Simplify the expression $z = ABC + AB\overline{C} + A\overline{B}C$.

Solution: We will look at two different ways to arrive at the same result.

Method 1: The first two terms in the expression have the variables AB in common. Thus,

$$z = AB(C + \bar{C}) + A\bar{B}C$$

$$= AB(1) + A\bar{B}C$$

$$= AB + A\bar{B}C$$

We can factor the variable A from both terms:

$$z = A(B + \bar{B}C)$$

Invoking theorem (15),

$$z = A(B + C)$$

Method 2: The original expression is $z = ABC + AB\bar{C} + A\bar{B}C$. The first two terms have the variables AB in common. The first and last terms have the variables AC in common. How do we know whether to factor AB from the first two terms or AC from the two end terms? Actually, we can do both by using the ABC term *twice*. In other words, we can rewrite the expression as

$$z = ABC + AB\bar{C} + A\bar{B}C + ABC$$

where we have added an extra term ABC. This is valid and will not change the value of the expression since $ABC + ABC = ABC$ [theorem (7)]. Now we can factor AB from the first two terms and AC from the last two terms:

$$z = AB(C + \bar{C}) + AC(\bar{B} + B)$$

$$= AB \cdot 1 + AC \cdot 1$$

$$= AB + AC = A(B + C)$$

This is, of course, the same result as method 1. This trick of using the same term twice can always be used. In fact, the same term can be used more than twice if necessary.

EXAMPLE 4.3

Simplify $z = \bar{A}C(\overline{\bar{A}BD}) + \bar{A}B\bar{C}\bar{D} + A\bar{B}C$.

Solution: First, use DeMorgan's theorem on the first term.

$$z = \bar{A}C(A + \bar{B} + \bar{D}) + \bar{A}B\bar{C}\bar{D} + A\bar{B}C$$

Multiplying out,

$$z = \bar{A}CA + \bar{A}C\bar{B} + \bar{A}C\bar{D} + \bar{A}B\bar{C}\bar{D} + A\bar{B}C$$

Since $\bar{A} \cdot A = 0$, the first term is eliminated.

$$z = \bar{A}\bar{B}C + \bar{A}C\bar{D} + \bar{A}B\bar{C}\bar{D} + A\bar{B}C$$

Factor $\bar{B}C$ from the first and last terms. Also factor $\bar{A}\bar{D}$ from the second and third terms.

$$z = \overline{B}C(\overline{A} + A) + \overline{A}\,\overline{D}(C + B\overline{C})$$

Now, since $\overline{A} + A = 1$, and $C + B\overline{C} = C + B$ [theorem (15)], we have

$$z = \overline{B}C + \overline{A}\,\overline{D}(C + B)$$

EXAMPLE 4.4

Simplify the expression $x = (\overline{A} + B)(A + B + D)\overline{D}$.

Solution: The expression can be put into sum-of-products form by multiplying out all the terms. The result is

$$x = \overline{A}A\overline{D} + \overline{A}B\overline{D} + \overline{A}D\overline{D} + BA\overline{D} + BB\overline{D} + BD\overline{D}$$

The first term can be eliminated, since $\overline{A}A = 0$. Likewise, the third and sixth terms can be eliminated, since $D\overline{D} = 0$. The fifth term can be simplified to $B\overline{D}$, since $BB = B$. This gives us

$$x = \overline{A}B\overline{D} + AB\overline{D} + B\overline{D}$$

We can factor $B\overline{D}$ from each term to obtain

$$x = B\overline{D}(\overline{A} + A + 1)$$

Clearly, the term inside the parentheses is always 1, so we finally have

$$x = B\overline{D}$$

EXAMPLE 4.5

Simplify the circuit of Figure 4.3(a).

Solution: The expression for output z is

$$z = (\overline{A} + B)(A + \overline{B})$$

Multiplying out to get the sum-of-products form, we obtain

$$z = \overline{A}A + \overline{A}\,\overline{B} + BA + B\overline{B}$$

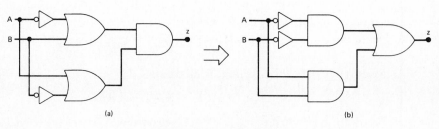

(a)   (b)

FIGURE 4.3   Example 4.5

We can eliminate $\bar{A}A = 0$ and $B\bar{B} = 0$ to end up with

$$z = \bar{A}\bar{B} + AB$$

This expression is implemented in Figure 4.3(b), and if we compare it to the original circuit, we see that both circuits contain the same number of gates and connections. In this case the simplification process produced an equivalent, but not simpler, circuit.

EXAMPLE 4.6

Simplify $x = A\bar{B}C + \bar{A}BD + \bar{C}D$.

Solution: You can try, but you will not be able to simplify this expression any further.

REVIEW QUESTIONS

1. Which of the following expressions are *not* in sum-of-products form? (a) $R\bar{S}T + \bar{R}S\bar{T} + \bar{T}$ (b) $A\bar{D}C + \bar{A}DC$ (c) $MN\bar{P} + (M + \bar{N})P$ (d) $AB + A\bar{B}C + AB\bar{C}D$ (*Ans.* b and c)
2. Simplify the circuit in Figure 4.1(a) to arrive at the circuit of Figure 4.1(b).
3. Change each AND gate in Figure 4.1(a) to a NAND gate. Determine the new expression for x and simplify it. (*Ans.* $x = \bar{A} + \bar{B} + \bar{C} = \overline{ABC}$)

4.4 DESIGNING COMBINATORIAL LOGIC CIRCUITS

When the desired output level of a logic circuit is given for all possible input conditions, the results can be conveniently displayed in a truth table. The Boolean expression for the required circuit can then be derived from the truth table. For example, consider Figure 4.4(a), where a truth table is shown for a circuit that has two inputs, A and B, and output x. The table shows that output x is to be at the 1 level *only* for the case where $A = 0$ and $B = 1$. It now remains to determine what logic circuit will produce this desired operation. It should be apparent that one possible solution is that shown in Figure 4.4(b). Here an AND gate is used with inputs $\bar{A}$ and B, so $x = \bar{A} \cdot B$. Obviously x will be 1 *only if* both inputs to the AND gate are 1, namely $\bar{A} = 1$ (which means $A = 0$) and $B = 1$. For all other values of A and B, the output x has to be 0.

A similar approach can be used for the other input conditions. For instance, if x were to be high only for the $A = 1$, $B = 0$ condition, the resulting circuit would be an AND gate with inputs A and $\bar{B}$. In other words, for any of the four possible input conditions we can generate a high x output by using an AND gate with appropriate

A	B	x
0	0	0
0	1	1
1	0	0
1	1	0

(a)　　　　　　　　　(b)　　　　　　　　**FIGURE 4.4**

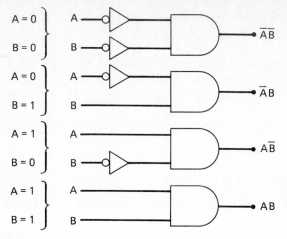

FIGURE 4.5 AND gates used to generate high outputs for each of the possible input conditions.

inputs. The four different cases are shown in Figure 4.5. Each of the AND gates shown generates an output that is 1 *only* for the one given input condition and is 0 for all other conditions. It should be noted that the AND inputs are inverted or not inverted depending on the values that the variables have for the given condition. If the variable is 0 for the given condition, it is inverted before entering the AND gate.

Let us now consider the case shown in Figure 4.6(a), where we have a truth table which indicates that the output x is to be 1 for two different cases: $A = 0, B = 1$ and $A = 1, B = 0$. How can this be implemented? We know that the AND term $\overline{A} \cdot B$ will generate a 1 only for the $A = 0$, $B = 1$ condition, and the AND term $A \cdot \overline{B}$ will generate a 1 for the $A = 1$, $B = 0$ condition. Since x has to be high for *either* condition, it should be clear that these terms should be ORed together to produce the desired output, x. This implementation is shown in Figure 4.6(b), where the resulting expression for the output is $x = \overline{A}B + A\overline{B}$.

In this example, an AND term is generated for each case in the table where the output x is to be a 1. The AND-gate outputs are then ORed together to produce the total output x, which will be high when either AND term is high. This same procedure can be extended to examples with more than two inputs. Consider the following truth table for a three-input circuit:

A	B	x
0	0	0
0	1	1
1	0	1
1	1	0

(a)

(b)

$$x = \overline{A}B + A\overline{B}$$

FIGURE 4.6

A	B	C	x
0	0	0	0
0	0	1	0
0	1	0	1 $\longrightarrow \overline{A}B\overline{C}$
0	1	1	1 $\longrightarrow \overline{A}BC$
1	0	0	0
1	0	1	0
1	1	0	0
1	1	1	1 $\longrightarrow ABC$

Here there are three cases where the output x is to be 1. The required AND term for each of these cases is shown. Again, note that for each case where a variable is 0, it appears complemented in the AND term. The final expression for x is obtained by ORing the three AND terms. Thus,

$$x = \overline{A}B\overline{C} + \overline{A}BC + ABC$$

This expression can be implemented with three AND gates feeding an OR gate.

In all the preceding examples the expression for the output x was derived from the truth table in sum-of-products form. The general procedure for obtaining the output expression from a truth table can be summarized as follows:

1. Write an AND term for each case in the table where the output is 1.
2. Each AND term contains each input variable in either inverted or noninverted form. If the variable is 0 for that particular case in the table, it is inverted in the AND term.
3. All the AND terms are then ORed together to produce the final expression for the output.

Complete Design Problem   Once the output expression has been determined from the truth table in sum-of-products form, it can easily be implemented using AND and OR gates. There will be one AND gate for each term in the expression and one OR gate, which is fed by the outputs of each AND gate. Usually, however, the expression can be simplified, thereby resulting in a more efficient circuit. The following example illustrates the complete design procedure.

EXAMPLE 4.7

Design a logic circuit that has three inputs, A, B, and C, and whose output will be high only when a majority of the inputs is high.

Solution: The first step is to set up the truth table based on the problem statement. The eight possible input cases are shown in Figure 4.7(a). Based on the problem statement, the output x should be a 1 whenever two or more inputs are 1. For all other cases the output should be 0. The next step is to write the AND terms for each case where $x = 1$, as shown in the figure. The expression for x can then be written as

$$x = \overline{A}BC + A\overline{B}C + AB\overline{C} + ABC$$

A	B	C	x	
0	0	0	0	
0	0	1	0	
0	1	0	0	
0	1	1	1	$\overline{A}BC$
1	0	0	0	
1	0	1	1	$A\overline{B}C$
1	1	0	1	$AB\overline{C}$
1	1	1	1	ABC

(a)

(b)   (c)

FIGURE 4.7 Example 4.7

This expression can be simplified in several ways. Perhaps the quickest way is to realize that the last term ABC has two variables in common with each of the other terms. Thus, we can use the ABC term to factor with each of the other terms. The expression is rewritten with the ABC term occurring three times (recall that this is legal in Boolean algebra).

$$x = \overline{A}BC + ABC + A\overline{B}C + ABC + AB\overline{C} + ABC$$

Factoring the appropriate pairs of terms, we have

$$x = BC(\overline{A} + A) + AC(\overline{B} + B) + AB(\overline{C} + C)$$

Since each term in parentheses is equal to 1, we have

$$x = BC + AC + AB$$

This expression is implemented in Figure 4.7(b). Since the expression is in sum-of-products form, the circuit consists of a group of AND gates working into a single OR gate.

This expression can be factored further, but it will not result in the elimination of any terms. It might, however, produce a slightly simpler circuit. For example, let's factor C from the first two terms to obtain

$$x = C(B + A) + AB$$

This expression is implemented in Figure 4.7(c). The resulting circuit requires the same number of gates as the circuit in Figure 4.7(b), but it requires one less connection because the final OR gate has only two inputs. In practice, a logic-circuit designer may still select the implementation in Figure 4.7(b) because of some other considerations, one of which is the fact that the *A* and *B* input signals in Figure 4.7(c) have to propagate through three gates (1, 2, and 3) before reaching the output. This might be an important consideration in a high-speed digital system.

EXAMPLE 4.8

Refer to 4.8(a), where four logic-signal lines *A*, *B*, *C*, *D* are being used to represent a 4-bit binary number with *A* as the MSB and *D* as the LSB. The binary inputs are fed to a logic circuit that produces a HIGH output only when the binary number is greater than $0110_2 = 6_{10}$. Design this circuit.

Solution: The truth table is shown in Figure 4.8(b). For each case in the truth table we have indicated the decimal equivalent of the binary number represented by the *ABCD* combination.

The output *z* is set equal to 1 for all those cases where the binary number is greater than 0110. For all other cases, *z* is set equal to 0. This truth table gives us the following sum-of-products expression

$$z = \overline{A}BCD + A\overline{B}\,\overline{C}\overline{D} + A\overline{B}\,\overline{C}D + A\overline{B}C\overline{D} + A\overline{B}CD + AB\overline{C}\,\overline{D}$$

$$+ AB\overline{C}D + ABC\overline{D} + ABCD$$

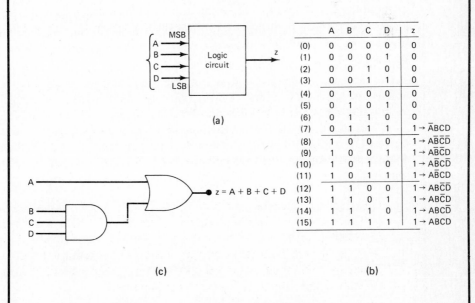

(a)

		A	B	C	D	z
(0)		0	0	0	0	0
(1)		0	0	0	1	0
(2)		0	0	1	0	0
(3)		0	0	1	1	0
(4)		0	1	0	0	0
(5)		0	1	0	1	0
(6)		0	1	1	0	0
(7)		0	1	1	1	1 → $\overline{A}BCD$
(8)		1	0	0	0	1 → $A\overline{B}\,\overline{C}\overline{D}$
(9)		1	0	0	1	1 → $A\overline{B}\,\overline{C}D$
(10)		1	0	1	0	1 → $A\overline{B}C\overline{D}$
(11)		1	0	1	1	1 → $A\overline{B}CD$
(12)		1	1	0	0	1 → $AB\overline{C}\,\overline{D}$
(13)		1	1	0	1	1 → $AB\overline{C}D$
(14)		1	1	1	0	1 → $ABC\overline{D}$
(15)		1	1	1	1	1 → $ABCD$

(b)

$z = A + B + C + D$

(c)

FIGURE 4.8   Example 4.8

Simplification of this expression will be a formidable task, but with a little care it can be accomplished. The step-by-step process involves factoring and eliminating terms of the form $A + \bar{A}$.

$$z = \bar{A}BCD + A\bar{B}\bar{C}(\bar{D} + D) + A\bar{B}C(\bar{D} + D) + AB\bar{C}(\bar{D} + D)$$
$$\qquad + ABC(\bar{D} + D)$$
$$= \bar{A}BCD + A\bar{B}\bar{C} + A\bar{B}C + AB\bar{C} + ABC$$
$$= \bar{A}BCD + A\bar{B}(\bar{C} + C) + AB(\bar{C} + C)$$
$$= \bar{A}BCD + A\bar{B} + AB$$
$$= \bar{A}BCD + A(\bar{B} + B)$$
$$= \bar{A}BCD + A$$

This can be reduced further by invoking theorem (15), which says that $x + \bar{x}y = x + y$. In this case $x = A$ and $y = BCD$. Thus,

$$z = \bar{A}BCD + A = BCD + A$$

This final expression is implemented in Figure 4.8(c).

As this example demonstrates, the algebraic simplification method can be quite lengthy when the original expression contains a large number of terms. This is a limitation that is not shared by the Karnaugh mapping method, as we will see later.

EXAMPLE 4.9

Suppose the design problem of Example 4.8 had been stated as follows: "The logic circuit output is to go LOW whenever the binary number is *less than* $0111_2 = 7_{10}$." Design the circuit.

Solution: Clearly, the statement above is a different, but equivalent, way of stating the same design problem as Example 4.8. Thus, the truth table will be exactly the same as in Figure 4.8(b). Likewise, the expression for z will be exactly the same, and it will produce the same circuit as in Figure 4.8(c).

The only difference is that the statement above implies that z is to be an active-LOW output. If we wish to convey this information, then the circuit diagram should be converted accordingly. This is done by changing the OR-gate symbol to its alternate symbol, as shown in Figure 4.9. This in turn requires

(a)   (b)

FIGURE 4.9   Example 4.9

changing the AND symbol to its alternate symbol, so that a bubble output is connected to a bubble input.

Remember, both circuits are physically the same AND and OR gates, but the circuit diagram in Figure 4.9(b) more accurately represents the circuit's operation as given in the statement of the design problem.

Implementing the Final Design   In the design examples we have shown, the final circuit was implemented using AND and OR gates. In fact, the sum-of-products form always yields a circuit containing one or more AND gates driving a single OR gate. One of the reasons for using the sum-of-products form is that it can be implemented using all NAND gates with little, if any, increase in circuit complexity over the AND/OR implementation. Since NAND gates are the most widely available logic gates in the TTL logic family, this is an important characteristic.

To illustrate, Figure 4.10 shows the equivalent NAND implementation for the circuits of Figures 4.7(b) and 4.8(c). You may wish to work out these conversions yourself as a review of the procedure covered in Chapter 3.

Comparing the NAND implementation with the circuit in Figure 4.10(a), we see that they are identical in structure; that is, *each* gate of the original circuit has been replaced by a single NAND gate. This characteristic is true only if the original circuit is in sum-of-products form. The only exception to this is when the sum-of-products form contains a single-variable term such as $z = A + BCD$ in Figure 4.10(b). Here the NAND implementation requires an extra NAND gate used as an INVERTER on the A input.

We can streamline the process of converting a *sum-of-products* circuit from AND/OR to NAND gates as follows:

1. Replace each AND gate, OR gate, and INVERTER by a *single* NAND gate.
2. Use a NAND gate to invert any single variable that is feeding the final OR gate.

REVIEW QUESTIONS

1. Write the sum-of-products expression for a circuit with four inputs and an output that is to be HIGH only when input A is LOW at the same time that exactly two other inputs are LOW. (*Ans.* $x = \overline{A}\,\overline{B}\,\overline{C}D + \overline{A}\,\overline{B}C\overline{D} + \overline{A}B\overline{C}\,\overline{D}$)
2. Implement the expression of question 1 using all NAND gates. How many are required? (*Ans.* Eight)

4.5   THE KARNAUGH MAP METHOD*

The Karnaugh map is a graphical device used to simplify a logic equation or to convert a truth table to its corresponding logic circuit in a simple, orderly process. Although a Karnaugh map (henceforth abbreviated *K map*) can be used for problems involving any number of input variables, its practical usefulness is limited to six variables. The following discussion will be limited to problems with up to four inputs, since even five- and six-input problems are too involved and are best done by a computer program.

*This topic may be omitted without affecting the continuity of the remainder of the text.

FIGURE 4.10 (a) Converting circuit of Figure 4.7(b) to NANDs; (b) converting circuit of Figure 4.8(c) to NANDs.

94

FIGURE 4.11 Karnaugh maps and truth tables for two, three, and four variables.

Karnaugh Map Format The K map, like a truth table, is a means for showing the relationship between logic inputs and the desired output. Figure 4.11 shows three examples of K maps for two, three, and four variables, together with the corresponding truth tables. These examples illustrate the following important points:

1. The truth table gives the value of output X for each combination of input values. The K map gives the same information in a different format. Each case in the truth table corresponds to a square in the K map. For example, in Figure 4.11(a), the $A = 0$, $B = 0$ condition in the truth table corresponds to the $\overline{A}\,\overline{B}$ square in the K map. Since the truth table shows $X = 1$ for this case, a 1 is placed in the $\overline{A}\,\overline{B}$ square in the K map. Similarly, the $A = 1$, $B = 1$ condition in the truth table corresponds to the AB square of the K map. Since $X = 1$ for this case, a 1 is placed in the AB square.

All other squares are filled with 0s. This same idea is used in the three- and four-variable maps shown in the figure.

2. The K-map squares are labeled so that horizontally adjacent squares differ only in one variable. For example, the upper-left-hand square in the four-variable map is $\overline{A}\,\overline{B}\,\overline{C}\,\overline{D}$, while the square immediately to its right is $\overline{A}\,\overline{B}\,\overline{C}\,D$ (only the D variable is different). Similarly, vertically adjacent squares differ only in one variable. For example, the upper-left-hand square is $\overline{A}\,\overline{B}\,\overline{C}\,\overline{D}$ while the square directly below it is $\overline{A}B\overline{C}\,\overline{D}$ (only the B variable is different).

3. Once a K map has been filled with 0s and 1s, the sum-of-products expression for the output X can be obtained by ORing together those squares that contain a 1. In the three-variable map of Figure 4.11(b), the $\overline{A}\,\overline{B}\,\overline{C}$, $\overline{A}\,\overline{B}C$, $\overline{A}BC$, and $AB\overline{C}$ squares contain a 1, so $X = \overline{A}\,\overline{B}\,\overline{C} + \overline{A}\,\overline{B}C + \overline{A}BC + AB\overline{C}$.

Looping   The expression for output X can be simplified by properly combining those squares in the K map which contain 1s. The process for combining these 1s is called *looping*.

Looping Groups of Two (Pairs)   Figure 4.12(a) is the K map for a particular three-variable truth table. This map contains a pair of 1s that are vertically adjacent to each other; the first one represents $\overline{A}B\overline{C}$ and the second one represents $AB\overline{C}$. Note that in these two terms only the A variable appears in both normal and complemented form (B and $\overline{C}$ remain unchanged). These two terms can be looped (combined) to give a resultant that eliminates the A variable since it appears in both uncomplemented and complemented forms. This is easily proved as follows:

$$X = \overline{A}B\overline{C} + AB\overline{C}$$
$$= B\overline{C}(\overline{A} + A)$$
$$= B\overline{C}(1) = B\overline{C}$$

This same principle holds true for any pair of vertically or horizontally adjacent 1s. Figure 4.12(b) shows an example of two horizontally adjacent 1s. These two can be looped and the C variable eliminated since it appears in both its uncomplemented and complemented forms to give a resultant of $X = \overline{A}B$.

Another example is shown in Figure 4.12(c). In a K map the top row and bottom row of squares are considered to be adjacent. Thus, the two 1s in this map can be looped to provide a resultant of $\overline{A}\,\overline{B}\,\overline{C} + A\overline{B}\,\overline{C} = \overline{B}\,\overline{C}$.

Figure 4.12(d) shows a K map that has two pairs of 1s which can be looped. The two 1s in the top row are horizontally adjacent. The two 1s in the bottom row are also adjacent since in a K map the leftmost column and rightmost column of squares are considered to be adjacent. When the top pair of 1s is looped, the D variable is eliminated (since it appears as both D and $\overline{D}$) to give the term $\overline{A}\,\overline{B}C$. Looping the bottom pair eliminates the C variable to give the term $A\overline{B}\,\overline{D}$. These two terms are ORed to give the final result for X.

To summarize: *Looping a pair of adjacent 1s in a K map eliminates the variable that appears in complemented and uncomplemented form.*

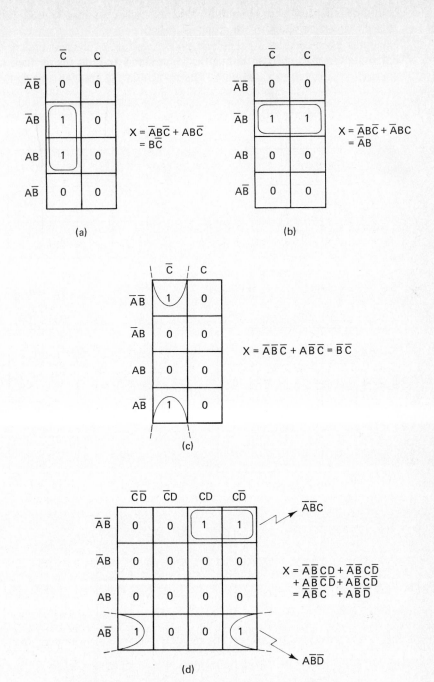

FIGURE 4.12 Examples of looping pairs of adjacent 1s.

Looping Groups of Four (Quads) A K map may contain a group of four 1s that are adjacent to each other. This group is called a *quad*. Figure 4.13 shows several examples of quads. In (a) the four 1s are vertically adjacent and in (b) they are horizontally adjacent. The K map in (c) contains four 1s in a square and they are considered adjacent to each other. The four 1s in (d) are also adjacent, as are those

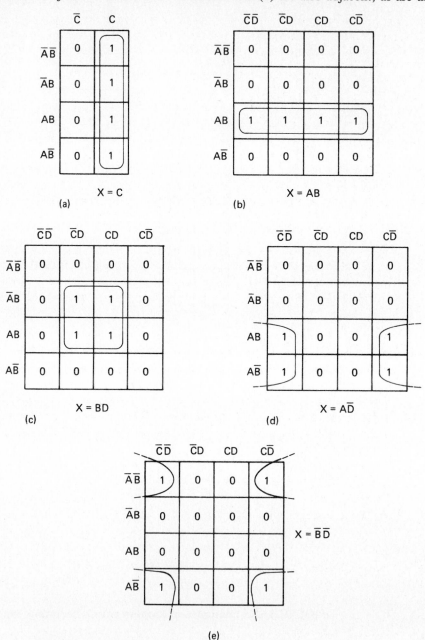

FIGURE 4.13 Example of looping groups of four 1s (quads).

in (e) because, as pointed out earlier, the top and bottom rows and the leftmost and rightmost columns are considered to be adjacent to each other.

When a quad is looped, the resultant term will contain only the variables that do not change form for all the squares in the quad. For example, in (a) the four squares which contain a 1 are $\overline{A}\,\overline{B}C, \overline{A}BC, ABC$, and $A\overline{B}C$. Examination of these terms reveals that only the variable C remains unchanged (both A and B appear in complemented and uncomplemented form). Thus, the resultant expression for X is simply $X = C$. This can be proved as follows:

$$X = \overline{A}\,\overline{B}C + \overline{A}BC + ABC + A\overline{B}C$$
$$= \overline{A}C(\overline{B} + B) + AC(B + \overline{B})$$
$$= \overline{A}C + AC$$
$$= C(\overline{A} + A) = C$$

As another example, consider Figure 4.13(d), where the four squares containing 1s are $AB\overline{C}\,\overline{D}, A\overline{B}\,\overline{C}\,\overline{D}, ABC\overline{D}$, and $A\overline{B}C\overline{D}$. Examination of these terms indicates that only the variables A and $\overline{D}$ remain unchanged, so the simplified expression for X is

$$X = A\overline{D}$$

This can be proved in the same manner that was used above.

The reader should check each of the other cases in Figure 4.13 to verify the indicated expressions for X. To summarize: *Looping a quad of 1s eliminates the two variables that appear in both complemented and uncomplemented form.*

Looping Groups of Eight (Octets) A group of eight 1s that are adjacent to each other is called an *octet*. Several examples of octets are shown in Figure 4.14. When an octet is looped in a four-variable map, three of the four variables are eliminated because only one variable remains unchanged. For example, examination of the eight looped squares in (a) shows that only the variable B is in the same form for all eight squares; the other variables appear in complemented and uncomplemented form. Thus, for this map, $X = B$. The reader can verify the results for the other examples in Figure 4.14.

To summarize: *Looping an octet of 1s eliminates the three variables that appear in both complemented and uncomplemented form.*

Complete Simplification Process We have seen how looping of pairs, quads, and octets on a K map can be used to obtain a simplified expression. We can summarize the rule for loops of *any* size: *When a variable appears in both complemented and uncomplemented form within a loop, that variable is eliminated from the expression. Variables that are the same for all squares of the loop must appear in the final expression.*

It should be clear that a larger loop of 1s eliminates more variables. To be exact, a loop of two eliminates one variable, a loop of four eliminates two, and a loop of eight eliminates three. This principle will now be used to obtain a simplified logic expression from a K map that contains any combination 1s and 0s.

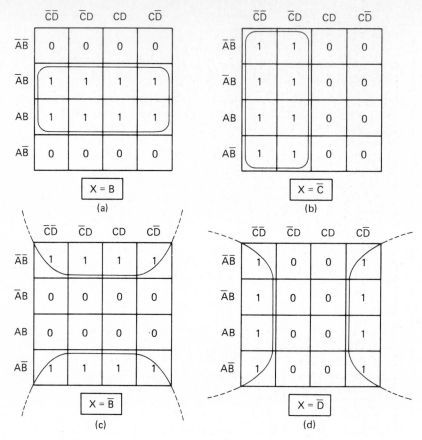

FIGURE 4.14   Examples of looping groups of eight 1's (octets).

The procedure will first be outlined and then applied to several examples. The steps below are followed in using the K-map method for simplifying a Boolean expression:

1. Construct the K map and place 1s in those squares corresponding to the 1s in the truth table. Place 0s in the other squares.
2. Examine the map for adjacent 1s and loop those 1s which are *not* adjacent to any other 1s. These are called *isolated* 1s.
3. Next, look for those 1s which are adjacent to only one other 1. Loop *any* pair containing such a 1.
4. Loop any octet even if some of the 1s have already been looped.
5. Loop any quad that contains one or more 1s which have not already been looped.
6. Loop any pairs necessary to include any 1s that have not yet been looped, making sure to use the minimum number of loops.
7. Form the OR sum of all the terms generated by each loop.

These steps will be followed exactly and referred to in the following examples. In each case, the resulting logic expression will be in its simplest sum-of-products form.

EXAMPLE 4.10

Figure 4.15(a) shows the K map for a four-variable problem. We will assume that the map was obtained from the problem truth table (step 1). The squares are numbered for convenience in identifying each loop.

(a)

(b)

(c)

FIGURE 4.15   Examples 4.10–4.12

Step 2: Square 4 is the only square containing a 1 that is not adjacent to any other 1. It is looped and is referred to as loop 4.

Step 3: Square 15 is adjacent *only* to square 11. This pair is looped and referred to as loop 11, 15.

Step 4: There are no octets.

Step 5: Squares 6, 7, 10, and 11 form a quad. This quad is looped (loop 6, 7, 10, 11). Note that square 11 is used again, even though it was part of loop 11, 15.

Step 6: All 1s have already been looped.

Step 7: Each loop generates a term in the expression for X. Loop 4 is simply $\overline{A}\,\overline{B}C\overline{D}$. Loop 11, 15 is ACD (the B variable is eliminated). Loop 6, 7, 10, 11, is BD (A and C are eliminated).

EXAMPLE 4.11

Consider the K map in Figure 4.15(b). Once again we can assume that step 1 has already been performed.

Step 2: There are no 1s which are not adjacent to other 1s.

Step 3: The 1 in square 3 is adjacent *only* to the 1 in square 7. Looping this pair (loop 3, 7) produces the term $\overline{A}CD$.

Step 4: There are no octets.

Step 5: There are two quads. Squares 5, 6, 7, and 8 form one quad. Looping this quad produces the term $\overline{A}B$. The second quad is made up of squares 5, 6, 9, and 10. This quad is looped because it contains two squares that have not been looped previously. Looping this quad produces $B\overline{C}$.

Step 6: All 1s have already been looped.

Step 7: The terms generated by the three loops are ORed together to obtain the expression for X.

EXAMPLE 4.12

Consider the K map in Figure 4.15(c).

Step 2: There are no isolated 1s.

Step 3: The 1 in square 2 is adjacent only to the 1 in square 6. This pair is looped to produce $\overline{A}\,\overline{C}D$. Similarly, square 9 is adjacent only to square 10. Looping this pair produces $B\overline{C}D$. Likewise, loop 7, 8, and loop 11, 15 produce the terms $\overline{A}BC$ and ACD, respectively.

Step 4: There are no octets.

Step 5: There is one quad formed by squares 6, 7, 10, and 11. This quad, however, is *not* looped because all the 1s in the quad have been included in other loops.

Step 6: All 1s have already been looped.

Step 7: The expression for X is shown in the figure.

EXAMPLE 4.13

Consider the K map in Figure 4.16(a).

$$X = \overline{A}\,\overline{C}\,D + \overline{A}BC + A\,\overline{B}\,\overline{C} + AC\overline{D}$$

(a)

$$X = \overline{A}BD + BC\overline{D} + \overline{B}C D + A\,\overline{B}\,\overline{D}$$

(b)

FIGURE 4.16 The same K map with two equally good solutions.

Step 2: There are no isolated 1s.

Step 3: There are no 1s that are adjacent to only one other 1.

Step 4: There are no octets.

Step 5: There are no quads.

Steps 6 and 7: There are many possible pairs. The looping must use the minimum number of loops to account for all the 1s. For this map there are *two* possible loopings, which require only four looped pairs. Figure 4.16(a) shows one solution and its resultant expression. Figure 4.16(b) shows the other. Note that both expressions are of the same complexity, so neither is better than the other.

EXAMPLE 4.14

Use the K map to simplify the expression $x = \overline{A}\,\overline{B}\,\overline{C} + \overline{B}C + A\overline{B}$.

Solution: In this problem we are not given the truth table from which to fill in the K map. Instead, we must fill in the K map by taking each of the product terms in the expression and placing 1s in the corresponding squares.

The first term, $\overline{A}\,\overline{B}\,\overline{C}$, tells us to enter a 1 in the $\overline{A}\,\overline{B}\,\overline{C}$ square of the map (see Figure 4.17). The second term, $\overline{B}C$, tells us to enter a 1 in each square that contains a $\overline{B}C$ in its label. In Figure 4.17 this would be the $A\overline{B}C$ and $\overline{A}\,\overline{B}C$ squares. Likewise, the $A\overline{B}$ term tells us to place a 1 in the $A\overline{B}C$ and $A\overline{B}\,\overline{C}$ squares. All other squares will be filled with 0s.

Now the K map can be looped for simplification. The result is $x = \overline{A} + \overline{B}C$, as shown in the figure.

$$x = \overline{A} + \overline{B}C$$

FIGURE 4.17 Example 4.14

"Don't Care" Conditions   Some logic circuits can be designed so that there are certain input conditions for which there are no specified output levels, usually because these input conditions will never occur. In other words, there will be certain combinations of input levels where we "don't care" whether the output is HIGH or LOW. This is illustrated in the truth table of Figure 4.18(a).

Here the output z is not specified as either 0 or 1 for the conditions $A, B, C = 1$, 0, 0 and $A, B, C = 0, 1, 1$. Instead, an x is shown for these conditions. The x represents the "don't care" condition. A "don't care" condition can come about for several reasons, the most common being that in some situations certain input combinations can never occur, and so it is not necessary to specify the output for these conditions.

A circuit designer is free to make any "don't care" condition either a 0 or a 1 in order to produce the simplest output expression. For example, the K map for this truth table is shown in Figure 4.18(b) with an x placed in the $A\overline{B}\,\overline{C}$ and $\overline{A}BC$ squares. The designer here would be wise to change the x in the $A\overline{B}\,\overline{C}$ square to a 1 and the x in the $\overline{A}BC$ square to a 0, since this would produce a quad that can be looped to produce $z = A$, as shown in Figure 4.18(c).

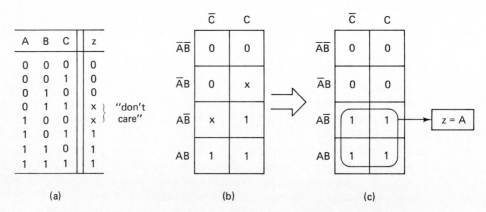

FIGURE 4.18 "Don't care" conditions should be changed to 0 or 1 to produce K-map looping that yields the simplest expression.

Whenever "don't care" conditions occur, we have to decide which ones to change to 0 and which to 1 to produce the best K-map looping (i.e., the simplest expression). This decision is not always an easy one.

SUMMARY

The K-map process has several advantages over the algebraic method. K mapping is a more orderly process with well-defined steps as compared to the trial-and-error process sometimes used in algebraic simplification. K mapping usually requires fewer steps, especially for expressions containing many terms, and it always produces a minimum expression.

Nevertheless, some instructors prefer the algebraic method because it requires a thorough knowledge of Boolean algebra and is not simply a mechanical procedure. Each method has its advantages, and though most logic designers are adept at both, being proficient in one method is all that is necessary to produce acceptable results.

REVIEW QUESTIONS

1. Use K mapping to simplify the expression of Example 4.7.
2. Use K mapping to simplify the expression of Example 4.8. This should emphasize the advantage of K mapping for expressions containing many terms.
3. What is meant by a "don't care" condition?

4.6 EXCLUSIVE-OR AND EXCLUSIVE-NOR CIRCUITS

Two special logic circuits that occur quite often in digital systems are the *exclusive-OR* and *exclusive-NOR* circuits. Their description and operation will now be investigated.

Exclusive-OR Consider the logic circuit of Figure 4.19(a). The output expression of this circuit is

$$x = \bar{A}B + A\bar{B}$$

The accompanying truth table shows that $x = 1$ for two cases: $A = 0, B = 1$ (the $\bar{A}B$ term) and $A = 1, B = 0$ (the $A\bar{B}$ term). In other words, *this circuit produces a HIGH output whenever the two inputs are at opposite levels*. This is the exclusive-OR circuit, which will hereafter be abbreviated EX-OR.

This particular combination of logic gates occurs quite often and is very useful in certain applications. In fact, the EX-OR circuit has been given a symbol of its own, shown in Figure 4.19(b). This symbol is assumed to contain all the logic contained in the EX-OR circuit and therefore has the same logic expression and truth table. This EX-OR circuit is commonly referred to as an EX-OR *gate,* and we can consider it as another type of logic gate.

An EX-OR gate has only *two* inputs; there are no three-input or four-input EX-OR gates. The two inputs are combined such that $x = \bar{A}B + A\bar{B}$. A shorthand way that is sometimes used to indicate the EX-OR output expression is

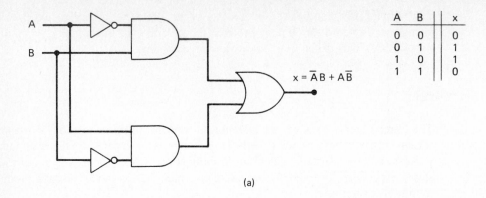

A	B	x
0	0	0
0	1	1
1	0	1
1	1	0

(a)

EX-OR

(b)

FIGURE 4.19 **(a) Exclusive-OR circuit and truth table; (b) exclusive-OR gate symbol.**

$$x = A \oplus B$$

where the symbol $\oplus$ represents the EX-OR gate operation.

The characteristics of an EX-OR gate are summarized as follows:

1. It has only two inputs and its output is

$$x = \overline{A}B + A\overline{B} = A \oplus B$$

2. Its output is *HIGH* only when the two inputs are at *different* levels.

Several ICs are available that contain EX-OR gates. Those listed below are *quad* EX-OR chips containing four EX-OR gates.

- **7486**   QUAD EX-OR (TTL family)
- **74C86** QUAD EX-OR (CMOS family)
- **4070**   QUAD EX-OR (CMOS)

Exclusive-NOR   The exclusive-NOR circuit (abbreviated EX-NOR) operates completely opposite to the EX-OR circuit. Figure 4.20(a) shows an EX-NOR circuit and its accompanying truth table. The output expression is

$$x = AB + \overline{A}\,\overline{B}$$

which indicates along with the truth table that x will be 1 for two cases: $A = B = 1$

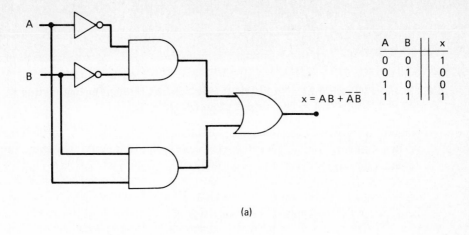

A	B	x
0	0	1
0	1	0
1	0	0
1	1	1

$x = AB + \overline{A}\,\overline{B}$

(a)

$x = \overline{A \oplus B} = AB + \overline{A}\,\overline{B}$

EX-NOR

(b)

FIGURE 4.20 (a) Exclusive-NOR circuit; (b) exclusive-NOR gate symbol.

(the AB term) and $A = B = 0$ (the $\overline{A}\,\overline{B}$ term). In other words, *this circuit produces a high output whenever the two inputs are at the same level.**

It should be apparent that the output of the EX-NOR circuit is the exact inverse of the output of the EX-OR circuit. The symbol for an EX-NOR gate is obtained by simply adding a small circle at the output of the EX-OR symbol [Figure 4.20(b)].

The EX-NOR gate also has only *two* inputs, and it combines them such that its output is

$$x = AB + \overline{A}\,\overline{B}$$

A shorthand way to indicate the output expression of the EX-NOR is

$$x = \overline{A \oplus B}$$

which is simply the inverse of the EX-OR operation. The EX-NOR gate is summarized as follows:

   1. It has only two inputs and its output is

$$x = AB + \overline{A}\,\overline{B} = \overline{A \oplus B}$$

   2. Its output is HIGH only when the two inputs are at the same level.

*For this reason it is often called a "coincidence gate."

Sec. 4.6   Exclusive-OR and Exclusive-NOR Circuits   **107**

FIGURE 4.21 An EX-NOR gate can be obtained by connecting an INVERTER to the output of an EX-OR gate.

Not too many ICs are available that contain EX-NOR gates. The 74LS266 is a TTL IC that contains four EX-NORs. Its special output circuitry, however, limits it to certain types of applications. The 4077 is a CMOS QUAD EX-NOR, but it is becoming obsolete. Typically, logic designers will obtain the EX-NOR function by combining an EX-OR and an INVERTER. This is shown in Figure 4.21, where a 7486 EX-OR is followed by an INVERTER to produce an EX-NOR. Incidentally, the 7404 is a TTL HEX INVERTER, which means that it contains *six* INVERTERs on one chip.

EXAMPLE 4.15

Determine the output waveform for the input waveforms given in Figure 4.22.

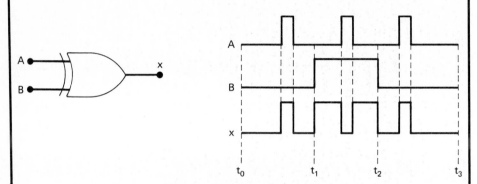

FIGURE 4.22 Example 4.15

Solution: The output waveform is obtained using the fact that the EX-OR output will go HIGH only when its inputs are at different levels. The resulting output waveform reveals several interesting points:

1. The x waveform matches the A input waveform during those time intervals when $B = 0$. This occurs during the $t_0 - t_1$ and $t_2 - t_3$ time intervals.

2. The x waveform is the *inverse* of the A input waveform during those time intervals when $B = 1$. This occurs during the $t_1 - t_2$ interval.

3. These observations show that an EX-OR gate can be used as a *controlled INVERTER;* that is, one of its inputs can be used to control whether the signal at the other input will be inverted or not. This property will be useful in certain applications.

EXAMPLE 4.16

$x_1 x_0$ represents a 2-bit binary number that can have any value (00, 01, 10, or 11); for example, when $x_1 = 1$ and $x_0 = 0$, the binary number is 10, and so on. Similarly, $y_1 y_0$ represents another 2-bit binary number. Design a logic circuit, using x_1, x_0, y_1, and y_0 inputs, whose output will be HIGH only when the two binary numbers $x_1 x_0$ and $y_1 y_0$ are *equal*.

Solution: The first step is to construct a truth table for the 16 input conditions.

x_1	x_0	y_1	y_0	z (output)
0	0	0	0	1
0	0	0	1	0
0	0	1	0	0
0	0	1	1	0
0	1	0	0	0
0	1	0	1	1
0	1	1	0	0
0	1	1	1	0
1	0	0	0	0
1	0	0	1	0
1	0	1	0	1
1	0	1	1	0
1	1	0	0	0
1	1	0	1	0
1	1	1	0	0
1	1	1	1	1

The output z has to be high whenever the $x_1 x_0$ values match the $y_1 y_0$ values; that is, whenever $x_1 = y_1$ and $x_0 = y_0$. The table shows that there are four such cases.

We could now continue with the normal procedure, which would be to obtain a sum-of-products expression for z, attempt to simplify it, and then implement the result. However, the nature of this problem makes it ideally suited for implementation using EX-NOR gates, and a little thought will produce a simple solution with minimum work. Refer to Figure 4.23; in this logic diagram x_1 and y_1 are fed to one EX-NOR gate and x_0 and y_0 are fed to another EX-NOR gate. The output of each EX-NOR will be HIGH only when its inputs are equal. Thus, for $x_0 = y_0$ and $x_1 = y_1$ both EX-NOR outputs will be high. This is the condition we are looking for, because it means that the two 2-bit numbers are equal. The AND-gate output will be HIGH only for this case, thereby producing the desired output.

REVIEW QUESTIONS

1. Use Boolean algebra to prove that the EX-NOR output expression is the exact inverse of the EX-OR output expression.

2. What is the output of an EX-NOR gate when a logic signal and its exact inverse are connected to its inputs? (*Ans.* A constant LOW)
3. A logic designer needs an INVERTER, and all that is available is one EX-OR gate from a 7486 chip. Does he need another chip? (*Ans.* No, he can use the EX-OR gate as an INVERTER by connecting one of its inputs permanently HIGH)

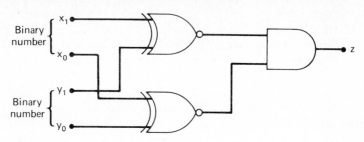

FIGURE 4.23 Circuit for detecting equality of two 2-bit binary numbers.

4.7 LOGIC CIRCUITS WITH MULTIPLE OUTPUTS

Many logic design problems involve more than one output. In other words, several outputs are generated from the same inputs. For these situations the outputs can be treated separately, with a complete design procedure followed for each output. Sometimes when this is done and the final output expressions are obtained, there may be some common terms among the expressions that may be shared. To illustrate, suppose that we have a two-output problem, where the output expressions are

$$x = AB + \bar{B}C$$

$$y = AB\bar{C} + \bar{A}B$$

Note that x contains an AB term. The y expression has an $AB\bar{C}$ term, which is the same as $(AB) \cdot \bar{C}$. Thus if AB is generated for the x output, it can be used to generate $AB\bar{C}$ for the y output.

This is illustrated in Figure 4.24. For this simple example, the savings realized by sharing common terms is not great (one gate-input connection is saved); however, in a more complex multiple-output problem, the savings can be considerable if the designer is clever in his use of "term sharing."

4.8 DESIGNING WITHOUT A TRUTH TABLE

Occasionally a logic design problem is simple enough to perform without constructing a truth table, because the expression or circuit diagram can be written directly from the problem statement. The following examples illustrate.

EXAMPLE 4.17

Design a logic circuit with inputs A, B, C, and D such that the output is HIGH *only* when both *C and D* are LOW *while* either *A or B* or both is HIGH.

Solution: The output expression can be written as

$$x = (A + B) \cdot \overline{C + D}$$

Let us examine how this was obtained. First, note that A is ORed with B, since either one has to be HIGH for a HIGH output. Next, note that the $\overline{C + D}$ term produces a HIGH *only* if C and D are both LOW (this should be recognized as a NOR-gate term). The two terms $A + B$ and $C + D$ are then ANDed since both conditions have to exist to produce a HIGH output.

FIGURE 4.24 Example of term sharing.

EXAMPLE 4.18

Design a circuit whose output goes HIGH only when either A or B or both is LOW while C and D are not both HIGH.

Solution:

$$x = \overline{AB} \cdot \overline{CD}$$

or

$$x = (\overline{A} + \overline{B}) \cdot (\overline{C} + \overline{D})$$

Inhibit Circuits Each of the basic logic gates can be used to control the passage of an input logic signal through to the output. This is depicted in Figure 4.25, where a logic signal, A, is applied to one input of each of the basic logic gates. The other input of each gate is the control input, B. The logic level at this control input will determine whether the input signal is enabled to reach the output, or inhibited from reaching the output.

FIGURE 4.25 Four basic gates can either enable or inhibit the passage of an input signal, A, under control of the logic level at control input, B.

Examine Figure 4.25 and you should notice that when the noninverting gates (AND, OR) are enabled, the output will follow the A signal exactly. Conversely, when the inverting gates (NAND, NOR) are enabled, the output will be the exact inverse of the A signal.

Also notice that AND and NOR gates produce a constant LOW output when they are in the inhibited condition. Conversely, the NAND and OR gates produce a constant HIGH output in the inhibited condition.

There will be many situations in digital-circuit design where the passage of a logic signal is to be enabled or inhibited, depending on conditions present at one or more control inputs. Several are shown in the following examples.

EXAMPLE 4.19

Design a logic circuit that will allow a signal to pass to the output only when control inputs *B* and *C* are both HIGH; otherwise the output will stay LOW.

Solution: An AND gate should be used because the signal is to be passed without inversion, and the inhibit output condition is a LOW. Since the enable

FIGURE 4.26 Examples 4.19 and 4.20

condition has to occur only when $B = C = 1$, a three-input AND gate is used, as shown in Figure 4.26(a).

EXAMPLE 4.20

Design a logic circuit that allows a signal to pass to the output only when one, but not both, of the control inputs are HIGH; otherwise, the output will stay HIGH.

Solution: The result is drawn in Figure 4.26(b). An OR gate is used because we want the output inhibit condition to be a HIGH, and we do not want to invert the signal. Control inputs B and C are combined in an EX-NOR gate. When B and C are different, the EX-NOR sends a LOW to enable the OR gate. When B and C are the same, the EX-NOR sends a HIGH to inhibit the OR gate.

EXAMPLE 4.21

Design a logic circuit with input signal A, control input B, and outputs X and Y to operate as follows:

    1. When $B = 0$, output X will follow input A, and output Y will be 0.

    2. When $B = 1$, output X will be 0, and output Y will follow input A.

Solution: The two outputs will be 0 when inhibited and will follow the input signal when enabled. Thus an AND gate should be used for each output. Since X is to be enabled when $B = 0$, its AND gate must be controlled by B, as shown in Figure 4.27. Since Y is to be enabled when $B = 0$, its AND gate is controlled by $\overline{B}$.

FIGURE 4.27 Example 4.21

This circuit is called a *pulse-steering* circuit because it steers the input pulse to one output or the other depending on B.

1. Design a logic circuit with three inputs A, B, C and an output that goes LOW only when A is HIGH while B and C are different. (*Ans.* $x = \overline{A[B \oplus C]}$)
2. What type of gate produces a 1 output when it is inhibited, and passes the inverse of the input signal when enabled?

PROBLEMS

4.1 Simplify the following expressions using Boolean algebra.
(a) $x = \overline{A}\,\overline{B}\,\overline{C} + \overline{A}BC + AB\underline{C} + A\overline{B}\,\overline{C} + A\overline{B}C$
(b) $z = (B + \overline{C})(\overline{B} + C) + \overline{A} + B + \overline{C}$
(c) $y = \overline{(C + D)} + \overline{A}C\overline{D} + A\overline{B}\,\overline{C} + \overline{A}BCD + AC\overline{D}$

4.2 Simplify the circuit of Figure 4.28 using Boolean algebra.

4.3 Change each gate in Figure 4.28 to a NOR gate and simplify the circuit using Boolean algebra.

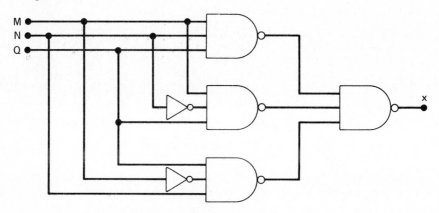

FIGURE 4.28

4.4 Design the logic circuit corresponding to the following truth table:

A	B	C	x
0	0	0	1
0	0	1	0
0	1	0	1
0	1	1	1
1	0	0	1
1	0	1	0
1	1	0	0
1	1	1	1

4.5 Design a logic circuit whose output is HIGH *only* when a majority of inputs A, B, and C are LOW.

4.6 Three photocells are being illuminated by three different flashing lights. The lights are supposed to be flashing in sequence so that at no time should all three lights be on at the same time or off at the same time. Each photocell is used to monitor one of the lights, and each photocell is in a circuit that produces a LOW output voltage when the photocell is dark and a HIGH output voltage when the photocell is illuminated. Design a logic circuit that has as its inputs the photocell-circuit outputs and which produces a HIGH output whenever the three lights are *all* on or *all* off at the same time.

4.7 A 4-bit binary number is represented as $A_3A_2A_1A_0$, where A_3, A_2, A_1, and A_0 represent the individual bits with A_0 equal to the LSB. Design a logic circuit that will produce a HIGH output whenever the binary number is greater than 0010 and less than 1000.

4.8 Implement the circuit of Problem 4.4 using all NAND gates.

4.9 Implement the circuit of Problem 4.5 using all NAND gates.

4.10 Implement the expression $z = \overline{D} + ABC + \overline{A}\,\overline{C}$ using AND/OR gates, then convert to all NAND gates.

4.11 Simplify the expression of Problem 4.1(a) using the K map.

4.12 Simplify the expression of Problem 4.1(c) using the K map.

4.13 Simplify the expression from Problem 4.7 using a K map.

4.14 Determine the minimum expressions for each K map in Figure 4.29.

(a)

	$\overline{CD}$	$\overline{C}D$	CD	$C\overline{D}$
$\overline{AB}$	1	1	1	1
$\overline{A}B$	1	1	0	0
AB	0	0	0	1
$A\overline{B}$	0	1	1	0

(b)

	$\overline{CD}$	$\overline{C}D$	CD	$C\overline{D}$
$\overline{AB}$	1	0	1	1
$\overline{A}B$	1	0	0	1
AB	0	0	0	0
$A\overline{B}$	1	0	1	1

(c)

	$\overline{C}$	C
$\overline{AB}$	1	1
$\overline{A}B$	0	0
AB	1	0
$A\overline{B}$	1	X

FIGURE 4.29

4.15 Determine the output waveform for the circuit of Figure 4.30.

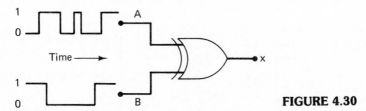

FIGURE 4.30

4.16 Repeat Problem 4.15 with the B input held LOW.

4.17 Repeat with B held HIGH.

4.18 Determine the input conditions needed to produce $x = 1$ in Figure 4.31.

FIGURE 4.31

4.19 Redo Problem 4.6 using EX-OR gates and one other gate.

4.20 Figure 4.32 represents a multiplier circuit that takes two 2-bit binary numbers $x_1 x_0$ and $y_1 y_0$ and produces an output binary number $Z_3 Z_2 Z_1 Z_0$ that is equal to the arithmetic product of the two input numbers. Design the logic circuit for the multiplier. (*Hint:* The logic circuit will have four inputs and four outputs.)

FIGURE 4.32

4.21 A BCD code is being transmitted to a remote receiver. The bits are A_3, A_2, A_1, A_0 with A_3 as the MSB. The receiver circuitry includes a *BCD error-detector* circuit that examines the received code to see if it is a legal BCD code (i.e., ≤ 1001). Design this circuit to produce a HIGH for any error condition.

4.22 Figure 4.33 represents a *relative-magnitude detector* that takes two 3-bit binary numbers $x_2 x_1 x_0$ and $y_2 y_1 y_0$ and determines whether they are equal and, if not, which one is larger. There are three outputs, defined as follows:
(a) $M = 1$ only if the two input numbers are equal.
(b) $N = 1$ only if $x_2 x_1 x_0$ is greater than $y_2 y_1 y_0$.
(c) $P = 1$ only if $y_2 y_1 y_0$ is greater than $x_2 x_1 x_0$.

FIGURE 4.33

116 Chap. 4 **Combinatorial Logic Circuits**

Design the logic circuitry for the comparator. The circuit has *six* inputs and *three* outputs and is therefore much too complex to handle using the truth-table approach. Refer to Example 4.16 as a hint to how you might start to solve this problem.

4.23 An EX-OR gate can be considered to be an *odd parity detector* because it will produce a HIGH output only if an odd number of inputs are HIGH. Show how several EX-OR gates can be combined in a circuit which examines four bits A, B, C, D and produces a HIGH output only if an odd number of bits are 1.

4.24 Design a logic circuit whose output is HIGH whenever A and B are both HIGH, while C and D are either both LOW or both HIGH. Try to do this without using a truth table. Then check your result by constructing a truth table for your circuit to see if it agrees with the problem statement.

4.25 Four large tanks at a chemical plant contain different liquids being heated. Liquid-level sensors are being used to detect whenever the level in tanks A and B rises above a predetermined level. Temperature sensors in tanks C and D detect when the temperature in these tanks drops below a prescribed temperature limit. Assume that the liquid-level sensor outputs A and B are LOW when the level is satisfactory and HIGH when the level is too high. Also, the temperature-sensor outputs C and D are LOW when the temperature is satisfactory and HIGH when the temperature is too low. Design a logic circuit that will detect whenever the level in tank A or tank B is too high at the same time that the temperature in either tank C or tank D is too low.

4.26 Figure 4.34 shows the intersection of a main highway with a secondary access road. Vehicle-detection sensors are placed along lanes C and D (main road) and lanes A and B (access road). These sensor outputs are LOW (0) when no vehicle is present and HIGH (1) when a vehicle is present. The intersection traffic light is to be controlled according to the following logic:

(a) The E-W traffic light will be green whenever *both* lanes C and D are occupied.

(b) The E-W light will be green whenever *either* C or D are occupied but lanes A and B are not *both* occupied.

(c) The N-S light will be green whenever *both* lanes A and B are occupied but C and D are not *both* occupied.

(d) The N-S light will also be green when *either* A or B is occupied while C and D are *both* vacant.

(e) The E-W will be green when *no* vehicles are present. Using the sensor outputs A, B, C, and D as inputs, design a logic circuit to control the traffic light. There should be two outputs, N/S and E/W, which go HIGH when the corresponding light is to be *green*. Simplify the circuit as much as possible and show *all* steps.

FIGURE 4.34

4.27 Design a logic circuit that will allow an input signal to pass through to the output only when control input B is LOW while control input C is HIGH; otherwise, the output is LOW.

4.28 Design a circuit that will *inhibit* the passage of an input signal only when control inputs B, C, and D are all HIGH; the output is to be HIGH in the inhibited condition.

4.29 Design a logic circuit that controls the passage of a signal A according to the following requirements:
  (a) Output X will equal A when control inputs B and C are the same.
  (b) X will remain HIGH when B and C are different.

4.30 Design a logic circuit that has two signal inputs A_1 and A_0 and a control input S so that it functions according to the requirements given in Figure 4.35. This type of circuit is called a *multiplexer* (covered in Chapter 9).

S	z
0	$= A_0$
1	$= A_1$

FIGURE 4.35

4.31 Use K mapping to design a circuit to meet the requirements of Example 4.16. Compare this circuit to the solution in Figure 4.23. This points out that the K-map method cannot take advantage of the EX-OR and EX-NOR gate logic. The designer has to be able to determine when these gates are applicable.

5

FLIP-FLOPS

The logic circuits considered thus far have been combinatorial circuits whose output levels at any instant of time are dependent upon the levels present at the inputs at that time. Any prior input-level conditions have no effect on the present outputs because combinatorial logic circuits have no memory. Most digital systems are made up of both combinatorial circuits and memory elements.

Figure 5.1 shows a block diagram of a general digital system that combines combinatorial logic gates with memory devices. The combinatorial portion accepts logic signals from external inputs and from the outputs of the memory elements. The combinatorial circuit operates on these inputs to produce various outputs, some of which are used to determine the binary values to be stored in the memory elements. The outputs of some of the memory elements, in turn, go to the inputs of logic gates in the combinatorial circuits. This process indicates that the external outputs of a digital system are a function of both its external inputs and the information stored in its memory elements.

The most widely used memory element is the flip-flop, which we shall study thoroughly in this chapter. The *flip-flop* (abbreviated FF) is a logic circuit with two outputs, which are the inverse of each other. Figure 5.2 indicates these outputs as Q and $\overline{Q}$ (actually any letter could be used, but Q is the most common). The Q output is called the *normal* FF output and $\overline{Q}$ is the *inverted* FF output. When a FF is said to be in the HIGH (1) state or the LOW (0) state, this is the condition at the Q output. Of course, the $\overline{Q}$ output is always the inverse of Q.

There are two possible operating states for the FF: (1) $Q = 0$, $\overline{Q} = 1$; and (2) $Q = 1$, $\overline{Q} = 0$. The FF has one or more inputs, which are used to cause the FF to switch back and forth between these two states. As we shall see, when an input is pulsed to send the FF to a given state, the FF will remain in that state even after the input returns to normal. This is its *memory* characteristic.

The flip-flop, incidentally, is known by several other names, including *bistable multivibrator, latch,* and *binary,* but we will generally use flip-flop because it is the most common designation in the digital field. Other memory elements, which we will

FIGURE 5.1 General digital system diagram.

FIGURE 5.2 General flip-flop symbol.

study in a later chapter, are also used in digital systems, but flip-flops are the most versatile because of their high speed of operation, the ease with which information can be stored into and read out of them, and the ease with which they can be interconnected with logic gates.

5.1 NAND GATE SET-CLEAR FLIP-FLOP

A basic FF circuit can be constructed from two cross-coupled NAND gates as shown in Figure 5.3(a). Notice that the output of NAND-1 is connected to one of the inputs of NAND-2 and vice versa. The gate outputs, Q and $\overline{Q}$, are the FF outputs, which, under normal conditions, will always be the inverse of each other. The two FF inputs are labeled SET and CLEAR for reasons that will be explained.

The SET and CLEAR inputs are normally resting in the HIGH state and one of them will be pulsed LOW whenever we want to change the FF output state. We begin our analysis by showing that there are two equally likely output states when SET = CLEAR = 1. One possibility is shown in Figure 5.3(a), where we have $Q = 0/\overline{Q} = 1$. With $Q = 0$, the inputs to NAND-2 are 0 and 1, which produce $\overline{Q} = 1$. The 1 from $\overline{Q}$ causes NAND-1 to have a 1 at both inputs to produce a 0 output at Q. In effect, what we have is the LOW at the NAND-1 output produces a HIGH at the NAND-2 output, which, in turn, keeps the NAND-1 output LOW.

The second possibility is shown in Figure 5.3(b), where $Q = 1/\overline{Q} = 0$. The HIGH from NAND-1 produces a LOW at the NAND-2 output, which, in turn, keeps the NAND-1 output HIGH. Thus, there are two possible output states when

(a)   (b)

FIGURE 5.3 NAND FF has two possible resting states when SET = CLEAR = 1.

SET = CLEAR = 1; as we shall soon see, the one that actually exists will depend on what has occurred previously at the inputs.

Now let's investigate what happens when the SET input is momentarily pulsed LOW while CLEAR is kept HIGH. Figure 5.4(a) shows what happens when $Q = 0$ prior to the occurrence of the pulse. As SET is pulsed LOW at time t_0, Q will go HIGH, and this HIGH will force $\overline{Q}$ to go LOW so that NAND-1 now has two LOW inputs. Thus, when SET returns to the 1 state at t_1, the NAND-1 output *remains* HIGH, which, in turn, keeps the NAND-2 output LOW.

Figure 5.4(b) shows what happens when $Q = 1/\overline{Q} = 0$ prior to the application of the SET pulse. Since $\overline{Q} = 0$ is already keeping the NAND-1 output HIGH, the LOW pulse at SET will not change anything. Thus, when SET returns HIGH, the FF outputs are still in the $Q = 1/\overline{Q} = 0$ state.

We can summarize Figure 5.4 by stating that a LOW pulse on the SET input will cause the FF to end up in the $Q = 1$ state. This operation is called *setting* the FF. In fact, the $Q = 1$ state is also called the *set state*.

Now let's consider what occurs when the CLEAR input is pulsed LOW while SET is kept HIGH. Figure 5.5(a) shows what happens when $Q = 0$ and $\overline{Q} = 1$ prior to the application of the pulse. Since $Q = 0$ is already keeping the NAND-2 output

(a)

(b)

FIGURE 5.4 Pulsing the SET input to the 0 state will always produce the $Q = 1/\overline{Q} = 0$ output state.

(a)

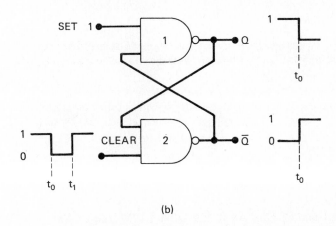

(b)

FIGURE 5.5 Pulsing the CLEAR input to the LOW state will always produce $Q = 0/\overline{Q} = 1$.

HIGH, the LOW pulse at CLEAR will not have any effect. When CLEAR returns HIGH, the FF outputs are still $Q = 0/\overline{Q} = 1$.

Figure 5.5(b) shows the situation where $Q = 1$ prior to the occurrence of the CLEAR pulse. As CLEAR is pulsed LOW at t_0, $\overline{Q}$ will go HIGH, and this HIGH forces Q to go LOW so that NAND-2 now has two LOW inputs. Thus, when CLEAR returns HIGH at t_1, the NAND-2 output *remains* HIGH, which, in turn, keeps the NAND-1 output LOW.

Figure 5.5 can be summarized by stating that a LOW pulse on the CLEAR input will cause the FF to end up in the $Q = 0$ state. This operation is called *clearing* the FF, and the $Q = 0$ state is also called the *cleared state*.

The last case to consider is the case where the SET and CLEAR inputs are simultaneously pulsed LOW. This will produce HIGH levels at both NAND outputs so that $Q = \overline{Q} = 1$. Clearly, this is an undesired condition, since the two outputs are supposed to be inverses of each other. Furthermore, when the SET and CLEAR inputs return HIGH, the resulting FF output state will depend on which input returns HIGH first. Simultaneous transitions back to the 1 state will produce unpredictable results. For these reasons the SET = CLEAR = 0 condition is never purposely used for the NAND FF.

Set	Clear	FF Output
1	1	No change
0	1	Q = 1
1	0	Q = 0
0	0	Ambiguous

(a)   (b)   **FIGURE 5.6 (a) NAND FF;
(b) truth table.**

Summary of NAND FF   The operation described above can be conveniently placed in a truth table (Figure 5.6) and is summarized as follows:

1. SET = CLEAR = 1: This condition is the normal resting state and it has no effect on the FF output state. The Q and $\overline{Q}$ outputs will remain in whatever state they were prior to this input condition.
2. SET = 0, CLEAR = 1: This will always cause the output to go to the $Q = 1$ state, where it will remain even after SET returns HIGH. This is called *setting* the FF.
3. SET = 1, CLEAR = 0: This will always produce the $Q = 0$ state, where the output will remain even after CLEAR returns HIGH. This is called *clearing* the FF.
4. SET = CLEAR = 0: This condition tries to set and clear the FF simultaneously and can produce ambiguous results. It should not be used.

Alternate Representations   From the description of the NAND FF operation, it should be clear that the SET and CLEAR inputs are active LOW. The SET input will set $Q = 1$ when SET goes LOW; the CLEAR input will clear $Q = 0$ when CLEAR goes LOW. For this reason, the NAND FF circuit is often drawn using the alternate representation for each NAND gate, as shown in Figure 5.7(a). The bubbles on the SET and CLEAR inputs emphasize the fact that these inputs are active LOW.

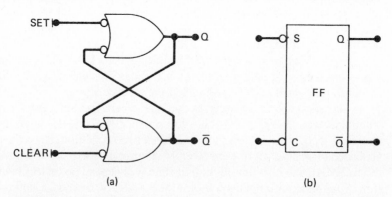

(a)   (b)

FIGURE 5.7 (a) NAND FF equivalent representation; (b) simplified block symbol.

Figure 5.7(b) shows a simplified block representation that we will sometimes use. The S and C labels represent the SET and CLEAR inputs, and the bubbles indicate the active LOW nature of these inputs. Whenever we use this block symbol, it represents a NAND-gate FF.

EXAMPLE 5.1

The waveforms of Figure 5.8 are applied to the inputs of a NAND-gate SET-CLEAR FF. Assume that initially $Q = 0$, and determine the Q waveform.

FIGURE 5.8   Example 5.1

Solution: Initially, SET = CLEAR = 1 so that Q will remain in the 0 state. The LOW pulse that occurs on the CLEAR input at time T_1 will have no effect, since Q is already in the cleared (0) state.

The only way that Q can go to the 1 state is by a LOW pulse on the SET input. This occurs at time T_2 when SET first goes LOW. When SET returns HIGH at T_3, Q will remain in its new HIGH state.

At time T_4 when SET goes LOW again, there will be no effect on Q because Q is already set to the 1 state.

The only way to bring Q back to the 0 state is by a LOW pulse on the CLEAR input. This occurs at time T_5. When CLEAR returns to 1 at time T_6, Q remains in its now LOW state.

This example shows that the FF output "remembers" the last input that was activated and will not change states until the opposite input is activated.

EXAMPLE 5.2

It is virtually impossible to obtain a single voltage transition from a mechanical switch because of the phenomenon of *contact bounce*. This is illustrated in Figure 5.9(a), where closing the switch produces several output-voltage transitions between 0 and 5 V as the switch contacts vibrate open and closed before finally staying closed. The duration of this switch bounce is generally only a couple of milliseconds or less, depending on the type of switch.

The multiple transitions on the output signal may not be acceptable in many applications, so some means has to be used to "debounce" the switch. A NAND-gate FF can be used to perform this function as shown in Figure 5.9(b). Describe the operation of this "debounce" circuit.

(a)

(b)

FIGURE 5.9 (a) **Mechanical contact bounce will produce multiple transitions; (b) NAND FF used to debounce a mechanical switch.**

Solution: Assume the switch is initially in position 1 so that the CLEAR input is LOW and $Q = 0$. When the switch is moved to position 2, it will put a LOW on the SET input when it first makes contact. This will set $Q = 1$ in a matter of a few nanoseconds, and it will stay there even if the contact bounces open (since that will produce SET = CLEAR = 1, which has no effect).

Likewise, when the switch is moved back to position 1, it will place a LOW on the CLEAR input upon first contact. This clears $Q = 0$, where it will stay even if the contact bounces open.

Thus, the Q output will consist of a *single* transition each time the switch is moved from one position to the other.

Resetting a FF The action of *clearing* a FF ($Q = 0$) is also called *resetting* a FF, and both terms are used interchangeably in the digital field. In fact, SET-CLEAR FFs are often called RESET-SET FFs (abbreviated as RS FF), where the RESET input is used in place of the CLEAR input. We will use both terms throughout the text.

1. What is the normal resting state of the SET and CLEAR inputs? What is the active state of each input?
2. What will be the states of Q and $\overline{Q}$ after a FF has been cleared (reset)?
3. True or false: The SET input can never be used to make $Q = 0$.
4. When power is first applied to any FF circuit, it is impossible to predict the initial states of Q and $\overline{Q}$. What would have to be done to ensure that a NAND FF always started off in the $Q = 1$ state? (*Ans.* A momentary LOW pulse has to be applied to the SET input).

5.2 NOR GATE SET-CLEAR FLIP-FLOP

Two cross-coupled NOR gates can be used as a SET-CLEAR FF. The arrangement, shown in Figure 5.10(a), is similar to the NAND FF except that the Q and $\overline{Q}$ outputs have reversed positions.

The analysis of the operation of the NOR FF can be performed in exactly the same manner as for the NAND FF. The results are given in the truth table in Figure 5.10(b) and are summarized as follows:

1. SET = CLEAR = 0: This is the normal resting state for the NOR FF and it has no effect on the output state. Q and $\overline{Q}$ will remain in whatever state they were prior to the occurrence of this input condition.
2. SET = 1, CLEAR = 0: This will always set $Q = 1$, where it will remain even after SET returns to 0.
3. SET = 0, CLEAR = 1: This will always clear $Q = 0$, where it will remain even after CLEAR returns to 0.
4. SET = 1, CLEAR = 1: This condition tries to set and clear the FF at the same time, and it produces $Q = \overline{Q} = 0$. If the inputs are returned to 0 simultaneously, the resulting output state is unpredictable. This input condition should not be used.

Set	Clear		FF Output
0	0		No change
1	0		$Q = 1$
0	1		$Q = 0$
1	1		Ambiguous

(b)

(a)

(c)

FIGURE 5.10 (a) NOR-gate FF; (b) truth table; (c) simplified block symbol.

The NOR-gate FF operates exactly like the NAND FF except that the SET and CLEAR inputs are active HIGH rather than active LOW, and the normal resting state is SET = CLEAR = 0. Q will be set HIGH by a HIGH pulse on the SET input, and it will be cleared LOW by a HIGH pulse on the CLEAR input. The simplified block symbol for the NOR FF in Figure 5.10(c) is shown with no bubbles on the S and C inputs; this indicates that these inputs are active HIGH.

EXAMPLE 5.3

Assume $Q = 0$ initially, and determine the Q waveform for the NOR FF inputs of Figure 5.11.

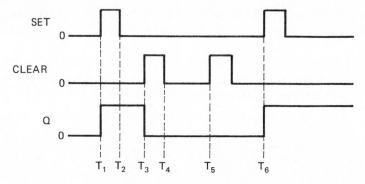

FIGURE 5.11   Example 5.3

Solution: Initially, SET = CLEAR = 0, which has no effect on Q, and Q stays LOW. When SET goes HIGH at time T_1, Q will be set to 1 and will remain there even after SET returns to 0 at T_2.

At T_3 the CLEAR input goes HIGH and clears Q to the 0 state, where it remains even after CLEAR returns LOW at T_4.

The CLEAR pulse at T_5 has no effect on Q, since Q is already LOW. The SET pulse at T_6 again sets Q back to 1, where it will stay.

This example shows that the FF "remembers" the last input that was activated, and it will not change states until the opposite input is activated.

EXAMPLE 5.4

Figure 5.12 shows a simple circuit that can be used to detect the interruption of a light beam. The light is focused on a phototransistor that is connected in the common-emitter configuration to operate as a switch. Assume that the FF has previously been cleared to the 0 state by momentarily opening switch S_1 and describe what happens if the light beam is momentarily interrupted.

Solution: With light on the phototransistor, we can assume that it is fully conducting so that the resistance between the collector and emitter is very small. Thus, v_0 will be close to 0 V. This places a LOW on the SET input of the FF so that SET = CLEAR = 0.

FIGURE 5.12   Example 5.4

When the light beam is interrupted, the phototransistor turns off and its collector-emitter resistance becomes very high (i.e., essentially an open circuit). This causes v_0 to rise to approximately 5 V; this activates the SET input and turns on the alarm.

Now, the alarm will remain on because Q will remain HIGH even if the light beam was only momentarily interrupted, and v_0 goes back to 0 V. The alarm can be turned off only by momentarily opening S1 to produce a HIGH on the CLEAR input.

In this application, the FF's memory characteristic is used to convert a momentary occurrence (beam interruption) into a constant output.

REVIEW QUESTIONS

1. What is the normal resting state of the NOR FF inputs? What is the active state?
2. When a FF is set, what are the states of Q and $\overline{Q}$?
3. What is the only way to cause the Q output of a NOR-gate FF to change from 1 to 0?
4. If the NOR FF in Figure 5.12 were replaced by a NAND FF, why wouldn't the circuit work properly?

5.3   CLOCK SIGNALS

Digital systems can operate either *asynchronously* or *synchronously*. In asynchronous systems, the outputs of logic circuits can change state any time one or more of the inputs change. An asynchronous system is difficult to design and troubleshoot.

In synchronous systems, the exact times at which any output can change states is determined by a signal commonly called the *clock*. This clock signal is generally a rectangular pulse train or squarewave as shown in Figure 5.13. The clock signal is distributed to all parts of the system, and most (if not all) of the system outputs can change state only when the clock makes a transition. The transitions (also called *edges*) are pointed out in Figure 5.13. When the clock changes from a 0 to a 1, this is called the *positive-going transition* (PGT); when the clock goes from 1 to 0, this is

FIGURE 5.13 Clock signals.

the *negative-going transition* (NGT). We will use the abbreviations PGT and NGT, since these terms appear so often throughout the text.

Most digital systems are principally synchronous (although there are always some asynchronous parts), since synchronous circuits are easier to design and troubleshoot. They are easier to troubleshoot because the circuit outputs can change only at specific instants of time. In other words, almost everything is synchronized to the clock-signal transitions.

The synchronizing action of the clock signals is accomplished through the use of *clocked flip-flops* that are designed to change states on one or the other of the clock's transitions.

5.4   CLOCKED FLIP-FLOPS

We will study several types of clocked flip-flops. First, however, we will discuss the principal ideas that are common to all of them.

1. All clocked FFs have a clock input that is typically labeled *CLK*, *CK*, or *CP*. We will use *CLK* as shown in Figure 5.14. The clock input is also symbolized by a small triangle to indicate that the FF responds only to the *transitions* of the *CLK* input. The FF represented in Figure 5.14(a) will trigger only on the positive-going transitions (PGT). The FF in Figure 5.14(b) triggers only on the negative-going transitions (NGT), as indicated by the bubble on the *CLK* input.

2. All clocked FFs have one or more *control* inputs. These control inputs can have

FIGURE 5.14   General symbols for clocked FFs: (a) triggers on PGT; (b) triggers on NGT.

various labels, depending on their exact function. The control inputs will determine the output state of the FF (like the SET and CLEAR inputs discussed earlier), but their effect is synchronized to the appropriate clock transition. As such, they are also called *synchronous* inputs.

For example, the synchronous inputs to the FF of Figure 5.14(a) will *not* affect Q until the PGT of the clock signal. Likewise, the synchronous inputs of Figure 5.14(b) will have no effect on Q until the NGT of the clock.

3. In other words, the logic levels present at the synchronous inputs will control how the FF output will change, while the signal at *CLK* will actually trigger the change when the appropriate transition (edge) occurs.

Set-Up and Hold Times   Two timing requirements must be met if a clocked FF is to respond reliably to its control inputs when the active *CLK* transition occurs. These requirements are illustrated in Figure 5.15 for a FF that triggers on a PGT.

The *set-up time*, t_S, is the time interval immediately preceding the active transition of the *CLK* signal during which the synchronous input has to be maintained at the proper level. IC manufacturers usually specify the minimum allowable set-up time. If this time requirement is not met, the FF may not respond reliably when the clock edge occurs.

The *hold time*, t_H, is the time interval immediately following the active transition of the *CLK* signal during which the synchronous input has to be maintained at the proper level. IC manufacturers usually specify the minimum acceptable value of hold time. If this requirement is not met, the FF will not trigger reliably.

Thus, to ensure that a clocked FF will respond properly when the active clock transition occurs, the synchronous inputs must be stable (unchanging) for a time interval equal to t_S *prior* to the clock transition, and for a time interval equal to t_H *after* the clock transition.

IC flip-flops will have t_S and t_H values in the nanosecond range. Set-up times are usually in the 5- to 50-ns range while hold times are generally from 0 to 10 ns. Notice that these times are measured between the 50 percent points on the transitions.

These timing requirements are very important in synchronous systems, because, as we shall see, there will be many situations where the synchronous inputs to a FF are changing at approximately the same time as the *CLK* input.

FIGURE 5.15 Set-up and hold times.

1. Define set-up time.
2. Define hold time.
3. The control inputs to a clocked FF can change the output state even if *CLK* is held at a constant level. True or false?

5.5 THE CLOCKED S-C FLIP-FLOP

Figure 5.16(a) shows the logic symbol for a *clocked S-C flip-flop* that is triggered by the positive-going edge of the clock signal. This means that the FF can change states *only* when a signal applied to its clock input makes a transition from 0 to 1. The *S* and *C* inputs control the state of the FF in the same manner as described earlier for the NOR-gate FF, but the FF does not respond to these inputs until the occurrence of the PGT of the clock signal.

The truth table in Figure 5.16(b) shows how the FF output will respond to the PGT at the *CLK* input for the various combinations of *S* and *C* inputs. This truth table uses some new nomenclature to distinguish between the value of Q prior to the PGT (labeled as Q_N) and the value of Q after the occurrence of the PGT (labeled as Q_{N+1}). This nomenclature is widely used by IC manufacturers on their IC data sheets.

The waveforms in Figure 5.16(c) illustrate the operation of the clocked S-C FF. If we assume that the set-up and hold time requirements are being met in all cases, we can analyze these waveforms as follows:

1. Initially all inputs are 0 and the Q output is assumed to be 0.
2. When the PGT of the first clock pulse occurs (point *a*), the *S* and *C* inputs are both 0, so the FF is not affected and remains in the $Q = 0$ state (i.e., $Q_{N+1} = Q_N$).
3. At the occurrence of the PGT of the second clock pulse (point *c*), the *S* input is now high, with *C* still low. Thus, the FF sets to the 1 state at the rising edge of this clock pulse.
4. When the third clock pulse makes its positive transition (point *e*), it finds that $S = 0$ and $C = 1$, which causes the FF to clear to the 0 state.
5. The fourth pulse sets the FF once again to the $Q = 1$ state (point *g*) because $S = 1$ and $C = 0$ when its positive edge occurs.
6. The fifth pulse also finds tht $S = 1$ and $C = 0$ when it makes its positive-going transition. However, Q is already high, so it remains in that state.
7. The $S = C = 1$ condition should not be used, because it results in an ambiguous condition.

It should be noted from these waveforms that the FF is not affected by the negative-going transitions of the clock pulses. Also, note that the *S* and *C* levels have no effect on the FF, except upon the occurrence of a positive-going transition of the clock signal. The *S* and *C* inputs are synchronous *control* inputs; they control which state the FF will go to when the clock pulse occurs; the *CLK* input is the *trigger* input that causes the FF to change states according to what the *S* and *C* inputs are when the active clock transition occurs.

FIGURE 5.16 (a) Clocked S-C FF that responds to the positive-going edge of clock pulse; (b) truth table; (c) typical waveforms.

Figure 5.17 shows the symbol for a clocked S-C flip-flop that triggers on the *negative*-going transition at its *CLK* input. The small circle and triangle on the *CLK* input indicates that this FF will trigger only when the *CLK* input goes from 1 to 0. This FF operates in the same manner as the positive-edge FF except that the output can change states only on the falling edge of the clock pulses (points *b*, *d*, *f*, *h*, and *j* in Figure 5.16). Both positive-edge and negative-edge triggering FFs are used in digital systems.

Sec. 5.5   The Clocked S-C Flip-Flop   **133**

S	C	Q_{N+1}
0	0	Q_N (no change)
1	0	1
0	1	0
1	1	Ambiguous

Triggers on
negative edge

FIGURE 5.17   Clocked S-C FF that triggers on negative-going transitions.

NAND Implementation of Edge-Triggered S-C FF   We are not interested in a detailed analysis of the internal circuitry of any of the clocked FFs, since all of them are available as ICs, and our major interest is in their external operation. It will be helpful, however, to look at a simplified version of a clocked S-C FF implemented from NAND gates, as shown in Figure 5.18.

The circuit contains the following three sections:

1. A basic NAND-gate FF formed by NANDs 3 and 4.
2. A *pulse-steering* circuit formed by NANDs 1 and 2.
3. An *edge-detector* circuit.

The edge-detector circuit generates a narrow pulse ($CLK*$) that occurs coincident with the appropriate transition of the CLK input pulse. For a positive-edge triggered FF, the edge detector is designed to produce $CLK*$ when CLK makes a PGT; for a negative-edge triggered FF, $CLK*$ occurs when CLK makes a NGT.

The pulse-steering circuit allows $CLK*$ to pass through NAND-1 or NAND-2 depending on the levels at S and C. For example, when $S = 1$ and $C = 0$, NAND-1 will pass $CLK*$ (inverted) and NAND-2 will inhibit $CLK*$. This produces a LOW pulse at the SET input of the NAND FF to set $Q = 1$. When $S = 0$ and $C = 1$,

FIGURE 5.18   Simplified version of the internal circuitry for an edge-triggered S-C FF.

NAND-2 will steer *CLK* ∗ (inverted) to the CLEAR input of the NAND FF to clear $Q = 0$.

The edge detector ensures that the S and C inputs have no effect on Q and $\overline{Q}$ except for a few nanoseconds when *CLK* ∗ is HIGH.

REVIEW QUESTIONS

1. True or false: A negative-edge triggered S-C FF will change states when the S or C inputs make a NGT.
2. Why is an edge-detector circuit used in the clocked S-C FF?

5.6 THE CLOCKED J-K FLIP-FLOP

Figure 5.19(a) shows a *clocked J-K flip-flop* that is triggered by the positive-going edge of the clock signal. The J and K inputs control the state of the FF in the same way as the S and C inputs do for the clocked S-C FF except for one major difference:

(a)

(b)

FIGURE 5.19 (a) Clocked J-K FF that responds to positive edge of clock; (b) waveforms.

the J = K = 1 condition does not result in an ambiguous output. For this 1,1 condition, the FF will always go to its opposite state upon the positive transition of the clock signal. This is called the **toggle** mode of operation. In this mode, if both *J* and *K* are left HIGH, the FF will change states (toggle) for each clock pulse.

The truth table in Figure 5.19(a) summarizes how the *J-K* FF responds to the PGT for each combination of *J* and *K*. Notice that the truth table is the same as for the clocked S-C FF (Figure 5.15) except for the $J = K = 1$ condition. This condition results in $Q_{N+1} = \overline{Q}_N$, which means that the new value of *Q* will be the inverse of the value it had prior to the PGT; this is the toggle operation.

The operation of this FF is illustrated by the waveforms in Figure 5.19(b). Once again we assume that the set-up and hold time requirements are being met.

1. Initially all inputs are 0 and the *Q* output is assumed to be 1.
2. When the positive-going edge of the first clock pulse occurs (point *a*) the $J = 0$, $K = 1$ condition exists. Thus, the FF will be cleared to the $Q = 0$ state.
3. The second clock pulse finds $J = K = 1$ when it makes its positive transition (point *c*). This causes the FF to toggle to its opposite state, $Q = 1$.
4. At point *e* on the clock waveform, *J* and *K* are both 0, so the FF does not change states on this transition.
5. At point *g*, $J = 1$ and $K = 0$. This is the condition that sets *Q* to the 1 state. However, it is already 1, so it will remain there.
6. At point *i*, $J = K = 1$, so the FF toggles to its opposite state. The same thing occurs at point *k*.

It should also be noted from these waveforms that the FF is not affected by the negative-going edge of the clock pulses. Also, the *J* and *K* input levels have no effect except upon the occurrence of the PGT of the clock signal. The *J* and *K* inputs by themselves cannot cause the FF to change states.

Figure 5.20 shows the symbol for a clocked J-K flip-flop that triggers on the negative-going clock-signal transitions. The small circle on the *CLK* input indicates that this FF will trigger when the *CLK* input goes from 1 to 0. This FF operates in the same manner as the positive-edge FF of Figure 5.19 except that the output can change states only on negative-going clock-signal transitions (points *b, d, f, h,* and *j*). Both polarities of edge-triggered J-K FFs are in common usage.

The J-K FF is much more versatile than the S-C FF because it has no ambiguous states. The $J = K = 1$ condition, which produces the toggling operation, finds extensive use in all types of binary counters. In essence, the J-K FF can do anything the S-C FF can do *plus* operate in the toggle mode. Because of this, the J-K FF presently enjoys widespread use in almost all modern digital systems.

J	K	Q_{N+1}
0	0	Q_N (no change)
1	0	1
0	1	0
1	1	$\overline{Q}_N$ (toggles)

FIGURE 5.20   J-K FF that triggers on negative-going transitions.

NAND Implementation of Edge-Triggered J-K FF A simplified version of the internal circuitry of an edge-triggered J-K FF is shown in Figure 5.21. It contains the same three sections as the edge-triggered S-C FF (Figure 5.18). In fact, the only difference between the two circuits is that the Q and $\overline{Q}$ outputs are fed back to the pulse-steering NAND gates. This feedback connection is what gives the J-K FF its toggle operation for the $J = K = 1$ condition.

Let's examine this toggle condition more closely by assuming that $J = K = 1$ and that Q is sitting in the LOW state when a *CLK* pulse occurs. With $Q = 0/\overline{Q} = 1$, NAND gate 1 will steer *CLK* * (inverted) to the SET input of the NAND FF to produce $Q = 1$. If we assume that Q is HIGH when a *CLK* pulse occurs, NAND gate 2 will steer *CLK* * (inverted) to the CLEAR input of the NAND FF to produce $Q = 0$. Thus, Q always ends up in the opposite state.

In order for the toggle operation to work as described above, the *CLK* * pulse must be very narrow. It has to return to 0 before the Q and $\overline{Q}$ outputs toggle to their new values; otherwise the new values of Q and $\overline{Q}$ will cause the *CLK* * pulse to toggle the NAND FF again.

REVIEW QUESTIONS

1. True or false: A J-K FF can be used as a S-C FF, but a S-C FF cannot be used as a J-K FF.
2. Does a J-K FF have any ambiguous input conditions?
3. What J-K input condition will set Q upon the occurrence of the active *CLK* transition?

5.7 THE CLOCKED D FLIP-FLOP

Figure 5.22(a) shows the symbol for a clocked D flip-flop that triggers on a PGT at the *CLK* input. The *D* input is a synchronous input that controls the state of the FF according to the accompanying truth table. The operation is very simple; Q will go to the same state that is present on the *D* input when a PGT occurs at *CLK*. This is illustrated by the waveforms in Figure 5.22(b).

Note that each time a PGT occurs on the *CLK* input, Q takes on the value present at the *D* input. The negative-going transitions at *CLK* have no effect, and the *D* input has no effect except when a PGT occurs.

Negative-edge triggered D FFs are also available and they operate in exactly the same manner, except that they respond only to NGTs at the *CLK* input. The symbol for these FFs will have a bubble on the *CLK* input.

Implementation of D FF An edge-triggered D FF is easily implemented by adding a single INVERTER to the edge-triggered S-C FF as shown in Figure 5.23. If you try both values of D, you should see that $Q_{N+1} = D$.

EXAMPLE 5.5

How can a J-K FF be modified to operate as a D FF?

Solution: The modification, shown in Figure 5.24, is the same as was done for the S-C FF in Figure 5.22.

FIGURE 5.21 NAND-gate implementation of edge-triggered J-K FF.

D	Q_{N+1}
0	0
1	1

Q_{N+1} is value of Q after PGT

(a)

(b)

FIGURE 5.22 (a) D FF that triggers on positive-going transitions; (b) waveforms.

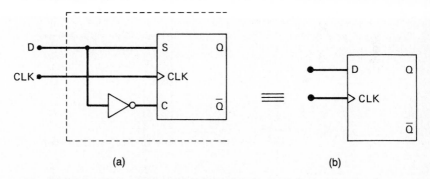

(a)   (b)

FIGURE 5.23 Edge-triggered D FF implemented from S-C FF.

FIGURE 5.24 D FF implemented from a J-K FF.

Parallel Data Transfer   At this point you may well be wondering about the usefulness of the D FF, since it appears that the Q output is the same as the D input. Not quite; remember, Q takes on the value of D only at certain time instances, and so it is not identical to D (e.g., see waveforms in Figure 5.22).

In most applications of the D FF, the Q output must take on the value at its D input only at precisely defined times. One example of this is illustrated in Figure 5.25. Logic-circuit outputs X, Y, Z are to be transferred to FFs Q_1, Q_2, and Q_3 for storage. Using the D FFs, the levels present at X, Y, and Z will be transferred to Q_1, Q_2, and Q_3 respectively, upon application of a TRANSFER pulse to the common *CLK* inputs. The FFs can store these values for subsequent processing. This is an example of *parallel* transfer of binary data; the three bits X, Y, and Z are all transferred simultaneously.

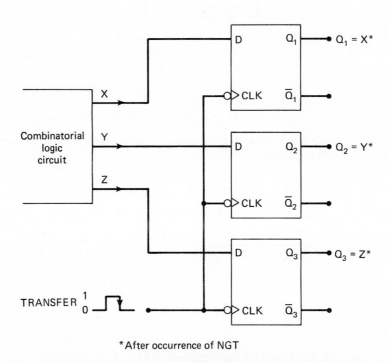

*After occurrence of NGT

FIGURE 5.25   Parallel transfer of binary data using D FFs.

REVIEW QUESTIONS

1. What will happen to the Q waveform in Figure 5.22(b) if the D input is held permanently LOW? (*Ans.* Q will go LOW and stay there)
2. True or false: The Q output will equal the level at the D input at all times.

---

140   **Chap. 5**   **Flip-Flops**

5.8 THE D LATCH

The edge-triggered D FF uses an edge-detector circuit to ensure that the output will take on the value at the D input *only* when the active clock transition occurs. If this edge detector is not used, the resultant circuit will operate somewhat differently. This circuit, called a *D latch*, is drawn in Figure 5.26.

The circuit contains a NAND SET-CLEAR FF and pulse-steering NAND gates 1 and 2. The operation is described as follows:

1. When *CLK* is LOW, the D input will have no effect, since this LOW keeps the NAND FF inputs both HIGH (no-change condition).
2. When *CLK* is HIGH, the D input will produce a LOW at either SET or CLEAR to cause Q to take on the same value as D. If D changes while *CLK* is HIGH, Q will follow this change exactly, and the latch is said to be "transparent."

This operation is somewhat different from the edge-triggered D FF. The absence of the edge-detector circuit allows Q to change states if D changes while *CLK* is HIGH.

The logic symbol for the D latch is shown in Figure 5.26(c). It does not have the small triangle on the *CLK* input since it is not edge-triggered. The reader should be aware, however, that the triangle symbol has not become standard yet, and some circuit diagrams and IC data books do not use it. In those cases, the symbols for the D latch and edge-triggered D FF would be the same, and unless you recognized the IC number, you would have to look it up in an IC data book.

CLK	D	Q
0	X	No change
1	0	0
1	1	1

"X" indicates "don't care"

(a) (b) (c)

FIGURE 5.26 D latch: (a) structure; (b) truth table; (c) logic symbol.

Some IC manufacturers have decided to use a different name for the *CLK* input of a *D* latch; they call it an ENABLE input and label it as *E* or *EN* on the logic symbol. They do this because, strictly speaking, a *CLK* input is supposed to affect *Q* only on its transition.

EXAMPLE 5.6

Determine the output waveform from a *D* latch for the input waveforms of Figure 5.27. Assume $Q = 0$ initially.

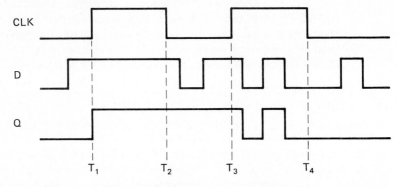

FIGURE 5.27   Example 5.6

Solution: Prior to time T_1, *CLK* is LOW so that *Q* cannot change. During the interval T_1–T_2, *CLK* is HIGH and *Q* will take on the value of $D = 1$. When *CLK* returns LOW at T_2, *Q* will hold (latch) at this HIGH level, since it cannot change while $CLK = 0$.

At T_3 when *CLK* goes HIGH again, *Q* will follow the *D* input exactly until T_4, when *CLK* returns LOW. During the T_3–T_4 interval, the latch is "transparent" because it allows *D* through to the output. At T_4 when *CLK* returns LOW, *Q* latches at the 0 level, since that was the last value of *D* while *CLK* was HIGH.

REVIEW QUESTIONS

1. Explain how a *D* latch operates differently from an edge-triggered D FF.
2. True or false: In a *D* latch the *Q* output "latches" to the last value of *D* prior to the NGT of *CLK*.

5.9   ASYNCHRONOUS INPUTS

For the clocked flip-flops that we have been studying, the *S*, *C*, *J*, *K*, and *D* inputs have been referred to as *control* inputs. These inputs are also called *synchronous inputs*, because their effect on the FF output is synchronized with the *CLK* input. As we have seen, the synchronous control inputs must be used in conjunction with a clock signal to trigger the FF.

Most clocked FFs also have one or more *asynchronous inputs* which operate independently of the synchronous inputs and clock input. These asynchronous inputs can be used to set the FF to the 1 state or clear the FF to the 0 state at any time, regardless of the conditions at the other inputs. Stated in another way, the asynchronous inputs are *override* inputs, which can be used to override all the other inputs in order to place the FF in one state or the other.

Figure 5.28 shows a J-K FF with two asynchronous inputs designated as DC SET and DC CLEAR. These are active-LOW inputs, as indicated by the bubbles on the FF symbol. The accompanying truth table summarizes how they affect the FF output. Let's examine the various cases.

- **DC SET = DC CLEAR = 1:** The asynchronous inputs are inactive and the FF is free to respond to the *J*, *K*, and *CLK* inputs; in other words, the clocked operation can take place.
- **DC SET = 0; DC CLEAR = 1:** The DC SET is activated and *Q* is *immediately* set to 1 no matter what conditions are present at the *J*, *K*, and *CLK* inputs. The *CLK* input cannot affect the FF while DC SET = 0.
- **DC SET = 1; DC CLEAR = 0:** The DC CLEAR is activated and *Q* is *immediately* cleared to 0 independent of the conditions on the *J*, *K*, or *CLK* inputs. The *CLK* input has no effect while DC CLEAR = 0.
- **DC SET = DC CLEAR = 0:** This condition should not be used, since it can result in an ambiguous response.

It is important to realize that these asynchronous inputs respond to dc levels. This means that if a constant 0 is held on the DC SET input, the FF will remain in the $Q = 1$ state regardless of what is occurring at the other inputs. Similarly, a constant low on the DC CLEAR input holds the FF in the $Q = 0$ state. Thus, the asynchronous inputs can be used to hold the FF in a particular state for any desired interval. Most often, however, the asynchronous inputs are used to set or clear the FF to the desired state by application of a momentary pulse.

Many clocked FFs that are available as ICs will have both of these asynchronous inputs; some will have only the DC CLEAR input. Some FFs will have asynchronous inputs that are active-HIGH rather than active-LOW. For these FFs the FF symbol would not have a bubble on the asynchronous inputs.

DC SET	DC CLEAR	FF response
1	1	Clocked operation
0	1	Q = 1
1	0	Q = 0
0	0	Not used

FIGURE 5.28 Clocked J-K FF with asynchronous inputs.

Unfortunately, IC manufacturers have not agreed on what designation to use for these asynchronous inputs. We will normally use the DC SET/DC CLEAR labels, but there are several other designations. The most common ones are listed below.

Asynchronous SET input	Asynchronous CLEAR input
DC SET	DC CLEAR
PRESET	CLEAR
SET	RESET
S_D (direct set)	C_D (direct clear)

While most IC FFs do have at least one asynchronous input, in many circuit applications these inputs are not used. In such cases they are permanently held at their inactive logic level. In our use of FFs throughout the text, we will not bother to show the asynchronous inputs unless they are being used.

The operation of the asynchronous inputs is illustrated in Figure 5.29. The synchronous inputs (J, K) are held HIGH as CLK pulses are applied. The FF will toggle on each negative-going transition as long as the PRESET and CLEAR inputs are both HIGH when the CLK edge occurs. This happens at points a, c, d, and f. The Q output can also change states in response to a LOW at PRESET or CLEAR. The LOW on PRESET causes Q to go to the 1 state *immediately*, regardless of the state of CLK (point b). Similarly, the LOW at CLEAR causes Q to go to 0 immediately (point e). Note that this LOW at CLEAR prevents the fourth CLK edge from toggling Q; this is the override feature.

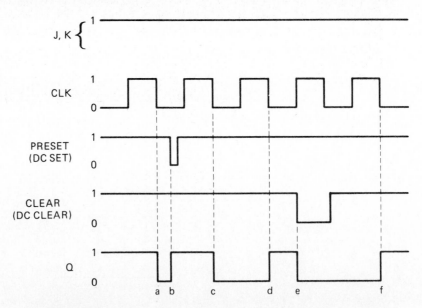

FIGURE 5.29 **Asynchronous inputs can cause FF to change states immediately without waiting for CLK transition, and they override the CLK input.**

1. True or false: If the DC CLEAR input is kept at its active level, the *CLK* input cannot affect *Q*.
2. A certain clocked D FF has active-HIGH asynchronous inputs. What conditions are necessary for the FF to operate in the clocked mode?
3. What are some of the other designations used for DC SET?
4. What are some of the other designations used for DC CLEAR?

5.10 FLIP-FLOP TIMING CONSIDERATIONS

Manufacturers of IC FFs will specify several important timing parameters and characteristics that must be considered before a FF is used in any circuit application. We will describe the most important of these and then give some actual examples of specific IC FFs from the TTL and CMOS logic families.

Set-Up and Hold Times These have already been discussed, and you may recall from Section 5.4 that they represent requirements that must be met for reliable FF triggering. The manufacturers' IC data sheet will always specify the *minimum* values of t_S and t_H.

Propagation Delays Whenever a signal is to change the state of a FF's output, there is a delay from the time the signal is applied to the time when the output makes its change. Figure 5.30 illustrates the propagation delays that occur in response to a positive transition on the *CLK* input. Note that these delays are measured between the 50 percent points on the input and output waveforms. The same types of delays occur in response to signals on a FF's asynchronous inputs (DC SET and CLEAR). The manufacturers' data sheets usually specify propagation delays in response to all inputs, and they usually specify the *maximum* values for t_{PLH} and t_{PHL}.

Modern IC flip-flops have propagation delays that range from a few nanoseconds to around 1 μs: The values of t_{PLH} and t_{PHL} are generally not the same and they increase in direct proportion to the number of loads being driven by the Q output. FF propagation delays play an important part in certain situations which we will encounter later.

FIGURE 5.30   FF propagation delays.

Maximum Clocking Frequency, f_{MAX}   This is the highest frequency that may be applied to the *CLK* input of a FF and still have it trigger reliably. The f_{MAX} limit will vary from FF to FF even with FFs having the same device number. For example, the manufacturer of the 7470 J-K FF chip tests many of these FFs and may find that the f_{MAX} values fall in a range from 20 to 35 MHz. He will then specify the *minimum f_{MAX}* as 20 MHz. This may seem confusing, but a little thought should make it clear that what the manufacturer is saying is that he cannot guarantee that the 7470 FF that you put in your circuit will work above 20 MHz; many of them will, but some of them will not. If you stay below 20 MHz, however, he guarantees that they will all work.

Clock Pulse HIGH and LOW Times   The manufacturer will also specify the *minimum* time duration that the *CLK* signal must remain LOW before it goes HIGH, sometimes called $t_W(L)$, and the *minimum* time that *CLK* must be kept HIGH before it returns LOW, sometimes called $t_W(H)$. These times are defined in Figure 5.31. Failure to meet these minimum time requirements can result in unreliable triggering.

FIGURE 5.31   Definition of clock LOW and clock HIGH times.

Asynchronous Active Pulse Width   The manufacturer will also specify the *minimum* time duration that a DC SET or DC CLEAR input has to be kept in its active state in order to reliably set or clear the FF.

Clock Transition Times   For reliable triggering, the clock waveform transition times (rise and fall times) should be kept very short. If the clock signal takes too long to make the transitions from one level to the other, the FF may trigger erratically or not at all. Manufacturers usually do not list a maximum transition-time requirement for each FF IC. Instead, it is usually given as a general requirement for all ICs within a given logic family. For example, the transition times should generally be $\leq$ 50 ns for TTL devices and $\leq$ 200 ns for CMOS. These requirements will vary slightly among the different manufacturers and among the various subfamilies within the broad TTL and CMOS logic families.

Actual ICs   As practical examples of these timing parameters, let's take a look at several actual IC FFs. In particular, we will look at the following ICs:

- **7474**   a dual edge-triggered D FF (TTL family)
- **74LS112**   a dual edge-triggered J-K FF (TTL)
- **4013B**   a dual edge-triggered D FF (CMOS family)
- **4027B**   a dual edge-triggered J-K FF (CMOS)

Table 5.1 lists the various timing values for each of these FFs as they appear in the

TABLE 5.1 FLIP-FLOP TIMING VALUES (IN NANOSECONDS)

	TTL		CMOS	
	7474	74LS112	4013B	4027B
t_S	20	20	60	100
t_H	5	0	0	0
t_{PHL}—from CLK to Q	40	24	200	200
t_{PLH}—from CLK to Q	25	16	200	200
t_{PHL}—from DC SET/DC CLEAR to Q	40	24	225	350
t_{PLH}—from DC SET/DC CLEAR to Q	25	16	225	350
$t_W(L)$—CLK LOW time	37	15	100	150
$t_W(H)$—CLK HIGH time	30	20	100	150
$t_W(L$ or $H)$—at DC SET/DC CLEAR	30	15	60	150
f_{MAX}—in MHz	15	30	5	4

manufacturers' data books. All the listed values are *minimum* values except for the propagation delays, which are *maximum* values.

Examination of this table reveals two interesting points.

1. All the FFs have very low values of hold time; this is typical of most modern edge-triggered FFs.
2. The values of the timing parameters for the CMOS FFs are generally much greater than for the TTL FFs, and the f_{MAX} values are lower for CMOS. As we shall discuss later, CMOS devices cannot respond to input signals as fast as TTL devices.

EXAMPLE 5.7

From Table 5.1 determine the following:
(a) Assume $Q = 0$. How long can it take for Q to go HIGH when a PGT occurs at the CLK input of a 7474?
(b) Assume $Q = 1$. How long can it take for Q to go LOW in response to the DC CLEAR input of a 4027B FF?
(c) What is the narrowest pulse that should be applied to the DC CLEAR input of the 74LS112 FF to reliably clear Q?
(d) Which FF in Table 5.1 requires that the control inputs remain stable *after* the occurrence of the active clock transition?

Solution:
(a) The PGT will cause Q to go from LOW to HIGH. The delay from CLK to Q is listed as $t_{PLH} = 25$ ns for the 7474.
(b) For the 4027B the time required for Q to go from HIGH to LOW in response to the DC CLEAR input is listed at $t_{PHL} = 350$ ns.
(c) For the 74LS112 the narrowest pulse at the DC CLEAR input is listed as $t_W(L) = 15$ ns.
(d) The 7474 is the only FF in Table 5.1 that has a nonzero hold-time requirement.

1. Define t_{PLH} and t_{PHL}.
2. Explain the meaning of the f_{MAX} requirement.
3. True or false: As long as a clock waveform meets the f_{MAX} requirement, it can have any duty cycle and still produce reliable triggering.

5.11 MASTER/SLAVE FLIP-FLOPS

In digital systems the outputs of flip-flops are often connected, directly or through logic gates, to the inputs of other flip-flops. Figure 5.32 is an illustration of this. The output of FF Q_1 acts as the J input of FF Q_2, and both FFs are triggered by the same clock pulse at their *CLK* inputs. Let us assume that initially $Q_1 = 1$ and $Q_2 = 0$. Since Q_1 has both its J and K inputs high, it will toggle to the 0 state on the negative transition of the clock pulse. Q_2 has $J = 1$ (from Q_1) and $K = 0$ prior to the negative transition of the clock pulse. However, when the clock pulse goes LOW, Q_1 will also go LOW, so the J input of Q_2 will be changing from 1 to 0 while Q_2 is being clocked. This is called a *race condition* and can sometimes lead to unpredictable triggering of Q_2.

Recall from our discussion of set-up and hold times (t_S and t_H) that edge-triggered FFs require that the control inputs (S, C or J, K or D) remain stable for a

FIGURE 5.32   Example of the type of race condition that can occur in FF circuits.

time t_H after the clock makes its transition. If the t_H requirement for the FF is very small or zero, the race situation in Figure 5.32 might not cause a problem. This is because Q_1 will actually go low a short time after the clock goes low, owing to the inherent delay of FF Q_1 in responding to the clock pulse. Thus, Q_2 will respond properly by going to the 1 state on the falling edge of the clock.

While most of the newer edge-triggered FFs will have no trouble with this race condition because of their small t_H requirements, there may be some situations where race conditions similar to this one can produce unreliable triggering. There is another class of clocked FFs that does not have a hold-time requirement (i.e., $t_H = 0$). They are called *master/slave* FFs because of their internal structure, which actually contains two FFs—the master FF and the slave FF. All three types of clocked FFs (S-C, J-K, and D) can be implemented in a master/slave arrangement as well as an edge-triggered version.

The J-K Master/Slave Flip-Flop Figure 5.33(a) shows a simplified circuit for a J-K master/slave (hereafter abbreviated M/S) flip-flop. It contains two almost identical stages. The first stage consists of a master FF that is a NAND SET-CLEAR FF with active-LOW inputs, and pulse-steering NAND gates 1 and 2. The second stage

(a)

(b)

J	K	Q_{N+1}
0	0	Q_N (no change)
0	1	0
1	0	1
1	1	$\overline{Q}_N$ (toggles)

(c)

FIGURE 5.33 (a) Internal circuit for an M/S J-K FF; (b) circuit symbol; (c) truth table.

consists of a slave FF that is a NAND SET-CLEAR FF, and pulse-steering NAND gates 3 and 4. The Q_S and $\overline{Q}_S$ outputs of the slave FF serve as the external FF outputs.

The *CLK* input is directly connected to NAND gates 1 and 2. When *CLK* = 0, there will be 1s at the S and C inputs of the master FF so that Q_M and $\overline{Q}_M$ cannot change states.

When *CLK* = 1, NAND gates 1 and 2 will be free to respond to J, K, and to the current slave FF outputs Q_S and $\overline{Q}_S$. For example, if $Q_S = 0/\overline{Q}_S = 1$ and $J = K = 1$, then when *CLK* goes HIGH, NAND 1 will go LOW and will set the master ($Q_M = 1$). Thus, *the master can change states only while CLK = 1.*

The *CLK* is inverted and fed to NAND gates 3 and 4. When *CLK* = 1, there will be 1s at the S and C inputs of the slave so that Q_S and $\overline{Q}_S$ cannot change.

When *CLK* = 0, NAND gates 3 and 4 will respond to the master outputs and will set or clear the slave accordingly. For example, if the master had been previously set to the $Q_M = 1$ state, then when *CLK* goes LOW, the LOW at the NAND-3 output will set $Q_S = 1$. Conversely, if $Q_M = 0$ when *CLK* goes LOW, a LOW will occur at the NAND-4 output to clear $Q_S = 0$. Thus, *the slave can change states only when CLK goes LOW, and Q_S takes on the current value of Q_M.*

This operation can be summarized in two basic steps:

1. **When CLK = 1:** The J, K levels are transferred into the master, which will go to a state determined by J, K, and the current Q_S. The slave cannot change states while *CLK* = 1.
2. **When CLK = 0:** The outputs of the master are transferred to the slave. The master cannot change states while *CLK* = 0.

It is important to understand that the slave outputs can change states only when *CLK* goes from 1 to 0. As such, the external FF output Q can be considered to change states on the NGT of *CLK* in accordance with the states of the J-K truth table. Thus, the logic symbol and truth table for the M/S flip-flop (Figure 5.33) are the same as for the edge-triggered J-K FF, except for the M/S designation.

What, then, is the difference in operation between the M/S and edge-triggered J-K FFs? The M/S responds to J, K, and *CLK* in the two steps described above, while the edge-triggered FF does it all in one step when the active transition of the *CLK* occurs.

Advantage of Master/Slave FFs   The master essentially "stores" the levels that are on J and K while *CLK* = 1, so that it is not necessary for J and K to remain stable once *CLK* goes LOW. In other words, *there is no hold-time requirement.* The master will transfer the correct value to the slave when *CLK* goes from 1 to 0 even if J and K are changing.

This allows us to use M/S FFs in situations where the J, K inputs are changing at the same time as the NGT of the *CLK* signal. This is illustrated in Figure 5.34.

Here the Q waveform can be determined in the same manner as for the edge-triggered J-K FF. Q can change states only on the NGT at T_2, T_4, T_6, and T_8. When these transitions occur, Q will take on the state determined by what the J, K levels were during the previous *CLK* = 1 interval. For example, when the NGT occurs at T_4, Q will respond to the $J = 1$, $K = 0$ condition that existed in the T_3–T_4 interval.

FIGURE 5.34 Typical waveforms for J-K M/S FF.

Limitation of M/S FF   An *edge-triggered* FF is sensitive only to what the *J* and *K* inputs are at the time of the active clock transition. At all other times the *J* and *K* inputs are *disabled* (have no effect). In the M/S FF, the *J* and *K* inputs can affect the FF at any time while $CLK = 1$. This characteristic can cause problems if the *J*, *K* inputs are not stable or there is noise on either one of them, during the *CLK* HIGH interval. To illustrate, the waveforms of Figure 5.35 show what can happen if a noise glitch appears on the *J* input at T_1 while $CLK = 1$. This glitch will set the master ($Q_M = 1$); then when *CLK* goes LOW at T_2, the 1 will be transferred from Q_M to output *Q*.

Actual ICs   Most of the newer FF ICs are edge-triggered types with very small hold-time requirements. The M/S types have decreased in importance because they no longer have any advantage over the edge-triggered FFs, and they are susceptible to the problem described above.

Some M/S flip-flops are still in use. The most common are the 7473, 7476, and 74107, all of which are TTL dual J-K FFs. These FFs operate as described above. The

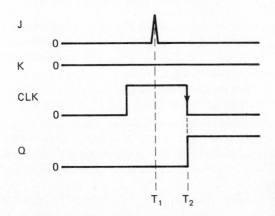

FIGURE 5.35 M/S FF can respond to noise or changes in *J*, *K* inputs while *CLK* = 1.

74110 and 74111 ICs are M/S J-K FFs that have been modified to prevent the J, K inputs from having any effect while CLK = HIGH.

─── **REVIEW QUESTIONS**
1. Describe the two steps by which the M/S FF responds to the J, K and CLK inputs.
2. True or false: The slave will respond to J and K while CLK = 0.
3. Refer to the diagram of Figure 5.33. Assume $Q_S = Q_M = 1$ and determine what happens to Q_M when CLK goes HIGH while J and K are at the following levels:
   (a) $J = 1$, $K = 0$ (b) $J = 0$, $K = 1$ (c) $J = 0$, $K = 0$ (d) $J = 1$, $K = 1$
   [*Ans.* (a) $Q_M = 1$; (b) $Q_M = 0$; (c) $Q_M = 1$; (d) $Q_M = 0$]
4. Why should J and K be kept stable and noise-free while CLK = 1 for a M/S FF?

5.12 FLIP-FLOP APPLICATIONS

Earlier in this chapter we saw some examples of how the simple NAND FF and NOR FF were used to perform switch debouncing (Example 5.2) and event storage (Example 5.4). These simple unclocked FFs are somewhat limited in their applications. Clocked FFs offer the logic designer a group of versatile devices that have numerous applications. We will briefly introduce the more common applications in the following sections and will expand on them in subsequent chapters.

5.13 FLIP-FLOP SYNCHRONIZATION

Most digital systems are principally synchronous in their operation in that most of the signals will change states in synchronism with the clock transitions. In many cases, however, there will be an external signal that is not synchronized to the clock; in other words, it is asynchronous. Asynchronous signals often occur as a result of a human operator's actuating an input switch at some random time relative to the clock signal. This randomness can produce unpredictable and undesirable results. The following example illustrates how a FF can be used to synchronize the effect of an asynchronous input.

EXAMPLE 5.8 ────────────────────────────────────

Figure 5.36(a) shows a situation where input signal A is generated from a debounced switch that is actuated by an operator (a debounced switch was first introduced in Example 5.2). A goes HIGH when the operator actuates the switch and goes LOW when the operator releases the switch. This A input is used to control the passage of the clock signal through the AND gate so that clock pulses appear at output X only as long as A is HIGH.

    The problem with this circuit is that A can change states at any time relative to the clock signal because the exact times when the operator actuates or releases the switch are essentially random. This can produce *partial* clock pulses at output X if either transition of A occurs while the clock signal is HIGH, as shown in the waveforms of Figure 5.36(b).

    This type of output is often not acceptable, so a method for preventing the appearance of partial pulses at X must be developed. One solution is shown in

FIGURE 5.36 Asynchronous signal A can produce partial pulses at X.

Figure 5.37(a). Describe how this circuit solves the problem, and draw the X waveform for the same situation as in Figure 5.36(b).

Solution: The A signal is connected to the D input of FF Q which is clocked by the NGT of the clock signal. Thus, when A goes HIGH, Q will not go HIGH until the next NGT of the clock at time T_1. This HIGH at Q will enable the AND gate to pass subsequent *complete* clock pulses to X, as shown in Figure 5.37(b).

FIGURE 5.37 A D FF is used to synchronize the enabling of the AND gate to the NGTs of the clock.

When *A* returns LOW, *Q* will not go LOW until the next NGT of the clock at T_2. Thus, the AND gate will not inhibit clock pulses until the clock pulse that ends at T_2 has been passed through to *X*.

Thus, output *X* contains only complete pulses.

5.14 DETECTING AN INPUT SEQUENCE

In many situations an output is to be activated only when the inputs are activated in a certain sequence. This cannot be accomplished using pure combinational logic but requires the storage characteristic of FFs.

For example, an AND gate can be used to determine when two inputs *A* and *B* are both HIGH, but its output will respond the same regardless of which input goes HIGH first. But suppose we want to generate a HIGH output *only* if *A* goes HIGH and then *B* goes HIGH some time later. One way to accomplish this is shown in Figure 5.38(a).

(a)   (b) A goes HIGH before B   (c) B goes HIGH before A

FIGURE 5.38   A clocked FF used to respond to a particular sequence of inputs.

The waveforms in (b) and (c) show that *X* will go HIGH only if *A* goes HIGH before *B* goes HIGH. This is because *A* must be HIGH in order for *X* to go HIGH on the PGT of *B*.

In order for this circuit to work properly, *A* has to go HIGH prior to *B* by at least an amount of time equal to the set-up time requirement of the FF.

5.15 DATA STORAGE AND TRANSFER

By far the most common use of flip-flops is for the storage of data or information. The data may represent numerical values (e.g., binary numbers, BCD-coded characters). These data are generally stored in groups of FFs called *registers*.

The operation most often performed on data that are stored in a FF or a register is the *transfer* operation. This involves the transfer of data from one FF or register to another. Figure 5.39 illustrates how data transfer can be accomplished between two FFs using clocked S-C, J-K, and D FFs. In each case, the logic value that is currently stored in FF *A* is transferred to FF *B* upon the NGT of the TRANSFER pulse. Thus, after this NGT, the *B* output will be the same as the *A* output.

FIGURE 5.39 Synchronous transfer operation performed by various types of FFs.

The transfer operations in Figure 5.39 are examples of *synchronous transfer*, since the synchronous and *CLK* inputs are used to perform the transfer. A transfer operation can also be obtained using the asynchronous inputs of a FF. Figure 5.40 shows how an *asynchronous transfer* can be obtained using the DC SET and DC CLEAR inputs of any type of FF. Here, the asynchronous inputs respond to LOW

FIGURE 5.40 Asynchronous transfer operation.

Sec. 5.15   **Data Storage and Transfer**   **155** levels (note the small circles on the symbol). When the TRANSFER ENABLE line is held LOW, the two NAND outputs are kept HIGH, with no effect on the FF outputs. When the TRANSFER ENABLE line is made HIGH, one of the NAND outputs will go LOW, depending on the state of the A and $\bar{A}$ outputs. This LOW will either set or clear FF B to the same state as FF A. This asynchronous transfer is done independently of the synchronous and CLK inputs of the FF. Asynchronous transfer is also called *jam transfer* because the data can be "jammed" into FF B even if the synchronous inputs are active.

Parallel Data Transfer   Figure 5.41 illustrates data transfer from one register to another using D-type FFs. Register X consists of FFs X_1, X_2, and X_3; register Y consists of FFs Y_1, Y_2, and Y_3. Upon application of the TRANSFER pulse, the value stored in X_1 is transferred to Y_1, X_2 to Y_2, and X_3 to Y_3. The transfer of the contents of the X register into the Y register is a synchronous transfer. It is also referred to as a *parallel* transfer, since the contents of X_1, X_2, and X_3 are transferred *simultaneously* into Y_1, Y_2, and Y_3. If a *serial* transfer were performed, the contents of the X register would be transferred to the Y register one bit at a time. This will be examined in the next section.

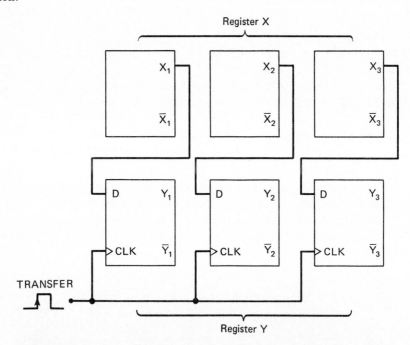

FIGURE 5.41   Parallel transfer of contents of register X into register Y.

1. True or false: Asynchronous data transfer uses the CLK input.
2. Which type of FF is best suited for synchronous transfer because it requires the fewest number of interconnections from one FF to the other?

3. If J-K FFs were used in the registers of Figure 5.41, how many interconnections would be required from each FF in register X to its corresponding FF in register Y?
4. True or false: Synchronous data transfer requires less circuitry than asynchronous transfer.

5.16 SHIFT REGISTERS

Shift registers are used to transfer the contents of one register into a second register one bit at a time. Before examining this operation of *serial* transfer, let us look at the operation of a basic shift register. Figure 5.42(a) shows four J-K FFs wired as a 4-bit shift register. Note that the FFs are connected so that the output of X_3 transfers into X_2, X_2 into X_1, and X_1 into X_0. What this means is that upon the occurrence of the shift pulse, each FF takes on the value stored previously in the FF on its left. FF X_3 takes on a value determined by the conditions present on its J and K inputs when the shift pulse occurs. For now, we will assume that X_3's J and K inputs are fed by the DATA IN waveform shown in Figure 5.42(b). We will also assume that all FFs are in the 0 state before shift pulses are applied.

The waveforms in Figure 5.42(b) show how the input data are shifted from left to right from FF to FF as shift pulses are applied. When the first NGT occurs at T_1, each of the FFs X_2, X_1, and X_0 will have the $J = 0$, $K = 1$ condition present at its inputs because of the state of the FF on its left. FF X_3 will have $J = 1$, $K = 0$ because of DATA IN. Thus, at T_1 only X_3 will go HIGH while all the other FFs remain LOW. When the second NGT occurs at T_2, FF X_3 will have $J = 0$, $K = 1$ because of DATA IN. FF X_2 will have $J = 1$, $K = 0$ because of the current HIGH at X_3. FFs X_1 and X_0 will still have $J = 0$, $K = 1$. Thus, at T_2 only FF X_2 will go HIGH, FF X_3 will go LOW, and FFs X_1 and X_0 will remain LOW.

Similar reasoning can be used to determine how the waveforms change at T_3 and T_4. Note that on each NGT of the shift pulses, each FF output takes on the level that was present at the output of the FF on its left *prior* to the NGT. Of course, X_3 takes on the level that was present at DATA IN prior to the NGT.

Hold-Time Requirement   In this shift-register arrangement it is necessary that the FFs have a very small hold-time requirement, because there are times when the J, K inputs are changing at about the same time as the *CLK* transition. For example, the X_3 output switches from 1 to 0 in response to the NGT at T_2, causing the J, K inputs of X_2 to change while its *CLK* input is changing. Actually, because of the propagation delay of X_3, the J, K inputs of X_2 won't change for a short time after the NGT. For this reason, a shift register should be implemented either from M/S FFs or from edge-triggered FFs that have a t_H value less than one FF propagation delay (*CLK*-to-output). This latter requirement is easily satisfied by most modern edge-triggered FFs.

Serial Transfer between Registers   Figure 5.43 shows two 3-bit shift registers connected so that the contents of the X register will be serially transferred (shifted) into register Y. We are using D FFs for each shift register, since this requires fewer connections than J-K FFs. Notice how X_0, the last FF of register X, is connected to the input of Y_2, the first FF of register Y. Thus, as the shift pulses are applied, the

(a)

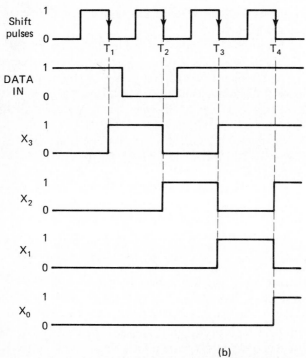

(b)

FIGURE 5.42   Four-bit shift register.

information transfer takes place as follows: $X_2 \rightarrow X_1 \rightarrow X_0 \rightarrow Y_2 \rightarrow Y_1 \rightarrow Y_0$. The X_2 FF will go to states determined by its D input. For now, D will be held LOW, so X_2 will go LOW on the first pulse and will remain there.

To illustrate, let us assume that before any shift pulses are applied, the contents of the X register is 1   0   1 (that is, $X_2 = 1$, $X_1 = 0$, $X_0 = 1$) and the Y register is at 0   0   0. Refer to the table in Figure 5.43(b), which shows how the states of each FF

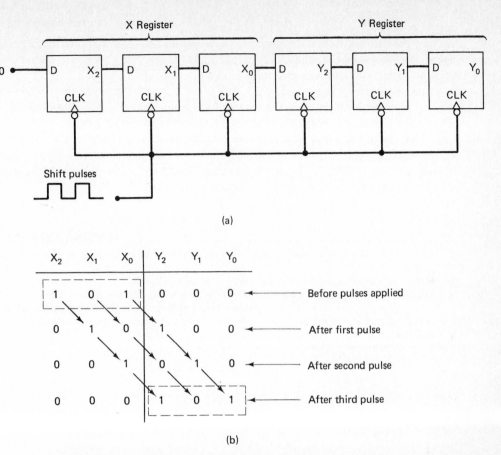

FIGURE 5.43 Serial transfer of information from X register into Y register.

change as shift pulses are applied. The following points should be noted:

1. On the falling edge of each pulse, each FF takes on the value that was stored in the FF on its left prior to the occurrence of the pulse.
2. After *three* pulses, the 1 that was initially in X_2 is in Y_2, the 0 initially in X_1 is in Y_1, and the 1 initially in X_0 is in Y_0. In other words, the 101 stored in the X register has now been shifted into the Y register. The X register is at 000.
3. The complete transfer of the *three* bits of data requires *three* shift pulses.

Parallel versus Serial Transfer   In parallel transfer, all the information is transferred simultaneously upon the occurrence of a *single* transfer command pulse (Figure 5.41), no matter how many bits are being transferred. In serial transfer, as exemplified by Figure 5.43, the complete transfer of N bits of information requires N clock pulses (3 bits requires three pulses, 4 bits requires four pulses, etc.). Parallel transfer, then, is obviously much faster than serial transfer using shift registers.

In parallel transfer, the output of each FF in register X is connected to a corresponding FF input in register Y. In serial transfer, only the last FF in register X is connected to register Y. In general, then, parallel transfer requires more interconnections between the sending register (X) and the receiving register (Y) than does serial transfer. This difference becomes more obvious when a greater number of bits of information are being transferred. This is an important consideration when the sending and receiving registers are remote from each other, since it determines how many lines (wires) are needed in the transmission of the information.

The choice of either parallel or serial transmission depends on the particular system application and specifications. Often, a combination of the two types is used to take advantage of the *speed* of parallel transfer and the *economy and simplicity* of serial transfer. More will be said later about information transfer.

── **REVIEW QUESTIONS**

1. True or false: The fastest method for transferring data from one register to another is parallel transfer.
2. What is the major advantage of serial transfer over parallel transfer?
3. Refer to Figure 5.43. Assume that the initial contents of the registers is: $X_2 = 0$, $X_1 = 1$, $X_0 = 0$, $Y_2 = 1$, $Y_1 = 1$, $Y_0 = 0$. Also assume that the D input of X_2 is held HIGH. Determine the value of each FF output after the occurrence of the fourth shift pulse. (*Ans.* $X_2 = X_1 = X_0 = 1$, $Y_2 = 1$, $Y_1 = 0$, $Y_0 = 1$)

5.17 FREQUENCY DIVISION AND COUNTING

Refer to Figure 5.44(a). Each FF has its J and K inputs at the 1 level, so it will change states (toggle) whenever the signal on its CLK input goes from HIGH to LOW. The train of clock pulses are applied only to the CLK input of FF X_0. Output X_0 is connected to the CLK input of FF X_1 and output X_1 is connected to the CLK input of FF X_2. The waveforms in Figure 5.44(b) show how the FFs change states as the pulses are applied. The following important points should be noted:

1. FF X_0 toggles on the negative-going transition of each input clock pulse. Thus, the X_0 output waveform has a frequency that is exactly $\frac{1}{2}$ of the clock pulse frequency.
2. FF X_1 toggles each time the X_0 output goes from HIGH to LOW. The X_1 waveform has a frequency equal to exactly $\frac{1}{2}$ the frequency of the X_0 output and therefore $\frac{1}{4}$ of the clock frequency.
3. FF X_2 toggles each time the X_1 output goes from HIGH to LOW. Thus, the X_2 waveform has $\frac{1}{2}$ the frequency of X_1 and therefore $\frac{1}{8}$ of the clock frequency.
4. Each FF output is a squarewave (50 percent duty cycle).

As described above, each FF divides the frequency of its input by 2. Thus, if we were to add a fourth FF to the chain, it would have a frequency equal to $\frac{1}{16}$ of the clock frequency; and so on. Using the appropriate number of FFs, this circuit could divide a frequency by any power of 2. Specifically, using N FFs would produce an output frequency from the last FF which is equal to $\frac{1}{2^N}$ of the input frequency.

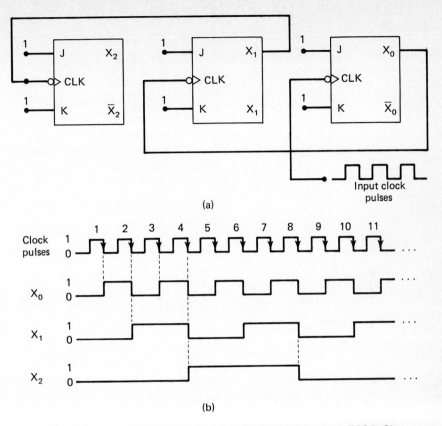

(a)

(b)

FIGURE 5.44   J-K FFs wired as a 3-bit binary counter (MOD-8).

Counting Operation   In addition to functioning as a frequency divider, the circuit of Figure 5.44 also operates as a *binary counter*. This can be demonstrated by examining the sequence of states of the FFs after the occurrence of each clock pulse. Figure 5.45 presents the results in tabular form. Let the $X_2X_1X_0$ values represent a binary number where X_2 is the 2^2 position, X_1 is the 2^1 position, and X_0 is the 2^0 position. The first eight $X_2X_1X_0$ states in the table should be recognized as the binary counting sequence from 000 to 111. After the first pulse the FFs are in the 001 state ($X_2 = 0$, $X_1 = 0$, $X_0 = 1$), which represents 001_2 (equivalent to decimal 1); after the second pulse the FFs represent 010_2, which is equivalent to 2_{10}; after three pulses, $011_2 = 3_{10}$; after four pulses, $100_2 = 4_{10}$; and so on until after seven pulses, $111_2 = 7_{10}$. On the eighth pulse the FFs return to the 000 state, and the binary sequence repeats itself for succeeding pulses.

Thus, for the first seven input pulses, the circuit functions as a binary counter in which the states of the FFs represent a binary number equivalent to the number of pulses that have occurred. This counter can count as high as $111_2 = 7_{10}$, and then it returns to 000.

2^2	2^1	2^0	
X_2	X_1	X_0	
0	0	0	Before applying clock pulses
0	0	1	After pulse #1
0	1	0	After pulse #2
0	1	1	After pulse #3
1	0	0	After pulse #4
1	0	1	After pulse #5
1	1	0	After pulse #6
1	1	1	After pulse #7
0	0	0	After pulse #8 recycles to 000
0	0	1	After pulse #9
0	1	0	After pulse #10
0	1	1	After pulse #11
.	.	.	. . .
.	.	.	. . .
.	.	.	. . .

FIGURE 5.45 Sequence of FF states shows binary counting sequence.

MOD Number   The counter of Figure 5.44 has $2^3 = 8$ different states (000 through 111). It would be referred to as a *MOD-8 counter*, where the MOD number indicates the number of states in the counting sequence. If a fourth FF were added, the sequence of states would count in binary from 0000 to 1111, a total of 16 states. This would be called a *MOD-16 counter*. In general, if N FFs are connected in the arrangement of Figure 5.44, the counter will have 2^N different states, so it is a MOD-2^N counter. It would be capable of counting up to $2^N - 1$ before returning to its zero state.

The MOD number of a counter also indicates the frequency division obtained from the last FF. For instance, a 4-bit counter has 4 FFs, each representing one binary digit (bit), so it is a MOD-2^4 = MOD-16 counter. It can therefore count up to 15 ($=2^4 - 1$). It can also be used to divide the input pulse frequency by a factor of 16 (the MOD number).

We have looked only at the basic FF binary counter. We will examine counters in much more detail in Chapter 7.

EXAMPLE 5.9

Consider a counter circuit that contains six FFs wired in the arrangement of Figure 5.44 (i.e., $X_5, X_4, X_3, X_2, X_1, X_0$).
(a) Determine the counter's MOD number.
(b) Determine the frequency at the output of the last FF (X_5) when the input clock frequency is 1 MHz.
(c) What is the range of counting states for this counter?

Solution:

(a) MOD number $= 2^6 = 64$.

(b) The frequency at the last FF will equal the input clock frequency divided by the MOD number. That is,

$$f \text{ (at } X_5) = \frac{1 \text{ MHz}}{64} = 15.625 \text{ kHz}$$

(c) The counter will count from 000000_2 to 111111_2 (0 to 63_{10}) for a total of 64 states. Note that the number of states is the same as the MOD number.

_____ **REVIEW QUESTIONS**

1. A 20-kHz clock signal is applied to a J-K FF with $J = K = 1$. What is the frequency of the FF output waveform?
2. How many FFs are required for a counter that will count from 0 to 255_{10}? (*Ans.* Eight)
3. What is the MOD number of this counter? (*Ans.* 256)
4. What is the frequency at the output of the eighth FF when the input clock frequency is 512 kHz? (*Ans.* 2 kHz)

5.18 THE ONE-SHOT

A digital circuit that is somewhat related to the FF is the *one-shot* (abbreviated OS). Like the FF, the OS has two outputs, Q and $\overline{Q}$, which are inverses of each other. Unlike the FF, the OS has only one *stable* output state (normally $Q = 0$, $\overline{Q} = 1$), where it remains until it is triggered by an input signal. Once triggered, the OS outputs switch to the opposite state ($Q = 1$, $\overline{Q} = 0$). It remains in this *quasi-stable* state for a fixed period of time, t_p, which is usually determined by an RC time constant which is connected to the OS. After a time t_p, the OS outputs return to their stable resting state until triggered again.

Figure 5.46(a) shows the logic symbol for a OS. The value of t_p is usually indicated somewhere on the OS symbol. In practice, t_p can vary from several nanoseconds to several tens of seconds. The exact value of t_p is essentially determined by the values of external components R_T and C_T.

Two types of one-shots are available in IC form: the *nonretriggerable* and the *retriggerable*.

Nonretriggerable One-Shot The waveforms in Figure 5.46 illustrate the operation of a nonretriggerable OS that triggers on positive-going transitions at its trigger (T) input. The important points to note are:

1. The PGTs at points a, b, c, and e will trigger the OS to its quasi-stable state for a time t_p, after which it returns to the stable state.
2. The PGTs at points d and f have no effect on the OS because it has already been triggered to the quasi-stable state. The OS must return to the stable state before it can be retriggered.

(a)

(b)

FIGURE 5.46 OS symbol and typical waveforms.

3. The OS output-pulse duration is always the same regardless of the duration of the input pulses. As stated above, t_p depends only on R_T and C_T and the internal OS circuitry. A typical OS may have a t_p given by $t_p = 0.7R_TC_T$.

The most common IC nonretriggerable OS is the 74121—a TTL IC containing one nonretriggerable OS that has three inputs that can be used to trigger the OS on either a PGT or a NGT.

The Retriggerable OS The retriggerable OS operates much like the non-retriggerable OS except for one major difference: *it can be retriggered while it is in the quasi-stable state, and it will begin a new t_p interval.* This characteristic is illustrated in Figure 5.47(a) for a retriggerable OS with $t_p = 2$ ms.

The first PGT at the T input occurs at $t = 1$ ms and triggers the Q output to its HIGH state for a time duration of $t_p = 2$ ms, after which it returns to 0. The second trigger pulse occurs at $t = 5$ ms and triggers Q to the HIGH state where it would normally stay until $t = 7$ ms; however, a third trigger pulse occurs at $t = 6$ ms and retriggers the OS to begin a new $t_p = 2$ ms interval. Thus, Q stays HIGH for 2 ms *after* this third trigger pulse.

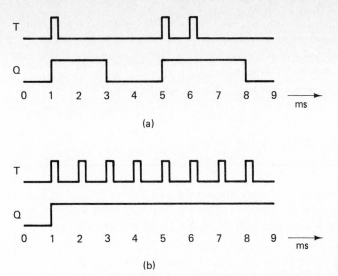

FIGURE 5.47 Waveforms for a
retriggerable OS with $t_p = 2$ ms.

In effect, then, a retriggerable OS will begin a new t_p interval *each* time a trigger pulse occurs, regardless of the state of Q. In fact, if a continuous train of trigger pulses having a frequency greater that $1/t_p$ is applied to the T input, the OS will trigger HIGH on the first pulse and remain HIGH as long as the pulses are applied. This is illustrated in Figure 5.47(b), where the trigger pulses occur every 1 ms ($f = 1$ kHz).

Two TTL retriggerable one-shots ICs are available. The 74122 contains one retriggerable OS that operates as described above. It also has an active-LOW CLEAR input that can be used to clear Q back to its stable 0 state at any time. The 74123 is a dual retriggerable OS IC containing two retriggerable one-shots that also have a CLEAR input. The CMOS 4528B is similar to the 74123.

Monostable Multivibrator One-shots are also called *monostable multivibrators* because they have only one stable state. The major applications of the OS are in timing circuits that utilize the predetermined t_p interval. Some of these applications will be introduced in the end-of-chapter problems and at various places throughout the text.

REVIEW QUESTIONS

1. In the absence of a trigger pulse, what will be the state of a OS output?
2. True or false: When a nonretriggerable OS is pulsed while it is in its quasi-stable state, the output is not affected.
3. What determines the t_p value for a OS?
4. Describe how a retriggerable OS operates differently from a nonretriggerable OS.

5.19 ANALYZING SEQUENTIAL CIRCUITS

Many logic circuits contain FFs, one-shots, and logic gates that are connected to perform a specific operation. Very often, a master clock signal is used to cause the logic levels in the circuit to go through a particular sequence of states. We can generally analyze these sequential circuits by following this step-by-step procedure:

1. Examine the circuit diagram and look for circuit structures such as counters or shift registers that you are familiar with. This can help to simplify the analysis.
2. Determine the logic levels that are present at the inputs of each FF *prior to* the occurrence of the first clock pulse.
3. Use these levels to determine how each FF output will change in response to the first clock pulse.
4. Repeat steps 2 and 3 for each successive clock pulse.

The following example will illustrate the procedure.

EXAMPLE 5.10

Consider the circuit of Figure 5.48. Initially, all the FFs and the OS are in the 0 state before the clock pulses are applied. These pulses have a repetition rate of 1 kHz. Determine the waveforms at $X, Y, Z, W, \bar{Q}, A$, and B for 16 cycles of the clock input.

Solution: Initially the FFs and OS are in the 0 state, so $X = Y = Z = W = Q = 0$. The inputs to the NAND gate are $X = 0$, $\bar{Y} = 1$, and $\bar{Z} = 1$, so its output $A = 1$. The inputs to the OR gate are $W = 0$ and $\bar{Q} = 1$, so its output is $B = 1$.

As long as B remains HIGH, FF Z will have its J and K inputs both HIGH, so it will operate in the toggle mode. FFs X and Y are kept in the toggle mode, since their J and K inputs are held at the 1 level permanently. It should be clear, then, that FFs X, Y, and Z will operate as a 3-bit counter as long as B stays HIGH. Thus, FF Z will toggle on the negative transition of each input clock pulse, FF Y will toggle on the negative transition of the Z output, and FF X will toggle on the negative transition of FF Y. This operation continues for clock pulses 1–4.

When clock pulse 4 makes its negative transition, Z goes LOW, Y goes LOW, and X goes HIGH. Therefore, the inputs to the NAND gate become $X = 1$, $\bar{Y} = 1$, and $\bar{Z} = 1$, causing the NAND output A to go LOW. This negative transition at A will trigger the OS to its quasi-stable state, so $\bar{Q}$ goes LOW for 3.5 ms. This negative transition of $\bar{Q}$ will not affect FF W, since this D-type FF triggers on positive transitions at its CLK input. With $\bar{Q}$ and W both LOW, the OR gate output B goes LOW, so FF Z now has $J = 1$, $K = 0$. This means that Z will be *set* to the 1 state on the *next* clock pulse and will remain in this state for each successive clock pulse as long as $J = 1$, $K = 0$. Thus, Z goes HIGH on clock pulse 5 and stays HIGH for clock pulses 6 and 7. (*Note:* A goes back HIGH when Z goes HIGH on pulse 5 because $\bar{Z} = 0$.)

After 3.5 ms the OS returns to its stable state with $Q = 0$, $\bar{Q} = 1$. This positive transition of $\bar{Q}$ causes FF W to go HIGH since $D = 1$. The 1 at W produces a 1 at B so that FF Z is back in the toggle mode with $J = K = 1$. The counter will now operate properly in response to all succeeding clock pulses.

On the negative transition of clock pulse 14, the NAND output A gain goes LOW, thereby triggering the OS. The OS, however, will not affect the K input of Z since W is keeping the OR output HIGH. Also, since its D input is HIGH, FF W will remain in the $W = 1$ state indefinitely, allowing the counter to count normally for all succeeding clock pulses.

FIGURE 5.48

5.20 FLIP-FLOP SUMMARY

1. NOR-gate FF

S	C	Q
0	0	No change
1	0	Q = 1
0	1	Q = 0
1	1	Ambiguous

(Alternate symbol)

FIGURE 5.49

2. NAND-gate FF

S	C	Q
0	0	Ambiguous
1	0	Q = 0
0	1	Q = 1
1	1	No change

(Alternate symbol)

FIGURE 5.50

3. Edge-triggered clocked S-C FF

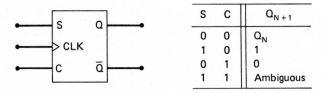

S	C	Q_{N+1}
0	0	Q_N
1	0	1
0	1	0
1	1	Ambiguous

FIGURE 5.51

4. Edge-triggered J-K FF

J	K	Q_{N+1}
0	0	Q_N (no change)
1	0	1
0	1	0
1	1	$\overline{Q}_N$ (toggles)

FIGURE 5.52

5. Edge-triggered D FF

D	Q_{N+1}
0	0
1	1

FIGURE 5.53

6. D-type latch

CLK	D	Q^*
0	X	No change
1	0	0
1	1	1

Q^* follows D input while CLK is HIGH

FIGURE 5.54

7. Asynchronous FF inputs

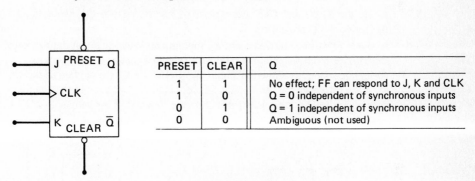

PRESET	CLEAR	Q
1	1	No effect; FF can respond to J, K and CLK
1	0	Q = 0 independent of synchronous inputs
0	1	Q = 1 independent of synchronous inputs
0	0	Ambiguous (not used)

FIGURE 5.55

8. Master/slave FFs Same as negative-edge-triggered FFs except that the control inputs have to remain stable during $CLK = 1$ interval.

PROBLEMS

5.1 Assume $Q = 0$ initially, apply the x and y waveforms of Figure 5.56 to the SET and CLEAR inputs of a NAND FF, and determine the Q and $\overline{Q}$ waveforms.

FIGURE 5.56

5.2 Invert the x and y waveforms of Figure 5.56, apply them to the SET and CLEAR inputs of a NOR FF, and determine the Q and $\bar{Q}$ waveforms. Assume $Q = 0$ initially.

5.3 The waveforms of Figure 5.56 are connected to the circuit of Figure 5.57. Assume $Q = 0$ initially, and determine the Q waveform.

FIGURE 5.57

5.4 Modify the circuit of Figure 5.9 to use a NOR-gate FF.

5.5 Modify the circuit of Figure 5.12 to use a NAND-gate FF.

5.6 A certain clocked FF is specified to have $t_S = 20$ ns and $t_H = 5$ ns. How long must the control inputs be stable prior to the active clock transition?

5.7 Apply the S, C, and CLK waveforms of Figure 5.16 to the FF of Figure 5.17 and determine the Q waveform.

5.8 A *toggle* FF is one that has a single input and operates such that the FF output changes state for each pulse applied to its input. The clocked S-C FF can be wired to operate in the toggle mode, as shown in Figure 5.58. The waveform applied to the CLK input is a 1-kHz squarewave. Verify that this arrangement operates in the toggle mode, then determine the Q output waveform. Assume that $Q = 0$ initially.

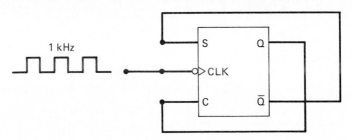

FIGURE 5.58

5.9 Apply the J, K, and CLK waveforms of Figure 5.19 to the FF of Figure 5.20. Assume that $Q = 1$ initially and determine the Q waveform.

5.10 Show how the J-K FF can be operated as a toggle FF. Apply a 10-kHz squarewave to its input and determine its output waveform.

5.11 Connect the Q output of the FF from Problem 5.10 to the CLK input of a second J-K FF that also has $J = K = 1$. Determine the frequency of the waveform at the second FF output.

5.12 A D-type FF is sometimes used to *delay* a binary waveform so that the binary information appears at the output a certain amount of time after it appears at the D input. Determine the Q waveform in Figure 5.59 and compare it to the input waveform. Note that it is delayed from the input by one clock period. How can a delay of two clock periods be obtained?

FIGURE 5.59

5.13 An edge-triggered D FF can be made to operate in the toggle mode by connecting it as shown in Figure 5.60. Assume $Q = 0$ initially and determine the Q waveform.

5.14 Change the circuit in Figure 5.60 so that Q is connected back to D. Then determine the Q waveform.

FIGURE 5.60   D FF connected to toggle.

5.15 Compare the operation of the D latch of Figure 5.26 with a negative-edge triggered D FF by applying the waveforms of Figure 5.61 to each and determining the Q waveforms.

FIGURE 5.61

5.16 In Problem 5.13 we saw how an edge-triggered D FF can be operated in the toggle mode. Explain why this same idea will not work for a D latch.

5.17 Determine the Q waveform for the FF in Figure 5.62. Assume that $Q = 0$ initially and remember that the asynchronous inputs override all other inputs.

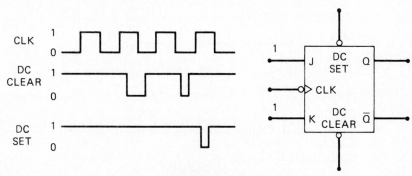

FIGURE 5.62

5.18 Apply the *CLK*, PRESET, and CLEAR waveforms of Figure 5.29 to a positive-edge triggered D FF with active-LOW asynchronous inputs. Assume that *D* is kept HIGH, and *Q* is initially LOW. Determine the *Q* waveform.

5.19 Use Table 5.1 in Section 5.10 to determine the following:
(a) How long can it take for the *Q* output of a 4013B to switch from 0 to 1 in response to an active *CLK* transition?
(b) Which FF in Table 5.1 requires its control inputs to remain stable for the longest time after the active *CLK* transition?
(c) What is the narrowest pulse that can be applied to the DC SET of a 7474 FF?

5.20 Refer to the circuit of Figure 5.63. It shows the two J-K FFs on a 74LS112 chip connected such that the Q_1 output serves as the *CLK* input for the Q_2 FF. Assume that $Q_1 = Q_2 = 1$ initially, and determine the *total* propagation delay between the NGT of the clock pulse and the output of Q_2.

FIGURE 5.63

5.21 The clock waveform of Figure 5.64 is a 25-MHz squarewave. Explain why this is the highest-frequency *squarewave* that can be used as the *CLK* input of a 74LS112 even though f_{MAX} is given as 30 MHz in Table 5.1. (*Hint:* Consider the minimum clock times given in the table.)

FIGURE 5.64

5.22 The waveforms shown in Figure 5.65 are to be applied to three different FFs: (a) positive-edge triggered J-K; (b) negative-edge triggered J-K; and (c) M/S J-K. Draw the *Q* waveform response for each of these FFs, assuming that *Q* = 0 initially. Assume each FF has $t_H = 0$.

FIGURE 5.65

5.23 Modify the circuit of Figure 5.37 to use a J-K FF.

5.24 In the circuit of Figure 5.66 the inputs A, B, and C are all initially LOW. Output Y is supposed to go HIGH only when A, B, and C go HIGH in a certain sequence.
  (a) Determine the sequence that will make Y go HIGH.
  (b) Explain why the START pulse is needed.

FIGURE 5.66

5.25 Draw the circuit diagram for the synchronous parallel transfer of data from one 3-bit register to another using J-K FFs.

5.26 Repeat Problem 5.25 for asynchronous parallel transfer.

5.27 A *recirculating* shift register is a shift register that keeps the binary information circulating through the register as clock pulses are applied. The shift register of Figure 5.42 can be made into a circulating register by connecting X_0 to the DATA IN line. No external inputs are used. Assume that this circulating register starts out with 1011 stored in it (that is, $X_3 = 1, X_2 = 0, X_1 = 1$, and $X_0 = 1$). List the sequence of states that the register FFs go through as eight shift pulses are applied.

5.28 Refer to Figure 5.43, where a 3-bit number stored in register X is serially shifted into register Y. How could the circuit be modified so that at the end of the transfer operation, the original number stored in X is present in both registers? (*Hint:* See Problem 5.27.)

5.29 Refer to the binary counter of Figure 5.44. Change it by connecting $\overline{X}_0$ to the *CLK* of FF X_1 and $\overline{X}_1$ to the *CLK* of FF X_2. Start out with all FFs in the 1 state and draw the various FF output waveforms (X_0, X_1, X_2) for 16 input pulses. Then list the sequence of FF states as was done in Figure 5.45. This counter is called a *down counter*. Why?

5.30 Show how clocked D FFs can be used in a counter such as that in Figure 5.44. (*Hint:* See Problem 5.13.)

5.31 (a) How many FFs are required to build a binary counter circuit that counts from 0 to 1023?
  (b) Determine the frequency at the output of the last FF of this counter for an input clock frequency of 2 MHz.
  (c) What is the counter's MOD number?

5.32 A certain counter is being pulsed by a 256-kHz clock signal. The output frequency from the last FF is 2-kHz.
  (a) Determine the MOD number.
  (b) Determine the counting range.

5.33 A photodetector circuit is being used to generate a pulse each time a customer walks into a certain establishment. The pulses are fed to an 8-bit counter. The counter is used to count these pulses as a means for determining how many customers have entered the store.

After closing the store, the proprietor checks the counter and finds that it shows a count of $00001001_2 = 9_{10}$. He knows that this is incorrect, because there were many more than nine people in his store.

Assuming that the counter circuit is working properly, what could be the reason for the discrepancy?

5.34 Determine the waveforms at Q_1, Q_2, and Q_3 in response to the single input pulse in Figure 5.67.

FIGURE 5.67

5.35 A retriggerable OS can be used as a pulse-frequency detector that detects when the frequency of a pulse input is below a predetermined value. A simple example of this application is shown in Figure 5.68. The operation begins by momentarily closing switch S_1.
(a) Describe how the circuit responds to input frequencies above 1 kHz.
(b) Describe how the circuit responds to input frequencies below 1 kHz.
(c) How would you modify the circuit to detect when the input frequency drops below 50 kHz?

FIGURE 5.68

5.36 Figure 5.69 shows the logic diagram for the 74121 IC. In addition to the non-retriggerable OS, this IC contains the two logic gates that allow different triggering possibilities.
(a) What input conditions are necessary if the OS is to be triggered by a PGT at the B input?
(b) What input conditions are necessary if the OS is to be triggered by a NGT at the A_1 input?

$$t_p \approx 0.7R_T C_T$$

FIGURE 5.69

5.37 The R_T value for a 74121 can be varied between 2 kΩ and 40 kΩ, and the C_T value can be as large as 1000 μF.

    (a) Show how a 74121 can be connected to produce a negative-going pulse of 5 ms duration whenever either of two logic inputs (E or F) goes from 1 to 0. Both E and F are normally HIGH signals.

    (b) Modify the circuit so that a LOW at control input, G, will inhibit the occurrence of any output pulse regardless of what happens at E or F.

5.38 Consider the circuit of Figure 5.70. Initially all FFs are in the 0 state. The circuit operation begins with a momentary start pulse applied to the DC SET inputs of FFs X and Y. Determine the waveforms at A, B, C, X, Y, Z, and W for 20 cycles of the clock pulses after the start pulse. State all assumptions.

FIGURE 5.70

6
DIGITAL ARITHMETIC: OPERATIONS and CIRCUITS

O U T L I N E

Digital computers and calculators perform the various arithmetic operations on numbers that are represented in binary form. The subject of digital arithmetic can be a very complex one if we want to understand all the various methods of computation and the theory behind them. Fortunately, this level of knowledge is not required by most technicians, at least not until they become experienced computer programmers. Our approach in this chapter will be to concentrate on those basic principles that are needed to understand how digital machines (i.e., computers) perform the basic arithmetic operations.

First we will see how the various arithmetic operations are performed on binary numbers using "pencil and paper," and then we will study the actual logic circuits that perform these operations in a digital system.

6.1 BINARY ADDITION

The addition of two binary numbers is performed in exactly the same manner as the addition of decimal numbers. In fact, binary addition is simpler, since there are fewer cases to learn. Let us first review decimal addition:

$$
\begin{array}{r}
3 \; 7 \; 6 \quad \text{LSD} \\
+ \; 4 \; 6 \; 1 \\
\hline
8 \; 3 \; 7
\end{array}
$$

The least-significant-digit (LSD) position is operated on first, producing a sum of 7. The digits in the second position are then added to produce a sum of 13, which produces a *carry* of 1 into the third position. This produces a sum of 8 in the third position.

The same general steps are followed in binary addition. However, only four cases can occur in adding the two binary digits (bits) in any position. They are:

$$0 + 0 = 0$$

$$1 + 0 = 1$$

$$1 + 1 = 10 = 0 + \text{carry of 1 into next position}$$

$$1 + 1 + 1 = 11 = 1 + \text{carry of 1 into next position}$$

The last case occurs when the two bits in a certain position are 1 and there is a carry from the previous position. Here are several examples of the addition of two binary numbers:

$$
\begin{array}{r}
011 \; (3) \\
+ \; 110 \; (6) \\
\hline
1001 \; (9)
\end{array}
\qquad
\begin{array}{r}
1001 \; (9) \\
+ \; 1111 \; (15) \\
\hline
11000 \; (24)
\end{array}
\qquad
\begin{array}{r}
11.011 \; (3.375) \\
+ \; 10.110 \; (2.750) \\
\hline
110.001 \; (6.125)
\end{array}
$$

It is not necessary to consider the addition of more than two binary numbers at a time, because in all digital systems the circuitry that actually performs the addition can handle only two numbers at a time. When more than two numbers are to be added, the first two are added together and then their sum is added to the third number; and so on. This is not a serious drawback, since modern digital machines can typically perform an addition operation in less than 1 μs.

Addition is the most important arithmetic operation in digital systems. As we shall see, the operations of subtraction, multiplication, and division as they are performed in most modern digital computers and calculators actually use only addition as their basic operation.

Binary Addition versus Logical OR Addition It is important to understand the difference between the OR-addition operation and binary addition. OR addition is the Boolean *logic* operation performed by an OR gate, which produces a 1 output whenever any one or more inputs are 1. Binary addition is an *arithmetic* operation that produces an arithmetic sum of two binary numbers. Although the + sign is used for both OR addition and binary addition, the meaning of the + sign will usually be apparent from the context in which it is being used. The major differences between OR addition and binary addition can be summarized as follows:

OR Addition	*Binary Addition*
1 + 1 = 1	1 + 1 = 0 + carry of 1
1 + 1 + 1 = 1	1 + 1 + 1 = 1 + carry of 1

6.2 REPRESENTING SIGNED NUMBERS

In binary machines, the binary numbers are represented by a set of binary storage devices (usually flip-flops). Each device represents one bit. For example, a 6-bit FF register could store binary numbers ranging from 000000 to 111111 (0 to 63 in decimal). This represents the *magnitude* of the number. Since most digital computers and calculators handle negative as well as positive numbers, some means is required for representing the *sign* of the number (+ or −). This is usually done by adding another bit to the number called the *sign bit*. In general, the common convention which has been adopted is that a 0 in the sign bit represents a positive number and a 1 in the sign bit represents a negative number. This is illustrated in Figure 6.1. Register A contains the bits 0110100. The 0 in the leftmost bit (A_6) is the sign bit that represents +. The other six bits are the magnitude of the number 110100_2, which is equal to 52 in decimal. Thus, the number stored in the A register is +52. Similarly, the number stored in the B register is −31, since the sign bit is 1, representing −.

The sign bit is used to indicate whether a stored binary number is positive or negative. For *positive* numbers, the rest of the bits are always used to represent the magnitude of the number in binary form. For *negative* numbers, however, there are three possible ways to represent the magnitude: true-magnitude form, 1's-complement form, and 2's-complement form.

True-Magnitude Form The numbers in Figure 6.1 contain a sign bit and 6 magnitude bits. The magnitude bits are the true binary equivalent of the decimal values being represented. We call this the *true-magnitude form* for representing signed binary numbers.

FIGURE 6.1 Representation of signed numbers in true magnitude form.

Although this true-magnitude system is straightforward and easy to understand, it is not as useful as two other systems for representing signed binary numbers. These other systems use the same true-magnitude form for *positive* numbers, but they use a different form for negative numbers.

1's-Complement Form The *1's-complement form* of any binary number is obtained simply by changing each 0 in the number to a 1 and each 1 in the number to a 0. In other words, change each bit to its complement. For example, the 1's complement of 101101 is 010010, and the 1's complement of 011010 is 100101.

When negative numbers are represented in 1's-complement form, the sign bit is made a 1 and the magnitude is converted from true binary form to its 1's complement. To illustrate, the number -57 would be represented as follows:

sign bits

$$-57 = 1 \quad 111001 \quad \text{(true-magnitude form)}$$
$$= 1 \quad 000110 \quad \text{(1's-complement form)}$$

Note that the sign bit is not complemented but is kept as a 1 to indicate a negative number.

Some additional examples of representing negative numbers in 1's-complement form are given below and should be verified by the reader:

$$-14 = 10001$$
$$-326 = 1010111001$$
$$-7 = 1000$$

2's-Complement Form The *2's-complement form* of a binary number is formed simply by taking the 1's complement of the number and adding 1 to the least-significant bit position. The procedure is illustrated below for converting 111001 (decimal 57) to its 2's-complement form.

```
1 1 1 0 0 1     complement each bit to form 1's complement
↓ ↓ ↓ ↓ ↓ ↓
0 0 0 1 1 0
        +1      add 1 to LSB to form 2's complement
─────────────
0 0 0 1 1 1
```

Thus, −57 would be written as 1000111 in its 2's-complement representation. Again, the leftmost bit is the sign bit. The other 6 bits are the 2's-complement form of the magnitude. As another example, the 2's complement of −14 is written 10010.

The three forms of representing signed numbers are summarized in Figure 6.2 for +57 and −57. All three forms are currently used in digital systems. Some digital machines store negative numbers in true-magnitude form but convert them to 1's- or 2's-complement form before performing any arithmetic operations. Other machines store the negative numbers in 1's- or 2's-complement form. In almost all modern digital machines, the negative numbers are in 1's- or 2's-complement form for the arithmetic operations. At present, the 2's-complement representation is the most frequently used.

Remember, in all three systems—true-magnitude, 1's-complement and 2's-complement—*positive* numbers are always in true binary form and a sign bit of 0. The differences are in the representation of negative numbers.

The 2's-complement form is used because, as we shall see, its use allows us to perform the operation of subtraction by actually using only the operation of addition. This is significant because it means that a digital machine can use the same circuitry to both add and subtract, thereby realizing a savings in hardware.

Converting from Complement Form to Binary   It is a relatively simple matter to take a number that is in its 1's- or 2's-complement form and convert it back to its true binary value. To go from 1's complement to true binary simply requires complementing each bit again. Similarly, to go from 2's complement to true binary requires complementing each bit and then adding 1 to the LSB. In both cases, the conversion back to binary is the same process that was used to produce the complement to begin with. The reader should verify this for a couple of binary numbers.

FIGURE 6.2   **Three systems for representing signed binary numbers.**

EXAMPLE 6.1

Represent each of the following signed decimal numbers as a signed binary number in the 2's-complement system. Use a total of five bits including the sign bit: (a) $+13$, (b) -9, (c) $+3$, (d) -3.

Solution: (a) Since the number is positive, the magnitude (13) will be represented in its true-magnitude form—that is, $13 = 1101_2$. Attaching the sign bit of 0, we have

$$+13 = 01101$$

$$\text{sign bit} \longrightarrow$$

(b) Since the number is negative, the magnitude (9) has to be represented in 2's-complement form:

$$9_{10} = 1001_2$$

$$
\begin{array}{ll}
0110 & \text{(1's complement)} \\
\underline{+1} & \text{(add 1 to LSB)} \\
0111 & \text{(2's complement)}
\end{array}
$$

When we attach the sign bit of 1, the complete signed number becomes

$$-9 = 10111$$

The procedure we have just followed required two steps. First we determined the 2's complement of the magnitude, then we attached the sign bit. This can be accomplished in one step if we include the sign bit in the 2's-complement process. For example, to find the representation for -9, we start with the representation for $+9$, *including the sign bit,* and we 2's-complement it in order to obtain the representation for -9.

$$+9 = 01001$$

$$
\begin{array}{ll}
10110 & \text{(1's comp. each bit)} \\
\underline{+1} & \text{(add 1 to LSB)} \\
10111 & \text{(2's-comp. representation of } -9)
\end{array}
$$

The result is, of course, the same as before. In general, we can say that

> *the 2's-complement operation on a signed number will change a positive number to a negative number and vice versa.*

(c) The decimal value 3 can be represented in binary using only 2 bits. However, the problem statement requires a 4-bit magnitude preceded by a sign bit. Thus, we have

$$+3_{10} = 00011$$

In many situations the number of bits is fixed by the size of the registers that will be holding the binary numbers, so that 0s may have to be added in order to fill the required number of bit positions.

(d) From above we have

$$+3 = 00011$$

11100	(1's comp.)
+1	(add 1)
11101	(2's-comp. representation of -3)

EXAMPLE 6.2

Each of the following numbers is a signed binary number in the 2's-complement system. Determine the decimal value in each case: (a) 01100, (b) 10110, (c) 10001.

Solution: (a) The sign bit is 0 so the number is *positive,* and the other 4 bits represent the true magnitude of the number. That is, $1100_2 = 12_{10}$. Thus, the decimal number is **+12.**

(b) The sign bit is a 1 so the number is *negative,* and the other 4 bits represent the 2's complement of the magnitude. To find out what this magnitude is, we must 2's-complement it again. Remember, the 2's-complement operation will change the polarity of a signed number, so if we 2's-complement a negative binary number (including the sign bit), we will get the positive number.

11010	(original negative number)
00101	(1's comp.)
+1	(add 1)
00110	(+6)

Since the result of 2's-complementing is a $+6$, the original number must have been -6. That is,

$$11010_2 = -\mathbf{6}_{10}$$

(c) Follow the same procedure as in (b):

10001	(original neg. number)
01110	(1's comp.)
+1	(add 1)
01111	(+15)

Thus, $10001 = -\mathbf{15}.$

Special Case in 2's-Complement Representation Whenever a signed number has a 1 in the sign bit and all 0s for the magnitude bits, then its decimal equivalent is -2^N, where N is the number of bits in the *magnitude*. For example,

$$1000 = -2^3 = -8$$

$$10000 = -2^4 = -16$$

$$100000 = -2^5 = -32$$

and so on.

Thus, we can state that the complete range of values that can be represented in the 2's-complement system having N magnitude bits is

$$-2^N \text{ to } +(2^N - 1)$$

EXAMPLE 6.3

What is the range of decimal values that can be represented in 8 bits including a sign bit?

Solution: Since one of the bits is a sign bit, there are 7 bits for the magnitude. The largest positive value occurs when the sign bit is 0 and all the magnitude bits are 1s. We know that

$$1111111_2 = 2^7 - 1 = 127$$

The largest negative value occurs when the sign bit is 1 and the magnitude bits are all 0s. From the special case described above, this is

$$10000000 = -2^7 = -128$$

Thus, the range of values is -128 to $+127$. This is a total of 256 possible values including zero (i.e., 00000000)

_____ **REVIEW QUESTIONS**

1. Represent each of the following values as a 5-bit signed number in the 2's-complement system: (a) +13, (b) −7, (c) −16. (*Ans.* 01101, 11001, 10000)
2. Each of the following is a signed binary number in the 2's-complement system. Determine the decimal equivalent for each: (a) 100011, (b) 1000000, (c) 0111111. (*Ans.* −29, −64, +63)
3. What range of decimal values can be represented in 12 bits (including the sign bit)? (*Ans.* −2048 to +2047)

6.3 ADDITION IN THE 2'S-COMPLEMENT SYSTEM

The 1's-complement system and 2's-complement system are very similar. However, the 2's-complement system is generally in use because of several advantages it enjoys in circuit implementation. We will now investigate how the operations of addition and subtraction are performed in digital machines that use the 2's-complement representation for negative numbers. In the various cases to be considered, it is important to note that the sign-bit of each number is operated on in the same manner as the magnitude portion.

Case I: two positive numbers. The addition of two positive numbers is straightforward. Consider the addition of +9 and +4:

$$
\begin{array}{r}
+9 \longrightarrow \quad 0 \; 1001 \quad \text{(augend)} \\
+4 \longrightarrow \quad 0 \; 0100 \quad \text{(addend)} \\
\hline
0 \; 1101 \quad \text{(sum} = +13)
\end{array}
$$

sign bits

Note that the sign bits of the *augend* and *addend* are both 0 and the sign bit of the sum is 0, indicating that the sum is positive. Also note that the augend and addend are made to have the same number of bits. This must *always* be done in the 2's complement system.

Case II: positive number and smaller negative number. Consider the addition of +9 and −4. Remember that the −4 will be in its 2's-complement form. Thus, +4(00100) must be converted to −4(11100)

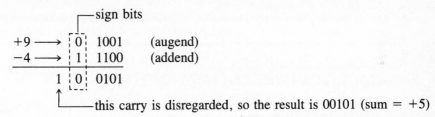

sign bits

$$
\begin{array}{r}
+9 \longrightarrow \quad 0 \; 1001 \quad \text{(augend)} \\
-4 \longrightarrow \quad 1 \; 1100 \quad \text{(addend)} \\
\hline
1 \; 0 \; 0101
\end{array}
$$

this carry is disregarded, so the result is 00101 (sum = +5)

In this case the sign bit of the addend is 1. Note that the sign bits also participate in the addition process. In fact, a carry is generated in the last position of addition. *This carry is always disregarded,* so the final sum is 00101, which is equivalent to +5.

Case III: positive number and larger negative number. Consider the addition of −9 and +4:

$$
\begin{array}{r}
-9 \longrightarrow \quad 10111 \\
+4 \longrightarrow \quad 00100 \\
\hline
11011 \quad \text{(sum} = -5)
\end{array}
$$

negative sign bit

The sum here has a sign bit of 1, indicating a negative number. Since the sum is negative, it is in 2's-complement form, so the last four bits, 1011, actually represent the 2's complement of the sum. To find the true magnitude of the sum, we must take the 2's complement of 1011; the result is 0101 (5). Thus, 11011 represents −5.

Case IV: two negative numbers.

$$
\begin{array}{r}
-9 \longrightarrow \quad 10111 \\
-4 \longrightarrow \quad 11100 \\
\hline
110011
\end{array}
$$

sign bit

this carry is disregarded, so the result is 10011 (sum = −13)

This final result is again negative and in 2's-complement form with a sign bit of 1. Note that 0011 is the 2's complement of 1101 (13).

Case V: equal and opposite numbers.

$$-9 \longrightarrow 10111$$
$$+9 \longrightarrow 01001$$
$$\overline{0 \quad 100000}$$

⤷disregard, so the result is 00000 (sum = +0)

The result is obviously +0, as expected.

————————————————————— REVIEW QUESTIONS

1. True or false: Whenever the sum of two signed binary numbers has a sign bit of 1, the magnitude of the sum is in 2's-complement form.
2. Add the following pairs of signed numbers. Express the sum as a signed binary number and as a decimal number: (a) 100111 + 111011, (b) 100111 + 011001. [*Ans.* 100010 (−30); 000000 (+0)]

6.4 SUBTRACTION IN THE 2'S-COMPLEMENT SYSTEM

The subtraction operation using the 2's-complement system actually involves the operation of addition and is really no different than the various cases considered in Section 6.3. When subtracting one binary number (the *subtrahend*) from another binary number (the *minuend*), the procedure is as follows:

1. Take the 2's complement of the subtrahend, *including* the sign bit. If the subtrahend is a positive number, this will change it to a negative number in 2's-complement form. If the subtrahend is a negative number, this will change it to a positive number in true binary form. In other words, we are changing the sign of the subtrahend.
2. After taking the 2's complement of the subtrahend, it is *added* to the minuend. The minuend is kept in its original form. The result of this addition represents the required *difference*. The sign bit of this difference determines whether it is + or − and whether it is in true binary form or 2's-complement form.
3. Remember, both numbers have to have same number of bits.

Let us consider the case where +4 is to be subtracted from +9.

$$\text{minuend } (+9) \longrightarrow 01001$$
$$\text{subtrahend } (+4) \longrightarrow 00100$$

Change the subtrahend to its 2's-complement form (11100), which represents −4. Now add this to the minuend.

$$01001 \quad (+9)$$
$$+ \ \underline{11100} \quad (-4)$$
$$1 \quad 00101 \quad (+5)$$

⤷disregard, so the result is 00101 = +5

When the subtrahend is changed to its 2's complement, it actually becomes -4, so we are *adding* $+9$ and -4, which is the same as subtracting $+4$ from $+9$. This is the same as case II of Section 6.3. Any subtraction operation, then, actually becomes one of addition when the 2's-complement system is used. This feature of the 2's-complement system has made it the most widely used of the methods available, since it allows addition and subtraction to be performed by the same circuitry.

The reader should verify the results of using the above procedure for the following subtractions: (a) $+9 - (-4)$; (b) $-9 - (+4)$; (c) $-9 - (-4)$; (d) $+4 - (-4)$. Remember that when the result has a sign bit of 1, it is negative and in 2's-complement form.

Arithmetic Overflow In each of the previous addition and subtraction examples, the numbers that were added consisted of a sign bit and 4 magnitude bits. The answers also consisted of a sign bit and 4 magnitude bits. Any carry into the sixth bit position was disregarded. In all of the cases considered, the magnitude of the answer was small enough to fit into 4 bits. Let's look at the addition of $+9$ and $+8$.

$$
\begin{array}{rclcl}
+9 & \longrightarrow & 0 & 1001 \\
+8 & \longrightarrow & 0 & 1000 \\
\hline
 & & 1 & 0001 \\
\end{array}
$$

$\uparrow$ sign bits

The answer has a negative sign bit, which is obviously incorrect. The answer should be $+17$, but the magnitude 17 requires more than 4 bits and therefore *overflows* into the sign-bit position. This overflow condition always produces an incorrect result, and its occurrence is detected by examining the sign bit of the result and comparing it to the sign bits of the numbers being added. In a computer, a special circuit is used to detect any overflow condition and to signal that the answer is erroneous. We will encounter such a circuit in one of the end-of-chapter problems.

REVIEW QUESTIONS

1. Perform the subtraction on the following pairs of signed numbers using the 2's-complement system. Express the results as signed binary numbers and as decimal values: (a) $01001 - 11010$, (b) $10010 - 10011$. (*Ans.* $01111 = +15$; $11111 = -1$)
2. How can arithmetic overflow be detected when signed numbers are being added?

6.5 MULTIPLICATION OF BINARY NUMBERS

The multiplication of binary numbers is done in the same manner as the multiplication of decimal numbers. The process is actually simpler, since the multiplier digits are either 0 or 1, so we are always multiplying by 0 or 1 and no other digits. The following example illustrates for unsigned binary numbers.

$$
\begin{array}{r}
1001 \quad \longleftarrow \text{ multiplicand } = 9_{10} \\
1011 \quad \longleftarrow \text{ multiplier } = 11_{10} \\
\hline
1001 \\
1001 \\
0000 \\
1001 \\
\hline
1100011\}
\end{array}
$$

$\left.\begin{array}{l}1001 \\ 1001 \\ 0000 \\ 1001\end{array}\right\}$ partial products

$1100011\}$ final product $= 99_{10}$

In this example the multiplicand and multiplier are in true binary form and no sign bits are used. The steps followed in the process are exactly the same as in decimal multiplication. First, the LSB of the multiplier is examined; in our example it is a 1. This 1 multiplies the multiplicand to produce 1001, which is written down as the first partial product. Next, the second bit of the multiplier is examined. It is a 1, so 1001 is written for the second partial product. Note that this second partial product is *shifted* one place to the left relative to the first one. The third bit of the multiplier is 0, so 0000 is written as the third partial product; again, it is shifted one place to the left relative to the previous partial product. The fourth multiplier bit is 1, so the last partial product is 1001 shifted again one position to the left. The four partial products are then summed to produce the final product.

Most digital machines can add only two binary numbers at a time. For this reason, the partial products formed during multiplication cannot all be added together at the same time. Instead, they are added together two at a time; that is, the first is added to the second, their sum is added to the third, and so on. This process is now illustrated for the example above:

Add $\left\{\begin{array}{l}1001 \quad \longleftarrow \text{ first partial product} \\ 1001 \quad \longleftarrow \text{ second partial product shifted left}\end{array}\right.$

Add $\left\{\begin{array}{l}11011 \quad \longleftarrow \text{ sum of first two partial products} \\ 0000 \quad \longleftarrow \text{ third partial product shifted left}\end{array}\right.$

Add $\left\{\begin{array}{l}011011 \quad \longleftarrow \text{ sum of first three partial products} \\ 1001 \quad \longleftarrow \text{ fourth partial product shifted left}\end{array}\right.$

$\quad\quad 1100011 \quad \longleftarrow$ sum of four partial products which equals final total product

Multiplication in the 2's-Complement System In machines that use the 2's-complement representation, multiplication is carried on in the manner described above provided that both the multiplicand and multiplier are put in true binary form. If the two numbers to be multiplied are positive, they are already in true binary form and are multiplied as they are. The resulting product is, of course, positive and is given a sign bit of 0. When the two numbers are negative, they will be in 2's-complement form. Each one is 2's-complemented to convert it to a positive number and then they are multiplied. The product is kept as a positive number and given a sign bit of 0.

When one of the numbers is positive and the other is negative, the negative number is first converted to a positive magnitude by taking its 2's complement. The product will be in true-magnitude form. However, the product has to be negative since the original numbers are of opposite sign. Thus, the product is then changed to 2's-complement form and given a sign bit of 1.

_____ **REVIEW QUESTION**

1. Multiply 0111 and 1110. (*Ans.* 1100010)

_____

6.6 BINARY DIVISION

The process for dividing one binary number (the *dividend*) by another (the *divisor*) is the same as that which is followed for decimal numbers, that which we usually refer to as "long division." The actual process is simpler in binary because when we are checking to see how many times the divisor "goes into" the dividend, there are only two possibilities, 0 or 1. To illustrate, consider the following division examples:

$$
\begin{array}{r}
0011 \\
11\overline{)1001} \qquad (9 \div 3 = 3) \\
\underline{011} \\
0011
\end{array}
$$

$$
\begin{array}{r}
0010.1 \\
100\overline{)1010.0} \qquad (10 \div 4 = 2.5) \\
\underline{100} \\
100 \\
\underline{100} \\
0
\end{array}
$$

In most modern digital machines the subtractions that are part of the division operation are usually carried out using 2's-complement subtraction—that is, 2's-complementing the subtrahend and then adding.

The division of signed numbers is handled in the same way as multiplication. Negative numbers are made positive by complementing and the division is then carried out. If the dividend and divisor are of opposite sign, the resulting quotient is changed to a negative number by 2's-complementing and is given a sign bit of 1. If the dividend and divisor are of the same sign, the quotient is left as a positive number and given a sign bit of 0.

6.7 BCD ADDITION

In Chapter 2 we stated that many computers and calculators use the BCD code to represent decimal numbers. Recall that this code takes *each* decimal digit and represents it by a 4-bit code ranging from 0000 to 1001. The addition of decimal numbers that are in BCD form can be best understood by considering the two cases that can occur when two decimal digits are added.

Sum Equals Nine or Less Consider adding 5 and 4 using BCD to represent each digit:

$$
\begin{array}{rll}
5 & 0101 & \longleftarrow \text{BCD for 5} \\
+\ 4 & +\ 0100 & \longleftarrow \text{BCD for 4} \\ \hline
9 & 1001 & \longleftarrow \text{BCD for 9}
\end{array}
$$

The addition is carried out as in normal binary addition and the sum is 1001, which is the BCD code for 9. As another example, take 45 added to 33:

$$
\begin{array}{rll}
45 & 0100\ \ 0101 & \longleftarrow \text{BCD for 45} \\
+33 & +\ 0011\ \ 0011 & \longleftarrow \text{BCD for 33} \\ \hline
78 & 0111\ \ 1000 & \longleftarrow \text{BCD for 78}
\end{array}
$$

In this example the 4-bit codes for 5 and 3 are added in binary to produce 1000, which is BCD for 8. Similarly, adding the second-decimal-digit positions produces 0111, which is BCD for 7. The total is 0111 1000, which is the BCD code for 78.

In the examples above, none of the sums of the pairs of decimal digits exceeded 9; therefore, *no decimal carries were produced.* For these cases the BCD addition process is straightforward and is actually the same as binary addition.

Sum Greater than Nine Consider the addition of 6 and 7 in BCD:

$$
\begin{array}{rll}
6 & 0110 & \longleftarrow \text{BCD for 6} \\
+\ 7 & +\ 0111 & \longleftarrow \text{BCD for 7} \\ \hline
+13 & 1101 & \longleftarrow \text{invalid code group for BCD}
\end{array}
$$

The sum 1101 does not exist in the BCD code; it is one of the six forbidden or invalid 4-bit code groups. This has occurred because the sum of the two digits exceeds 9. Whenever this occurs the sum has to be corrected by the addition of six (0110) to take into account the skipping of the six invalid code groups:

$$
\begin{array}{rll}
6 & 0110 & \longleftarrow \text{BCD for 6} \\
+\ 7 & +\ 0111 & \longleftarrow \text{BCD for 7} \\ \hline
13 & 1101 & \longleftarrow \text{invalid sum} \\
\quad 1 & 0110 & \longleftarrow \text{add 6 for correction} \\ \hline
0001 & 0011 & \longleftarrow \text{BCD for 13} \\
\underbrace{}_{1} & \underbrace{}_{3} &
\end{array}
$$

As shown above, 0110 is added to the invalid sum and produces the correct BCD result. Note that a carry is produced into the second decimal position. This addition of 0110 has to be performed whenever the sum of the two decimal digits is greater than 9.

As another example, take 47 plus 35 in BCD:

```
 47        0100  0111  ←——— BCD for 47
+ 35      + 0011  0101  ←——— BCD for 35
 82        0111  1100  ←——— invalid sum in first digit
              1↖ 0110  ←——— add 6
           1000 ⌞0010  ←——— correct BCD sum
           ‿‿‿  ‿‿‿
            8     2
```

The addition of the 4-bit codes for the 7 and 5 digits results in an invalid sum and is corrected by adding 0110. Note that this generates a carry of 1, which is carried over to be added to the BCD sum of the second-position digits.

Consider the addition of 59 and 38 in BCD:

```
                      1
                      ↧
 59        0101 │ 1001  ←——— BCD for 59
+ 38      + 0011 │ 1000  ←——— BCD for 38
 97        1001 ⌞0001  ←——— perform addition
                  0110  ←——— add 6 to correct
           1001  0111       BCD for 97
           ‿‿‿  ‿‿‿
            9     7
```

Here, the addition of the least significant digits (LSDs) produces a sum of $17 = 10001$. This generates a carry into the next digit position to be added to the codes for 5 and 3. Since $17 > 9$, a correction factor of 6 has to be added to the LSD sum. Addition of this correction does not generate a carry; the carry was already generated in the original addition.

To summarize the BCD addition procedure:

1. Add, using ordinary binary addition, the BCD code groups for each digit position.
2. For those positions where the sum is 9 or less, no correction is needed. The sum is in proper BCD form.
3. When the sum of two digits is greater than 9, a correction of 0110 should be added to that sum to get the proper BCD result. This case always produces a carry into the next digit position, either from the original addition (step 1.) or from the correction addition.

The procedure for BCD addition is clearly more complicated than straight binary addition. This is also true of the other BCD arithmetic operations. Readers should perform the addition of $275 + 641$. Then check the correct procedure below.

```
 275       0010 0111 0101  ←——— BCD for 275
+ 641     + 0110 0100 0001  ←——— BCD for 641
 916       1000 1011 0110  ←——— perform addition
       +        0110        ←——— add 6 to correct 2nd digit
           1001 0001 0110  ←——— BCD for 916
```

1. How can you tell when a correction is needed in BCD addition?
2. True or false: A correction has to be added to a BCD sum only when the sum overflows into the fifth bit.

6.8 HEXADECIMAL ARITHMETIC*

Hex numbers are used extensively in machine-language computer programming and in conjunction with computer memories (i.e., addresses). When working in these areas, there will be situations where hex numbers have to be added or subtracted.

Hex Addition   This can be done in much the same manner as decimal addition, as long as we remember that the largest hex digit is F instead of 9. To illustrate, let's add the hex numbers 58 and 24.

$$
\begin{array}{r}
58 \\
+ \ 24 \\
\hline
7C
\end{array}
$$

Adding the LSDs (8 and 4) produces 12, which is C in hex. There is no carry into the next digit position.

Now let's add 58_{16} and $4B_{16}$:

$$
\begin{array}{r}
58 \\
+ \ 4B \\
\hline
A3
\end{array}
$$

Here the addition of 8 and B can be thought of as $8 + 11 = 19_{10}$ (remember that $B_{16} = 11_{10}$). Since $19_{10} = 1 \times 16^1 + 3 \times 16^0 = 13_{16}$, we write down the 3 digit and carry a 1 into the next position. This carry is added to the 5 and 4 digits to produce A (remember that $A_{16} = 10_{10}$).

Here's another example:

$$
\begin{array}{r}
3AF \\
+ \ 23C \\
\hline
5EB
\end{array}
$$

The sum of F and C can be considered to be $15 + 12 = 27_{10}$. Since $27_{10} = 1 \times 16^1 + 11 \times 16^0 = 1B_{16}$, the sum digit is B and there is a carry of 1 into the second digit position.

Hex Subtraction   Remember that hex numbers are just an efficient way to represent binary numbers. Thus we can subtract hex numbers using the same method we used for binary numbers. The hex subtrahend will be 2's-complemented and then *added* to the minuend, and any carry out of the MSD position will be disregarded.

---

*This section may be omitted until needed.

How do we 2's-complement a hex number? One way is to convert it to binary, 2's-complement the binary equivalent, and then convert it back to hex. This process is illustrated below.

$$73A \longleftarrow \text{hex number}$$

$$0111\ 0011\ 1010 \longleftarrow \text{convert to binary}$$

$$1000\ 1100\ 0110 \longleftarrow \text{2's-complement it}$$

$$8C6 \longleftarrow \text{convert back to hex}$$

There is a quicker procedure. Subtract *each* hex digit from F, then add 1. Let's try this for the same hex number from the example above.

$$
\left.
\begin{array}{ccc}
F & F & F \\
- 7 & - 3 & - A \\
\hline
8 & C & 5
\end{array}
\right\} \longleftarrow \text{subtract each digit from F}
$$

$$
\begin{array}{c}
+ 1 \longleftarrow \text{add 1} \\
\hline
8 \quad C \quad 6 \longleftarrow \text{hex. equivalent of 2's comp.}
\end{array}
$$

Try either of the above procedures on the hex number E63. The correct result for the 2's complement is 19D.

EXAMPLE 6.4

Subtract $3A5_{16}$ from 592_{16}.

Solution: First, convert the subtrahend (3A5) to its 2's-complement form by using either method presented above. The result is C5B. Then add this to the minuend (592):

$$
\begin{array}{r}
592 \\
+ \quad C5B \\
\hline
11ED
\end{array}
$$

$\quad\quad\quad$ ↑—disregard carry

Ignoring the carry out of the MSD addition, the result is 1ED. We can prove that this is correct by adding 1ED to 3A5 and checking to see that it equals 592.

REVIEW QUESTIONS

1. Add 67F + 2A4. (*Ans.* 923)
2. Subtract 67F − 2A4. (*Ans.* 3DB)

6.9 ARITHMETIC CIRCUITS

One essential function of most computers and calculators is the performance of arithmetic operations. These operations are all performed in the arithmetic unit of a computer, where logic gates and flip-flops are combined so that they can add, subtract, multiply, and divide binary numbers. These circuits perform arithmetic operations at speeds that are not humanly possible. Typically, an addition operation will take less than 1 μs.

We will now study some of the basic arithmetic circuits that are used to perform the arithmetic operations discussed earlier. In some cases we will go through the actual design process, even though the circuits may be commercially available in integrated-circuit form, to provide more practice in the use of the techniques learned in Chapter 4.

Arithmetic Unit All arithmetic operations take place in the *arithmetic unit* of a computer. Figure 6.3 is a block diagram showing the major elements included in a typical arithmetic unit. The main purpose of the arithmetic unit is to accept binary data that are stored in the memory and to execute arithmetic operations on these data according to instructions from the control unit.

The arithmetic unit contains at least two flip-flop registers: the *B register* and the *accumulator register*. It also contains combinatorial logic, which performs the arithmetic operations on the binary numbers that are stored in the *B* register and the accumulator. A typical sequence of operations may occur as follows:

1. The control unit receives an instruction (from the memory unit) specifying that a number stored in a particular memory location (address) is to be added to the number presently stored in the accumulator register.

2. The number to be added is transferred from memory to the *B* register.

3. The number in the *B* register and the number in the accumulator register are added together in the logic circuits (upon command from control unit). The resulting sum is then sent to the accumulator to be stored.

4. The new number in the accumulator can remain there so that another number can be added to it or, if the particular arithmetic process is finished, it can be transferred to memory for storage.

These steps should make it apparent how the accumulator register derives its name. This register "accumulates" the sums that occur when performing successive additions between new numbers acquired from memory and the previously accumu-

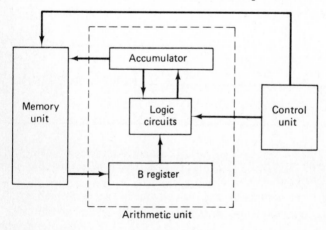

FIGURE 6.3 Functional parts
of an arithmetic unit.

lated sum. In fact, for any arithmetic problem containing several steps, the accumulator always contains the results of the intermediate steps as they are completed as well as the final result when the problem is finished.

6.10 THE PARALLEL BINARY ADDER

Computers and calculators perform the addition operation on two binary numbers at a time, where each binary number can have several binary digits. Figure 6.4 illustrates the addition of two 5-bit numbers. The *augend* is stored in the accumulator register; that is, the accumulator contains five FFs, storing the values 10101 in successive FFs. Similarly, the *addend,* the number that is to be added to the augend, is stored in the *B* register (in this case, 00111).

The addition process starts by adding the least significant bits (LSBs) of the augend and addend. Thus, $1 + 1 = 10,$* which means that the *sum* for that position is 0 with a *carry* of 1.

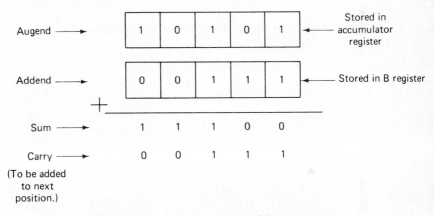

FIGURE 6.4 Typical binary addition process.

This carry has to be added to the next position along with the augend and addend bits in that position. Thus, in the second position, $1 + 0 + 1 = 10$, which is again a sum of 0 and a carry of 1. This carry is added to the next position together with the augend and addend bits in that position, and so on for the remaining positions.

At each step in this addition process we are performing the addition of 3 bits; the augend bit, the addend bit, and a carry bit from the previous position. The result of the addition of these 3 bits produces 2 bits: a *sum* bit, and a *carry* bit which is to be added to the next position. It should be clear that the same process is followed for each bit position. As such, if we can design a logic circuit that can duplicate this process, then we simply have to use identical circuits for each of the bit positions. This is illustrated in Figure 6.5.

In this diagram variables A_4, A_3, A_2, A_1, and A_0 represent the bits of the augend that are stored in the accumulator (which is also called the *A* register). Variables B_4, B_3, B_2, B_1, and B_0 represent the bits of the addend stored in the *B* register Variables

*Remember: Now we are talking about *arithmetic* addition, not OR addition.

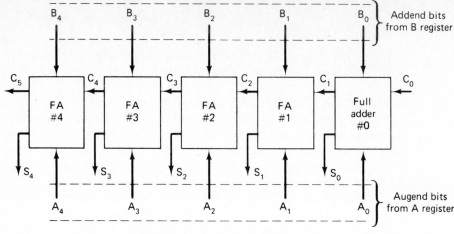

Sum appears at S_4, S_3, S_2, S_1, S_0 outputs.

FIGURE 6.5 Block diagram of 5-bit parallel adder circuit using full adders.

C_4, C_3, C_2, C_1, and C_0 represent the carry bits into the corresponding positions. Variables S_4, S_3, S_2, S_1, S_0 are the sum output bits for each position. Corresponding bits of the augend and addend are fed to a logic circuit called a *full adder,* along with a carry bit from the previous position. For example, bits A_1 and B_1 are fed into full adder 1 along with C_1, which is the carry bit produced by the addition of the A_0 and B_0 bits. Bits A_0 and B_0 are fed into full adder 0 along with C_0. Since A_0 and B_0 are the LSBs of the augend and addend, it appears that C_0 would always have to be 0, since there can be no carry into that position. However, we shall see that there will be situations when C_0 can also be 1.

The full-adder circuit used in each position has three inputs: an A bit, a B bit, and a C bit, and it produces two outputs: a sum bit and a carry bit. For example, full adder 0 has inputs A_0, B_0, and C_0, and it produces outputs S_0 and C_1. Full adder 1 has A_1, B_1, and C_1 as inputs and S_1 and C_2 as outputs; and so on. This arrangement is repeated for as many positions as there are in the augend and addend. Although this illustration is for 5-bit numbers, in modern computers the numbers usually range from 8 to 64 bits.

The arrangement in Figure 6.5 is called a *parallel adder* because all the bits of the augend and addend are present and are fed into the adder circuits *simultaneously.* This means that the additions in each position are taking place at the same time. This is different from how we add on paper, taking each position one at a time starting with the LSB. Clearly, parallel addition is extremely fast. More will be said about this later.

_____ **REVIEW QUESTIONS**

1. How many inputs does a full adder have? How many outputs?
2. Assume the following input levels in Figure 6.5: $A_4 A_3 A_2 A_1 A_0 = 01001$; $B_4 B_3 B_2 B_1 B_0 = 00111$; $C_0 = 0$.
   (a) What are the logic levels at the outputs of FA #2? (*Ans.* $S_2 = 0$, $C_3 = 1$)
   (b) What is the logic level at the C_5 output? (*Ans.* $C_5 = 0$)

6.11 DESIGN OF A FULL ADDER

Now that we know the function of the full adder, we can proceed to design a logic circuit that will perform this function. First, we must construct a truth table showing the various input and output values for all possible cases. Figure 6.6 shows the truth table having three inputs A, B, and C_{IN}, and two outputs, S and C_{out}. There are eight possible cases for the three inputs, and for each case the desired output values are listed. For example, consider the case $A = 1$, $B = 0$, and $C_{IN} = 1$. The full adder (hereafter abbreviated FA) must add these bits to produce a sum (S) of 0 and a carry (C_{out}) of 1. The reader should check the other cases to be sure they are understood.

Since there are two outputs, we will design the circuitry for each output individually, starting with the S output. The truth table shows that there are four cases where S is to be a 1. Using the sum-of-products method, we can write the expression for S as

$$S = \overline{A}\overline{B}C_{IN} + \overline{A}B\overline{C}_{IN} + A\overline{B}\overline{C}_{IN} + ABC_{IN} \tag{6.1}$$

We can now try to simplify this expression by factoring. Unfortunately, none of the terms in the expression has two variables in common with any of the other terms. However, $\overline{A}$ can be factored from the first two terms and A can be factored from the last two terms:

$$S = \overline{A}(\overline{B}C_{IN} + B\overline{C}_{IN}) + A(\overline{B}\,\overline{C}_{IN} + BC_{IN})$$

The first term in parentheses should be recognized as the exclusive-OR combination of B and C_{IN}, which can be written as $B \oplus C_{IN}$. The second term in parentheses should be recognized as the exclusive-NOR of B and C_{IN}, which can be written as $\overline{B \oplus C_{IN}}$. Thus, the expression for S becomes

$$S = \overline{A}(B \oplus C_{IN}) + A(\overline{B \oplus C_{IN}})$$

Augend bit input	Addend bit input	Carry bit input	Sum bit output	Carry bit output
A	B	C_{IN}	S	C_{OUT}
0	0	0	0	0
0	0	1	1	0
0	1	0	1	0
0	1	1	0	1
1	0	0	1	0
1	0	1	0	1
1	1	0	0	1
1	1	1	1	1

FIGURE 6.6 Truth table for a full-adder circuit.

If we let $X = B \oplus C_{IN}$, this can be written as

$$S = \overline{A} \cdot X + A \cdot \overline{X} = A \oplus X$$

which is simply the EX-OR of A and X. Replacing the expression for X, we have

$$S = A \oplus [B \oplus C_{IN}] \qquad (6.2)$$

Consider now the output C_{OUT} in the truth table of Figure 6.6. We can write the sum-of-products expression for C_{OUT} as follows:

$$C_{OUT} = \overline{A}BC_{IN} + A\overline{B}C_{IN} + AB\overline{C}_{IN} + ABC_{IN}$$

This expression can be simplified by factoring. We will employ the trick introduced in Chapter 3, whereby we will use the ABC_{IN} term *three* times since it has common factors with each of the other terms. Hence,

$$
\begin{aligned}
C_{OUT} &= BC_{IN}(\overline{A} + A) + AC_{IN}(\overline{B} + B) + AB(\overline{C}_{IN} + C_{IN}) \\
&= BC_{IN} + AC_{IN} + AB
\end{aligned}
\qquad (6.3)
$$

This expression cannot be simplified further.

Expressions (6.2) and (6.3) can be implemented as shown in Figure 6.7. There are several other implementations that can be used to produce the same expressions for S and C_{OUT}, none of which has any particular advantage over those shown. The complete circuit with inputs A, B, and C_{IN} and outputs S and C_{OUT} represents the full adder. Each of the FAs in Figure 6.5 contains this same circuitry (or its equivalent).

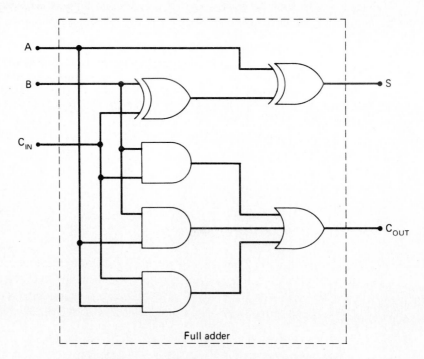

FIGURE 6.7 Complete circuitry for a full adder.

FIGURE 6.8   K mappings for the full-adder outputs.

K-Map Simplification   We simplified the expressions for S and C_{OUT} using algebraic methods. The K-map method can also be used. Figure 6.8(a) shows the K map for the S output. This map has no adjacent 1s, so there are no pairs or quads to loop. Thus, the expression for S cannot be simplified using the K map. This points out a limitation of the K-map method as compared to the algebraic method. We were able to simplify the expression for S through factoring and the use of EX-OR and EX-NOR operations.

The K map for the C_{OUT} output is shown in Figure 6.8(b). The three pairs that are looped will produce the same expression obtained from the algebraic method.

The Half Adder   The FA operates on three inputs to produce a sum and carry output. In some cases a circuit is needed that will add only 2 input bits, to produce a sum and carry output. An example would be the addition of the LSB position of two binary numbers where there is no carry input to be added. A special logic circuit can be designed to take *two* input bits, A and B, and to produce sum (S) and carry (C_{OUT}) outputs. This circuit is called a *half adder* (*HA*). Its operation is similar to a FA except that it operates on only 2 bits. We shall leave the design of the HA as an exercise at the end of the chapter, since it is relatively simple. However, we will use the HA symbol whenever it is appropriate.

6.12   COMPLETE PARALLEL ADDER WITH REGISTERS

In a computer, the numbers that are to be added are stored in FF registers. Figure 6.9 shows the complete diagram of a 4-bit parallel adder including the storage registers. The augend bits A_3–A_0 are stored in the accumulator (A register); the addend bits B_3–B_0 are stored in the B register. Each of these registers is made up of D FFs for easy transfer of data.

The contents of the A register (i.e., the binary number stored in A_3–A_0) is added to the contents of the B register by the four FAs, and the sum is produced at outputs

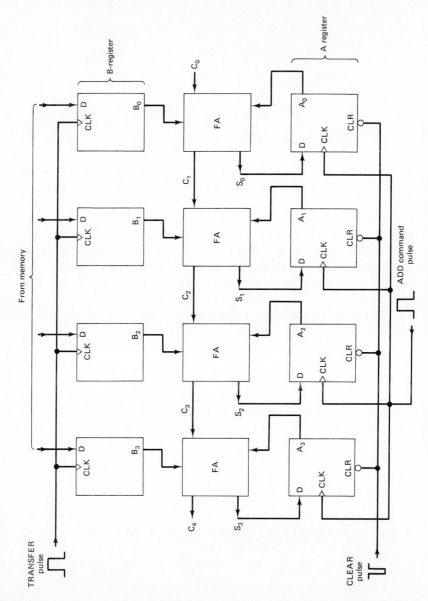

FIGURE 6.9 Complete 4-bit parallel adder with registers.

S_3–S_0. C_4 is the carry out of the fourth FA and it can be used as the carry input to a fifth FA, or as an *overflow* bit to indicate that the sum exceeds 1111.

Note that the sum outputs are connected to the D inputs of the A register. This will allow the sum to be parallel-transferred into the A register on the PGT of the ADD pulse. In this way, the sum can be stored in the A register.

Also note that D inputs of the B register are coming from the computer's memory, so that numbers from memory will be parallel-transferred into the B register on the PGT of the TRANSFER pulse. For now we will not be concerned with how the numbers come from memory.

Register Notation Before we go through the complete process of how this circuit adds two binary numbers, it will be helpful to introduce some notation that makes it easy to describe the contents of a register and data transfer operations.

Whenever we want to give the levels that are present at each FF in a register or at each output of a group of outputs, we will use brackets as illustrated below:

$$[A] = 1011$$

This is the same as saying that $A_3 = 1, A_2 = 0, A_1 = 1, A_0 = 1$. In other words, think of $[A]$ as representing "the contents of register A."

Whenever we want to indicate the transfer of data to or from a register, we will use an arrow as illustrated below:

$$[B] \longrightarrow [A]$$

This means that the contents of the B register has been transferred to the A register. The old contents of the A register will be lost as a result of this operation, and the B register will be unchanged.

Sequence of Operations We will now describe the process by which the circuit of Figure 6.9 will add the binary numbers 1001 and 0101. Assume $C_0 = 0$; that is, there is no carry into the LSB position.

1. **$[A] = 0000$.** A CLEAR pulse is applied to the asynchronous inputs (CLR) of each FF in register A.
2. **$[M] \rightarrow [B]$.** The first binary number is transferred from memory (M) to the B register. In this case, the binary number 1001 is transferred into register B on the occurrence of a TRANSFER pulse.
3. **$[S]^* \rightarrow [A]$.** With $[B] = 1001$ and $[A] = 0000$, the full adders produce a sum of 1001; that is, $[S] = 1001$. These sum outputs are transferred into the A register when the ADD pulse occurs. This makes $[A] = 1001$.
4. **$[M] \rightarrow [B]$.** The second binary number, 0101, is transferred from memory into the B register on the occurrence of a second TRANSFER pulse. This makes $[B] = 0101$.
5. **$[S] \rightarrow [A]$.** With $[B] = 0101$ and $[A] = 1001$, the FAs produce $[S] = 1110$. These sum outputs are transferred into the A register when the second ADD pulse occurs. Thus, $[A] = 1110$.

*Even though S is not a register, we will use $[S]$ to represent the group of S outputs.

6. At this point, the sum of the two binary numbers is stored in the accumulator. In most computers the contents of the accumulator, $[A]$, will usually be transferred to the computer's memory so that the adder circuit can be used for a new set of numbers. The circuitry that performs this $[A] \rightarrow [M]$ transfer is not shown in the diagram.

REVIEW QUESTIONS

1. Suppose four different 4-bit numbers are to be taken from memory and added by the circuit of Figure 6.9. How many CLEAR pulses will be needed? How many TRANSFER pulses? How many ADD pulses? (*Ans.* One, four, four)
2. Determine the contents of the A register after the following sequence of operations: $[A] = 0000$, $[0110] \rightarrow [B]$, $[S] \rightarrow [A]$, $[1110] \rightarrow [B]$, $[S] \rightarrow [A]$. (*Ans.* 0100)

6.13 CARRY PROPAGATION

The parallel adder of Figure 6.9 performs additions at a relatively high speed, since it adds the bits from each position simultaneously. However, its speed is limited by an effect called *carry propagation* or *carry ripple,* which can best be explained by considering the following addition:

$$0111$$
$$+ \ 0001$$
$$\overline{1000}$$

Addition of the LSB position produces a carry into the second position. This carry, when added to the bits of the second position, produces a carry into the third position. This latter carry, when added to the bits of the third position, produces a carry into the last position. The key thing to notice in this example is that the sum bit generated in the *last* position (MSB) depended on the carry that was generated by the addition in the *first* position (LSB).

Looking at this from the viewpoint of the circuit of Figure 6.9, S_3 out of the last full adder depends on C_1 out of the first full adder. But the C_1 signal must pass through three FAs before it produces S_3. What this means is that the S_3 output will not reach its correct value until C_1 has propagated through the intermediate FAs. This represents a time delay that depends on the propagation delay produced in a FA. For example, if each FA is considered to have a propagation delay of 40 ns, then S_3 will not reach its correct level until 120 ns after C_1 is generated. This means that the add command pulse cannot be applied until 160 ns after the augend and addend numbers are present in the FF registers (the extra 40 ns is due to the delay of the LSB FA, which generates C_1).

Obviously, the situation becomes much worse if we extend the adder circuitry to add a greater number of bits. If the adder were handling 32-bit numbers, the carry propagation delay could be 1280 ns = 1.28 μs. The add pulse could not be applied until at least 1.28 μs after the numbers were present in the registers.

This magnitude of delay is prohibitive for high-speed computers. Fortunately, logic designers have come up with several ingenious schemes for reducing this delay. One of the schemes, called *look-ahead carry,* utilizes logic gates to look at the lower-order bits of the augend and addend to see if a higher-order carry is to be generated. For example, it is possible to build a logic circuit with B_2, B_1, B_0, A_2, A_1, and A_0 as inputs and C_3 as an output. This logic circuit would have a shorter delay than is obtained by the carry propagation through the FAs. This scheme requires a large amount of extra circuitry but is necessary to produce high-speed adders. The extra circuitry is not a significant consideration with the present use of integrated circuits. Many high-speed adders available in integrated-circuit form utilize the look-ahead carry or a similar technique for reducing overall propagation delays.

6.14 INTEGRATED-CIRCUIT PARALLEL ADDER

Several parallel adder circuits are available as ICs. One of the most common is a 4-bit parallel adder package that contains four interconnected FAs and the look-ahead carry circuitry needed for high-speed operation. The TTL version of this IC is the 7483 (also the 74283), and the CMOS version is the 4008.

Figure 6.10(a) shows the logic symbol for the 7483 4-bit parallel adder. The inputs to this IC are two 4-bit numbers $A_3A_2A_1A_0$ and $B_3B_2B_1B_0$, and the carry, C_0, into the LSB position. The outputs are the sum bits $S_3S_2S_1S_0$ and the carry, C_4, out of the MSB.

Cascading Parallel Adders Two or more parallel adder blocks may be connected in cascade to accommodate the addition of larger binary numbers. To illustrate, Figure 6.10(b) shows how two 7483 adders can be connected to add two 8-bit numbers. The adder on the right adds the four least significant bits of the numbers. The C_4 output of this adder is connected as the input carry to the first position of the second adder, which adds the four most significant bits of the numbers. The eight sum outputs represent the resultant sum of the two 8-bit numbers. C_8 is the carry out of the last position (MSB) of the second adder. C_8 can be used as an overflow bit or as a carry into another adder stage if larger binary numbers are to be handled.

6.15 THE 2's-COMPLEMENT SYSTEM

Most modern computers use the 2's-complement system to represent negative numbers and to perform subtraction. The operations of addition and subtraction of signed numbers can be performed using only the addition operation if we use the 2's-complement form to represent negative numbers.

Addition Positive and negative numbers, including the sign bits, can be added together in the basic parallel adder circuit when the negative numbers are in 2's-complement form. This is illustrated in Figure 6.11 for the addition of -3 and $+6$. The -3 is represented in its 2's-complement form as 1101, where the first 1 is the sign bit; the $+6$ is represented as 0110, with the first zero as the sign bit. These numbers are stored in their corresponding registers. The 4-bit parallel adder produces sum outputs of 0011, which represents $+3$. The C_4 output is 1 but is disregarded in the 2's-complement method.

(a)

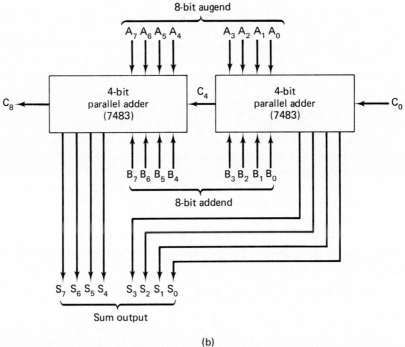

(b)

FIGURE 6.10 (a) **Block symbol for the 7483 4-bit parallel adder;**
(b) **cascading two 7483s.**

Subtraction When the 2's-complement system is used, the number to be subtracted
(the subtrahend) is 2's-complemented and then *added* to the minuend (the number the
subtrahend is being subtracted from). For example, we can assume that the minuend
is already stored in the accumulator (*A* register). The subtrahend is then placed in the
B register (in a computer it would be transferred here from memory) and is changed
to its 2's-complement form before it is added to the number in the *A* register. The sum

FIGURE 6.11   Parallel adder used to add + and − numbers in 2's-complement system.

outputs of the adder circuit now represent the *difference* between the minuend and subtrahend.

The parallel-adder circuit that we have been discussing can be adapted to perform the subtraction described above if we provide a means for taking the 2's complement of the *B* register number. The 2's complement of a binary number is obtained by complementing (inverting) each bit and then adding 1 to the LSB. Figure 6.12 shows how this can be accomplished. The *inverted* outputs of the *B* register are used rather than the normal outputs; that is, $\bar{B}_0$, $\bar{B}_1$, $\bar{B}_2$, and $\bar{B}_3$ are fed to the adder inputs (remember, B_3 is the sign bit). This takes care of complementing each bit of the *B*

FIGURE 6.12   Parallel adder used to perform subtraction using 2's-complement system.

number. Also, C_0 is made a logical 1, so it adds an extra 1 into the LSB of the adder; this accomplishes the same effect as adding 1 to the LSB of the B register for forming the 2's complement.

The S_3–S_0 outputs represent the results of the subtraction operation. Of course, S_3 is the sign bit of the result and indicates whether the result is $+$ or $-$. The carry output C_4 is again disregarded.

To help clarify this operation, study the following steps for subtracting $+6$ from $+4$:

1. $+4$ is stored in the A register as 0100.
2. $+6$ is stored in the B register as 0110.
3. The inverted outputs of the B-register FFs are fed to the adder—that is, 1001.
4. The 1001 is added to 0100 by the parallel adder along with a 1 added to the LSB position by making $C_0 = 1$. This produces sum output bits 1110 and a $C_4 = 1$, which is disregarded. This 1110 represents the required difference. Since the sign bit $= 1$, it is a negative result and is in 2's-complement form. We can verify that 1110 represents -2_{10} by 2's-complementing it and obtaining $+2_{10}$:

$$
\begin{array}{r}
1110 \\
0001 \\
+ \quad 1 \\
\hline
0010 = +2_{10}
\end{array}
$$

Combined Addition and Subtraction It should now be clear that the basic parallel adder circuit can be used to perform addition or subtraction depending on whether the B number is left unchanged or is 2's-complemented. A complete circuit that can perform *both* addition and subtraction in the 2's-complement system is shown in Figure 6.13.

This adder/subtractor circuit is controlled by the two control signals ADD and SUB. When the ADD level is HIGH, the circuit performs addition of the numbers stored in the A and B registers. When the SUB level is HIGH, the circuit subtracts the B-register number from the A-register number. The operation is described as follows:

1. Assume that ADD $= 1$ and SUB $= 0$. The SUB $= 0$ *disables* (inhibits) AND gates 2, 4, 6, and 8, holding their outputs at 0. The ADD $= 1$ *enables* AND gates 1, 3, 5, and 7, allowing their outputs to pass the B_0, B_1, B_2, and B_3 levels, respectively.
2. The B_0–B_3 levels pass through the OR gates into the 4-bit parallel adder to be added to the A_0–A_3 bits. The *sum* appears at the S_0–S_3 outputs.
3. Note that SUB $= 0$ causes $C_0 = 0$ into the adder.
4. Now assume that ADD $= 0$ and SUB $= 1$. The ADD $= 0$ inhibits AND gates 1, 3, 5, and 7. The SUB $= 1$ enables AND gates 2, 4, 6, and 8, so their outputs pass the $\bar{B}_0$, $\bar{B}_1$, $\bar{B}_2$, and $\bar{B}_3$ levels, respectively.
5. The $\bar{B}_0$–$\bar{B}_3$ levels pass through the OR gates into the adder to be added to the A_0–A_3 bits. Note also that C_0 is now 1. Thus, the B-register number has essentially been 2's-complemented.
6. The *difference* appears at the S_0–S_3 outputs.

The configuration of Figure 6.13 is one of the most widely used in computers because it provides a relatively simple method for adding and subtracting positive or

FIGURE 6.13   Parallel adder/subtractor using 2's-complement system.

negative numbers. In most computers the outputs present at the S output lines are usually transferred into the A register (accumulator), so the results of the addition or subtraction always end up stored in the A register. This is accomplished by applying a pulse to the *CLK* inputs of register A.

Sec. 6.15   The 2's-Complement System **207**

1. Why does C_0 have to be a 1 in order to use the adder circuit in Figure 6.12 as a subtractor?
2. Assume that $[A] = 0011$ and $[B] = 0010$ in Figure 6.13. If ADD = 1 and SUB = 0, determine the logic levels at the OR-gate outputs. (*Ans.* 0010)
3. Repeat for ADD = 0, SUB = 1. (*Ans.* 1101)
4. True or false: When the adder/subtractor circuit is used for subtraction, the 2's complement of the subtrahend appears at the input of the adder.

6.16 THE BCD ADDER

The BCD addition process was discussed in Section 6.7 and is reviewed below:

1. Add the BCD code groups for each decimal digit position; use ordinary binary addition.
2. For those positions where the sum is 9 or less, the sum is in proper BCD form and no correction is needed.
3. When the sum of two digits is greater than 9, a correction of 0110 should be added to that sum to produce the proper BCD result. This will produce a carry to be added to the next decimal position.

A BCD adder circuit must be able to operate in accordance with the above steps. In other words, the circuit must be able to do the following:

1. Add two 4-bit BCD code groups, using straight binary addition.
2. Determine if the sum of this addition is greater than 1001 (decimal 9); if it is, add 0110 (6) to this sum and generate a carry to the next decimal position.

The first requirement is easily met by using a 4-bit binary parallel adder such as the 7483 IC. For example, if the two BCD code groups represented by $A_3A_2A_1A_0$ and $B_3B_2B_1B_0$, respectively, are applied to a 4-bit parallel adder, the adder will perform the following operation:

$$
\begin{array}{rl}
A_3A_2A_1A_0 & \longleftarrow \text{BCD code group} \\
+\ \ B_3B_2B_1B_0 & \longleftarrow \text{BCD code group} \\
\hline
S_4S_3S_2S_1S_0 & \longleftarrow \text{straight binary sum}
\end{array}
$$

S_4 is actually C_4, the carry out of the MSB.

The sum outputs $S_4\,S_3\,S_2\,S_1\,S_0$ can range anywhere from 00000 to 10010 (when both BCD code groups are $1001 = 9$). The circuitry for a BCD adder must include the logic needed to detect whenever the $S_4\,S_3\,S_2\,S_1\,S_0$ is greater than 01001, so that the correction can be added in. These cases where the sum is greater than 01001 are listed below:

Let's define X as a logic output that will go HIGH only when the sum is greater than 01001 (i.e., for the cases listed above). If we examine these cases, it can be reasoned that X will be HIGH for either of the following conditions:

1. Whenever $S_4 = 1$ (sums greater than 15).
2. Whenever $S_3 = 1$ and either S_2 or S_1 or both are 1 (sums 10–15).

S_4	S_3	S_2	S_1	S_0	
0	1	0	1	0	(10)
0	1	0	1	1	(11)
0	1	1	0	0	(12)
0	1	1	0	1	(13)
0	1	1	1	0	(14)
0	1	1	1	1	(15)
1	0	0	0	0	(16)
1	0	0	0	1	(17)
1	0	0	1	0	(18)

This can be expressed as

$$X = S_4 + S_3(S_2 + S_1)$$

Whenever $X = 1$, it is necessary to add the correction 0110 to the sum bits and to generate a carry. Figure 6.14 shows the complete circuitry for a BCD adder, including the logic-circuit implementation for X.

The circuit consists of three basic parts. The two BCD code groups $A_3 - A_0$ and $B_3 - B_0$ are added together in the upper 4-bit adder to produce the sum S_4 S_3 S_2 S_1 S_0. The logic gates implement the expression for X. The lower 4-bit adder will add the correction 0110 to the sum bits *only* when $X = 1$, producing the final BCD sum output represented by $\Sigma_3\Sigma_2\Sigma_1\Sigma_0$ (the Greek letter Σ, sigma, is a common mathematical symbol for *sum*). X is also the carry output that is produced when the sum is greater than 01001. Of course, when $X = 0$, there is no carry and no addition of 0110. In such cases, $\Sigma_3\Sigma_2\Sigma_1\Sigma_0 = S_3S_2S_0S_0$.

To help in the understanding of the BCD adder, the reader should try several cases by following them through the circuit. The following cases would be particularly instructive:

Inputs
(a) $[A] = 0101$, $[B] = 0011$, $C_0 = 0$
(b) $[A] = 0111$, $[B] = 0110$, $C_0 = 0$

Outputs
(a) $[S] = 01000$, $X = 0$, $[\Sigma] = 1000$, CARRY $= 0$
(b) $[S] = 01101$, $X = 1$, $[\Sigma] = 0011$, CARRY $= 1$

Cascading BCD Adders   The circuit of Figure 6.14 is used for adding two decimal digits that have been encoded in BCD code. When decimal numbers with several digits are to be added together, it is necessary to use a separate BCD adder for each digit position. Figure 6.15 illustrates for the addition of two three-digit decimal numbers. The A register contains 12 bits, which are the three BCD code groups for one of the three-digit decimal numbers; similarly, the B register contains the BCD representation of the other three-digit decimal number. The A_3–A_0 and B_3–B_0 code groups representing the least significant digits are fed to the first BCD adder. Each BCD adder block is assumed to contain the circuitry of Figure 6.14. This first BCD adder produces sum outputs $\Sigma_3\Sigma_2\Sigma_1\Sigma_0$, which is the BCD code for the least significant digit of the sum.

FIGURE 6.14   A BCD adder contains two 4-bit adders and a correction detector circuit.

It also produces a carry output that is sent to the second BCD adder, which is adding A_7–A_4 and B_7–B_4, the BCD code groups for the second-decimal-digit position. The second BCD adder produces $\Sigma_7\Sigma_6\Sigma_5\Sigma_4$, the BCD code for the second digit of the sum, and so on. This arrangement can, of course, be extended to decimal numbers of any size by simply adding more FFs to the registers and including a BCD adder for each digit position.

REVIEW QUESTIONS

1. What are the three basic parts of a BCD adder circuit?
2. Describe how the BCD adder circuit detects the need for a correction and executes it.

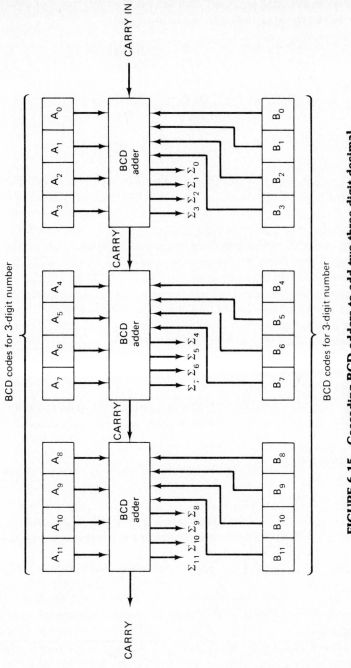

FIGURE 6.15 Cascading BCD adders to add two three-digit decimal numbers.

6.17 BINARY MULTIPLIERS

The multiplication of two binary numbers is done with paper and pencil by performing successive additions and shifting. To illustrate:

$$
\begin{array}{r}
1011 \\
\times\ 1101 \\
\hline
1011 \\
0000 \\
1011 \\
1011 \\
\hline
10001111
\end{array}
\quad
\begin{array}{l}
\text{multiplicand (11)} \\
\text{multiplier (13)} \\
\\
\\
\\
\\
\text{product (143)}
\end{array}
$$

This process consists of examining the successive bits of the multiplier, beginning with the LSB. If the multiplier bit is a 1, the multiplicand is copied down; if it is a 0, zeros are written down. The numbers written down in successive lines are shifted one position to the left relative to the previous line. When all the multiplier bits have been examined, the various lines are *added* to produce the final product.

In digital machines this process is modified somewhat because the binary adder is designed to add only two binary numbers at a time. Instead of adding all the lines at the end, they are added two at a time and their sum is accumulated in a register (the accumulator register). In addition, when the multiplier bit is 0, there is no need to write down and add zeros since it does not affect the final result. The previous example is redone here showing the modified process.

multiplicand:	1011	
multiplier:	1101	
	1011	1st multiplier bit = 1; write down multiplicand; shift multiplicand one position to the left (10110)
	1011	2nd multiplier bit = 0; write down previous result; shift multiplicand to the left again (101100)
	+ 101100	3rd multiplier bit = 1; write down multiplicand (101100)
	110111	add shift new multiplicand to the left (1011000)
	+ 1011000	4th multiplier bit = 1; write down new multiplicand (1011000)
	10001111	add to obtain final product

This multiplication process can be implemented as shown in Figure 6.16. It contains three registers. The X register is used to store the multiplier bits. The B register is used to store the multiplicand bits. These two registers are connected as shift registers (B shifts right and X shifts left on each clock pulse). The A register is the accumulator register, which is used to accumulate the partial products. Note that the adder outputs S_0–S_7 are connected to the D inputs of the A register so that the sum is transferred to the A register *only when a clock pulse gets through the AND gate*. Also

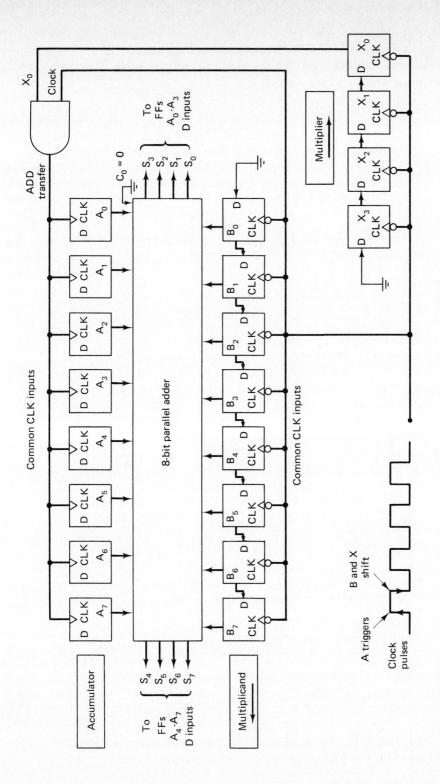

FIGURE 6.16 Binary multiplier circuit.

213

note that the B and X registers will shift on the negative-going edge of the clock pulses while the A register responds to the positive clock transitions.

The circuit operation can best be described with the aid of Figure 6.17, which shows the contents of all the registers and the adder outputs at each point during the operation. The step-by-step process takes place as follows:

1. Initially the A register is at 00000000. The B register contains the multiplicand 00001011. The X register contains the multiplier 1101. The adder outputs represent the sum of A and B, that is, 00001011. These conditions are shown in Figure 6.17(a).

2. The first clock pulse occurs. Since the LSB of the multiplier (X_0) is a 1, this clock pulse gets through the AND gate and its positive edge transfers the adder outputs into the A register (adds multiplicand to accumulator). The negative edge of this clock

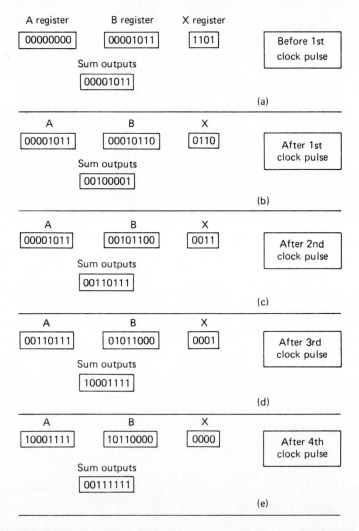

FIGURE 6.17 Contents of various registers during multiplication of 1011 × 1101.

pulse causes the B register to shift left and the X register to shift right. This situation is shown in Figure 6.17(b).

3. The second pulse occurs. Since X_0 is now at 0, this pulse will not affect the A register. Thus, the accumulator contents does not change. The negative edge of this pulse shifts the B and X registers. This situation is shown in Figure 6.17(c).

4. The third clock pulse occurs. X_0 is now a 1, so the adder outputs are transferred to the A register. Then, the B and X registers are shifted on the negative edge of the clock pulse.

5. The fourth clock pulse occurs. X_0 is at 1, so the adder outputs transfer to the A register. B and X both shift. After this fourth pulse the A register contains the final product, 10001111.

REVIEW QUESTIONS

1. Describe the sequence of events that occurs in the multiplier circuit during one clock cycle.
2. True or false: The outputs of the adder are transferred to the accumulator on *every* PGT of the clock.

PROBLEMS

6.1 Add the following groups of binary numbers using binary addition:
(a) 1010 + 1011   (b) 1111 + 0011   (c) 1011.1101 + 11.1
(d) 0.1011 + 0.1111   (e) 10011011 + 10011101

6.2 Represent each of the following signed decimal numbers in the 2's-complement system. Use a total of eight bits including sign bit.
(a) $+32$ (b) -14 (c) $+63$ (d) -104 (e) -1 (f) -128

6.3 Each of the following numbers represents a signed decimal number in the 2's-complement system. Determine the decimal value in each case.
(a) 01101   (b) 11101   (c) 01111011   (d) 10011001   (e) 01111111
(f) 100000   (g) 11111111   (h) 10000001

6.4 (a) What range of signed decimal values can be represented using 12 bits including sign bit?
(b) How many bits would be required to represent decimal numbers from $-32,768$ to $+32,767$?

6.5 Perform the following operations in the 2's-complement system. Use 8 bits (including the sign bit) for each number. Check your results by converting the binary result back to decimal.
(a) Add $+9$ to $+6$ (b) Add $+14$ to -17
(c) Add $+19$ to -24 (d) Add -48 to -80
(e) Subtract $+16$ from $+17$ (f) Subtract $+21$ from -13
(g) Subtract $+47$ from $+47$ (h) Subtract -36 from -15

6.6 Repeat Problem 6.5 for the following cases, and show that overflow occurs in each case.
(a) Add $+37$ to $+95$ (b) Subtract $+37$ from -95

6.7 Multiply the following pairs of binary numbers:
(a) 111×101 (b) 1011×1011 (c) 101.101×110.010
(d) $.1101 \times .1011$

6.8 Perform the following divisions:
(a) $1100 \div 100$ (b) $111111 \div 1001$ (c) $10111 \div 100$
(d) $10110.1101 \div 1.1$

6.9 Add the following decimal numbers after converting each to its BCD code:
(a) 74 + 23   (b) 58 + 37   (b) 147 + 380   (d) 385 + 118

6.10 Find the sum of each of the following pairs of hex numbers:
(a) 3E91 + 2F93   (b) 91B + 6F2

6.11 Perform the following subtractions on the pairs of hex numbers:
(a) 3E91 − 2F93   (b) 91B − 6F2   (c) 0300 − 005A

6.12 The owner's manual for a certain personal computer states that the computer has usable memory locations at the following hex addresses: 0200 through 03FF, and 4000 through 7FD0. What is the total number of available memory locations?

6.13 In the adder and subtractor circuits discussed in this chapter we gave no consideration to the possibility of *overflow*. Overflow occurs when the two numbers being added or subtracted produce a result that contains more bits than the capacity of the accumulator. For example, using 4-bit registers, including a sign bit, numbers ranging from +7 to −8 (in 2's complement) can be stored. Therefore, if the result of an addition or subtraction exceeds +7 or −8, we would say that an overflow had occurred. When an overflow occurs, the results are useless since they cannot be stored correctly in the accumulator register. To illustrate, add +5 (0101) and +4 (0100), which results in 1001. This 1001 would be interpreted incorrectly as a negative number since there is a 1 in the sign-bit position.

In computers and calculators there are usually circuits that are used to detect an overflow condition. There are several ways to do this. One method that can be used for the adder that operates in the 2's-complement system works as follows:

1. Examine the sign bits of the two numbers being added.
2. Examine the sign bit of the result.
3. Overflow occurs whenever the numbers being added are *both positive* and the sign bit of the result is 1 *or* when the numbers are *both negative* and the sign bit of the result is 0.

This method can be verified by trying several examples.

The reader should try the following cases for his own clarification: (1) 5 + 4; (2) −4 + (−6); (3) 3 + 2. Cases 1 and 2 will produce an overflow and case 3 will not. Thus, by examining the sign bits a logic circuit can be designed that will produce a 1 output whenever the overflow condition occurs. Design this overflow circuit for the adder of Figure 6.9.

6.14 Convert the FA circuit of Figure 6.7 to all NAND gates.

6.15 Write the truth table for a half adder (inputs A and B; outputs SUM and CARRY). From the truth table design a logic circuit that will act as a half adder.

6.16 A full adder can be implemented in many different ways. Figure 6.18 shows how one may be constructed from two half adders. Construct a truth table for this arrangement and verify that it operates as a FA.

6.17 Design a look-ahead carry circuit for the adder of Figure 6.9 which generates the carry C_3 to be fed to the FA of the MSB position based on the values of $A_0, B_0, C_0, A_1, B_1,$ $A_2,$ and B_2. In other words, derive an expression for C_3 in terms of $A_0, B_0, C_0, A_1, B_1,$ $A_2,$ and B_2. (*Hint:* Begin by writing the expression for C_1 in terms of $A_0, B_0,$ and C_0. Then write the expression for C_2 in terms of $A_1, B_1,$ and C_1. Substitute the expression for C_1 into the expression for C_2. Then write the expression for C_3 in terms of $A_2, B_2,$ and C_2. Substitute the expression for C_2 into the expression for C_3. Simplify the final expression for C_3 and put it in sum-of-products form. Implement the circuit.)

Full adder

FIGURE 6.18

6.18 Refer to Figure 6.9. Determine the contents of the A register after the following sequence of operations: $[A] = 0000$, $[0101] \longrightarrow [B]$, $[S] \longrightarrow [A]$, $[1011] \longrightarrow [B]$, $[S] \longrightarrow [A]$.

6.19 Refer to Figure 6.9. Assume that each FF has $t_{PLH} = t_{PHL} = 30$ ns, a set-up time of 10 ns, and each FA has a propagation delay of 40 ns. What is the minimum time allowed between the PGT of the TRANSFER pulse and the PGT of the ADD pulse for proper operation?

6.20 For the circuit of Figure 6.13, determine the sum outputs for the following cases:
(a) A register = 0101 (+5), B register = 1110 (−2); SUBTRACT = 1, ADD = 0.
(b) A register = 1100 (−4), B register = 1110 (−2); SUBTRACT = 0, ADD = 1.

6.21 Modify the circuit of Figure 6.13 so that a single control input, X, is used in place of ADD and SUB. The circuit is to function as an adder when $X = 0$, and as a subtractor when $X = 1$. Then simplify each set of gates. (*Hint:* Note that now each set of gates is functioning as a controlled inverter.)

6.22 Assume the following inputs in Figure 6.14: $[A] = 0101, [B] = 1001, C_0 = 0$. Determine the logic levels at $[S]$, X, $[\Sigma]$, and CARRY.

6.23 Would it make any difference in the BCD adder of Figure 6.14 if the C_0 of the upper adder was held LOW, while the C_0 of the lower adder was used as the carry input?

6.24 Assume the A register in Figure 6.15 holds the BCD code for 376, and the B register holds the BCD code for 469. Determine the outputs.

6.25 Refer to the circuit of Figure 6.16. Assume that the B register initially holds the multiplicand 0111, and the X register holds the multiplier 1001. Also assume $[A] = 00000000$ prior to the first clock pulse. Show the contents of registers A, B, and X after each clock pulse.

7
COUNTERS and REGISTERS

In Chapter 5 we saw how flip-flops could be connected to function as counters and registers. At that time we studied only the basic counter and register circuits. Digital systems employ many variations of these basic circuits, mostly in integrated-circuit form. In this chapter we will look at how FFs and logic gates can be combined to produce different types of counters and registers.

We will present several practical applications to illustrate the many ways in which these logic circuits can be used in digital systems. We will also look at several of the numerous counters and registers that are available as ICs.

7.1 ASYNCHRONOUS (RIPPLE) COUNTERS

Figure 7.1 shows a 4-bit binary counter circuit such as the one discussed in Chapter 5. Recall the following points concerning its operation:

1. The clock pulses are applied only to the *CLK* input of FF *A*. Thus, FF *A* will toggle

All J and K inputs assumed to be 1.

Clock pulses

D	C	B	A	Number of clock pulses
0	0	0	0	0
0	0	0	1	1
0	0	1	0	2
0	0	1	1	3
0	1	0	0	4
0	1	0	1	5
0	1	1	0	6
0	1	1	1	7
1	0	0	0	8
1	0	0	1	9
1	0	1	0	10
1	0	1	1	11
1	1	0	0	12
1	1	0	1	13
1	1	1	0	14
1	1	1	1	15
0	0	0	0	16 (recycles)
0	0	0	1	17
0	0	1	0	18
0	0	1	1	19
0	1	0	0	20
0	1	0	1	21

FIGURE 7.1   Four-bit asynchronous (ripple) counter.

(change to its opposite state) each time the clock pulses make a negative (HIGH to LOW) transition. Note that $J = K = 1$ for all FFs.

2. The normal output of FF A acts as the *CLK* input for FF B, so FF B will toggle each time the A output goes from 1 to 0. Similarly, FF C will toggle when B goes from 1 to 0 and FF D will toggle when C goes from 1 to 0.

3. The table in Figure 7.1 shows the sequence of binary states that the FFs will follow as clock pulses are continuously applied. If we let the FF outputs D, C, B, and A represent a binary number, with D being the MSB and A the LSB, then a binary counting sequence from 0000 to 1111 is produced.

4. After the 15th clock pulse has occurred, the counter FFs are in the 1111 condition. On the 16th clock pulse FF A goes from 1 to 0, which causes FF B to go from 1 to 0, and so on until the counter is in the 0000 state. In other words, the counter has gone through one complete cycle (0000 through 1111) and has *recycled* back to 0000, from where it will begin a new counting cycle as subsequent clock pulses are applied.

This type of counter, where each FF output serves as the *CLK* input signal for the next FF, is referred to as an *asynchronous counter*. This is because all the FFs do *not* change states in exact synchronism with the clock pulses; only FF A responds to the clock pulses. FF B has to wait for FF A to change states before it is triggered; FF C has to wait for FF B; and so on. Thus, there is a delay between the responses of each FF. In modern FFs this delay may be very small (typically 10 to 40 ns), but in some cases, as we shall see, it can be troublesome. Because of the manner in which this type of counter operates, it is also commonly referred to as a *ripple counter*. In the following discussions we will use the terms "asynchronous counter" and "ripple counter" interchangeably.

EXAMPLE 7.1

The counter in Figure 7.1 starts off in the 0000 state and then clock pulses are applied. Some time later the clock pulses are removed and the counter FFs read 0011. How many clock pulses occurred?

Solution: The apparent answer seems to be 3, since 0011 is the binary equivalent of 3. However, with the information given there is no way to tell whether the counter has recycled or not. This means that there could have been 19 clock pulses; the first 16 pulses bring the counter back to 0000 and the last 3 bring it to 0011. There could have been 35 pulses (two complete cycles and then 3 more), or 51 pulses, and so on.

MOD Number The counter in Figure 7.1 has 16 distinctly different states (0000 through 1111). Thus, it is a *MOD-16 ripple counter*. Recall that the MOD number is always equal to the number of states which the counter goes through in each complete cycle before it recycles back to its starting state. The MOD number can be increased simply by adding more FFs to the counter. That is,

$$\text{MOD number} = 2^N \tag{7.1}$$

where N is the number of FFs connected in the arrangement of Figure 7.1.

EXAMPLE 7.2

A counter is needed that will count the number of items passing on a conveyor belt. A photocell and light source combination is used to generate a single pulse each time an item crosses its path. The counter has to be able to count as many as one thousand items. How many FFs are required?

Solution: It is a simple matter to determine what value of N is needed so that $2^N \geq 1000$. Since $2^9 = 512$, 9 FFs will not be enough. $2^{10} = 1024$, so 10 FFs would produce a counter that could count as high as $1111111111_2 = 1023_{10}$. Therefore, we should use 10 FFs. We could use more than 10, but it would be a waste of FFs, since any FF past the tenth one will never be triggered.

Frequency Division In Chapter 5 we saw that in the basic counter each FF provides an output waveform that is exactly *half* the frequency of the waveform at its *CLK* input. To illustrate, suppose that the clock signal in Figure 7.1 is 16 kHz. Figure 7.2 shows the FF output waveforms. The waveform at output A is an 8-kHz *squarewave*, at output B it is 4 kHz, at output C it is 2 kHz, and at output D it is 1 kHz. Notice that the output of FF D has a frequency equal to the original clock frequency divided by 16. In general, for any counter the output from the last FF divides the input clock frequency by the MOD number of the counter. For example, a MOD-16 counter could also be called a *divide-by-16 counter*.

FIGURE 7.2 Counter waveforms showing frequency-division-by-two for each FF.

EXAMPLE 7.3

The first step involved in building a digital clock* is to take the 60-Hz power-line waveform and feed it into a shaping circuit to produce a squarewave as illustrated in Figure 7.3. The 60-Hz squarewave is then put into a MOD-60 counter, which is used to divide the 60-Hz frequency by exactly 60 to produce a 1-Hz waveform. This 1-Hz waveform is fed to a series of counters, which then count seconds, minutes, hours, etc. How many FFs are required for the MOD-60 counter?

*Here we are talking about a clock that indicates time in hours, minutes, etc.

FIGURE 7.3

Solution: There is no integer power of 2 that will equal 60. The closest is $2^6 = 64$. Thus, a counter using six FFs would act as a MOD-64. Obviously, this will not satisfy the requirement. It seems that there is no solution using a counter of the type shown in Figure 7.1. This is partly true; in the next section we will see how to modify this basic binary counter so that virtually *any* MOD number can be obtained and we will not be limited to values of 2^N.

REVIEW QUESTIONS

1. True or false: In an asynchronous counter, all FF's change states at the same time.
2. Assume that the counter in Figure 7.1 is holding the count 0101. What will be the count after 27 clock pulses? (*Ans.* 0000)

7.2 COUNTERS WITH MOD NUMBERS $< 2^N$

The basic ripple counter of Figure 7.1 is limited to MOD numbers that are equal to 2^N, where N is the number of FFs. This value is actually the maximum MOD number that can be obtained using N FFs. The basic counter can be modified to produce MOD numbers less than 2^N by allowing the counter to *skip states* that are normally part of the counting sequence. One of the most common methods for doing this is illustrated in Figure 7.4 where a 3-bit ripple counter is shown. Disregarding the NAND gate for a moment we can see that the counter is a MOD-8 binary counter which will count in sequence from 000 to 111. However, the presence of the NAND gate will alter this sequence as follows:

1. The NAND output is connected to the DC CLEAR inputs of each FF. As long as the NAND output is HIGH, it will have no affect on the counter. When it goes LOW, however, it will clear all the FFs so that the counter immediately goes to the 000 state.

2. The inputs to the NAND gate are the outputs of the B and C FFs, so the NAND output will go LOW whenever $B = C = 1$. This condition will occur when the counter goes from the 101 state to the 110 state (input pulse 6 on waveforms). The LOW at the NAND output will immediately (generally within a few nanoseconds) clear the counter to the 000 state. Once the FFs have been cleared, the NAND output goes back HIGH, since the $B = C = 1$ condition no longer exists.

FIGURE 7.4 MOD-6 counter produced by clearing a MOD-8 counter when count of six (110) occurs.

3. The counting sequence is, therefore,

CBA
000
001
010
011
100
101
110 (temporary state needed to clear counter)

Although the counter does go to the 110 state, it remains there for only a few nanoseconds before it recycles to 000. Thus, we can essentially say that this counter counts from 000 (zero) to 101 (five) and then recycles to 000. It essentially skips 110 and 111 so that it only goes through six different states; thus, it is a MOD-6 counter.

Notice that the waveform at the B output contains a *spike* or *glitch* caused by the momentary occurrence of the 110 state before clearing. This glitch is very narrow and so would not produce any visible indication on indicator lights or numerical displays. It could, however, cause a problem if the B output is being used to drive other circuitry outside the counter. It should also be noted that the C output has a frequency equal to $\frac{1}{6}$ of the input frequency; in other words, this MOD-6 counter has divided the input frequency by *six*. The waveform at C is *not* a symmetrical squarewave (50 percent duty cycle) because it is only HIGH for two clock cycles while it is LOW for four cycles.

Changing the MOD Number   The counter of Figure 7.4 is a MOD-6 because of the choice of inputs to the NAND gate. Any desired MOD number can be obtained by changing these inputs. For example, using a three-input NAND gate with inputs A, B, and C, the counter would function normally until the 111 condition was reached, at which point it would immediately reset to the 000 state. Ignoring the temporary excursion into the 111 state, the counter would go from 000 through 110 and then recycle back to 000, resulting in a MOD-7 counter (seven states).

EXAMPLE 7.4

Determine the MOD number of the counter in Figure 7.5(a). Also determine the frequency at the D output.

Solution: This is a 4-bit counter, which would normally count from 0000 through 1111. The NAND inputs are D, C, and B which means that the counter will immediately recycle to 0000 when the 1110 (decimal 14) count is reached. Thus, the counter actually has 14 stable states 0000 through 1101 and is therefore a **MOD-14** counter. Since the input frequency is 30 kHz, the frequency at output D will be

$$\frac{30 \text{ kHz}}{14} = 2.14 \text{ kHz}$$

General Procedure   To construct a counter that starts counting from all 0s and has a MOD number of X:

1. Find the smallest number of FFs such that $2^N \geq X$ and connect them as a counter.
2. Connect a NAND gate to the asynchronous CLEAR inputs of all the FFs.
3. Determine which FFs will be in the HIGH state at a count $= X$; then connect the normal outputs of these FFs to the NAND-gate inputs.

EXAMPLE 7.5

Construct a MOD-10 counter that will count from 0000 (zero) through 1001 (decimal 9).

*All J, K inputs are HIGH

(a)

*All J, K inputs are HIGH

(b)

FIGURE 7.5 **(a) MOD-14 ripple counter; (b) MOD-10 (decade) ripple counter.**

Solution: $2^3 = 8$ and $2^4 = 16$; thus, four FFs are required. Since the counter is to have stable operation up to the count of 1001, it must be reset to zero when the count of 1010 is reached. Therefore, FF outputs D and B must be connected as the NAND-gate inputs. Figure 7.5(b) shows the arrangement.

Decade Counters/BCD Counters   The MOD-10 counter of Example 7.5 is also referred to as a *decade counter*. In fact, a decade counter is any counter that has 10 distinct states, no matter what the sequence. A decade counter such as the one in Figure 7.5(b), which counts in sequence from 0000 (zero) through 1001 (decimal 9), is also commonly called a *BCD counter* because it uses only the 10 BCD code groups 0000, 0001, . . . , 1000, and 1001. To reiterate, any MOD-10 counter is a decade counter; and any decade counter that counts in binary from 0000 to 1001 is a BCD counter.

Decade counters, especially the BCD type, find widespread use in applications where pulses or events are to be counted and the results displayed on some type of decimal numerical readout. We shall examine this later in more detail. A decade counter is also often used for dividing a pulse frequency *exactly* by 10. The input pulses are applied to FF *A* and the output pulses are taken from the output of FF *D*, which has $\frac{1}{10}$ the frequency of the input.

EXAMPLE 7.6

In Example 7.3 a MOD-60 counter was needed to divide the 60-Hz line frequency down to 1 Hz. Construct an appropriate MOD-60 counter.

FIGURE 7.6   MOD-60 counter.

Solution: $2^5 = 32$ and $2^6 = 64$, so we need six FFs, as shown in Figure 7.6. The counter is to be cleared when it reaches the count of sixty (111100). Thus, the outputs of FFs *C, D, E,* and *F* must be connected to the NAND gate. The output of FF *F* will have a frequency of 1 Hz.

REVIEW QUESTIONS

1. What FF outputs should be connected to the clearing NAND gate to form a MOD-12 counter? (*Ans.* *D* and *C*)
2. True or false: All BCD counters are decade counters.
3. What is the output frequency of a decade counter that is clocked from a 50-kHz signal? (*Ans.* 5 kHz)

7.3 IC ASYNCHRONOUS COUNTERS

There are several TTL and CMOS asynchronous counter chips. One of the most popular is the TTL 7493 and its logical equivalent, the 74293. Figure 7.7(a) shows the logic diagram for the 7493 as it appears in the manufacturer's data book. Some of the nomenclature on this diagram is different from what we have been using, but it should be easy to figure out. By examining this diagram, we can ascertain the following points.

1. The 7493 contains four *J-K* FFs with outputs Q_0, Q_1, Q_2, Q_3 (instead of *A, B, C, D*). Each FF has a *CP* (clock pulse) input, which are just another name for the *CLK* input. The clock inputs to Q_0 and Q_1, labeled $\overline{CP}_0$ and $\overline{CP}_1$, respectively, are

FIGURE 7.7 (a) Logic diagram for 7493 and 74293 asynchronous counter IC; (b) simplified symbol. (Courtesy of Fairchild, a Schlumberger company)

externally accessible. The inversion bars over these inputs are used to indicate that they are acting on a HIGH-TO-LOW transition.

2. Each FF has a DC CLEAR input, C_D, connected to the output of a two-input NAND gate. The NAND inputs are MR_1 and MR_2, where MR stands for *master reset*.

3. FFs Q_1, Q_2, and Q_3 are already connected as a 3-bit ripple counter. FF Q_0 is not connected to anything internally. This allows the user the option of either connecting Q_0 to Q_1 to form a 4-bit counter, or using Q_0 separately if desired.

4. The FFs are arranged in the opposite order to what we have been using in our counter diagrams; that is, Q_0 is the LSB and Q_3 is the MSB of the counter. This is just an equivalent way of drawing the counter circuit; its logical operation is the same as that of the counters we have been using.

The following examples will illustrate some of the ways the 7493 can be wired to produce different counters. In these examples we will use the simplified logic symbol shown in Figure 7.7(b).

EXAMPLE 7.7

Show how the 7493 should be connected to operate as a MOD-16 counter with a 10-kHz clock input.

Solution: A MOD-16 requires four FFs, so we have to connect the Q_0 output to $\overline{CP_1}$, the clock input of FF Q_1 (see Figure 7.8). The 10-kHz pulses are applied to $\overline{CP_0}$, the clock input of Q_0.

FIGURE 7.8   A 7493 wired as a MOD-16.

In order for the counter to function as a MOD-16 counter, it must be allowed to count through its complete sequence 0000 to 1111. This means that the reset NAND gate has to be disabled; that is, its output should be kept permanently HIGH so that it has no effect on the counter operation. This can be accomplished by connecting MR_1 and MR_2 to a constant LOW.

EXAMPLE 7.8

Show how to wire the 7493 as a MOD-10 counter.

Solution: A MOD-10 requires four FFs, so again we need to connect Q_0 to $\overline{CP_1}$. This time, however, we want the counter to recycle back to 0000 when it tries to go to the count of 1010 (ten). Thus, the Q_3 and Q_1 outputs have to be connected to the master reset inputs; when they both go HIGH at the count of 1010, the NAND output will immediately reset the counter to 0000.

The circuit wiring is shown in Figure 7.9.

FIGURE 7.9   A 7493 wired as a MOD-10.

EXAMPLE 7.9

Show how to wire a 7493 as a MOD-14.

Solution: When the counter reaches the count of 1110 (14), the Q_3, Q_2, and Q_1 outputs are all HIGH. Unfortunately, the 7493's built-in reset NAND gate has only two inputs. Thus, we have to add some extra logic to ensure that the counter will reset back to 0000 when $Q_3 = Q_2 = Q_1 = 1$. In fact, all we need is a two-input AND gate as shown in Figure 7.10.

FIGURE 7.10   A 7493 wired as a MOD-14 requires an AND gate.

EXAMPLE 7.10

In Example 7.6 we divided the input frequency by 60 with a MOD-60 counter using six J-K FFs and a NAND gate. Another way to get a MOD-60 is shown in Figure 7.11. Explain how this circuit works.

FIGURE 7.11 Two 7493s can be combined to provide a frequency division of 60.

Solution: This circuit divides the input frequency by 60 in two steps. The 7493 counter on the right is wired as a MOD-10 so that its output Q_3 has a frequency $= f_{in}/10$. This signal is connected to the $\overline{CP}_1$ input of the second 7493 counter which is wired as a MOD-6 (note that Q_0 is not being used). Thus, the Q_3 output of the second counter will have a frequency

$$f_{out} = \frac{f_{in}/10}{6} = \frac{f_{in}}{60}$$

This last example shows that two (or more) counters can be cascaded to produce an overall MOD number equal to the *product* of their individual MOD numbers. This can be very useful in applications where a large amount of frequency division is required.

CMOS Asynchronous Counter There are several asynchronous counter ICs in the CMOS logic family. One of them, the 4024, is a 7-bit counter circuit with one asynchronous master reset input. The logic symbol for the 4024 is drawn in Figure 7.12. The seven FFs are internally connected in the ripple-counter configuration. The

FIGURE 7.12 CMOS 4024 counter

master reset input, *MR*, is active-HIGH, and it will reset all the counter FFs when it goes HIGH.

Another CMOS ripple counter is the 4020, a 14-bit counter with one asynchronous master reset input. In general, CMOS ICs will have more circuitry on a chip than TTL ICs. For this reason, you will find that CMOS counters will usually contain more FFs than TTL counters.

7.4 ASYNCHRONOUS DOWN COUNTER

All the counters we have looked at thus far have counted *upward* from zero; that is, they were *up counters*. It is a relatively simple matter to construct asynchronous (ripple) *down counters*, which will count downward from a maximum count to zero. Before looking at a ripple down counter, let us examine the count-down sequence for a 3-bit down counter:

	CBA	CBA	CBA
(7)	1 1 1	1 1 1	1 1 1
(6)	1 1 0	1 1 0	1 1 0
(5)	1 0 1	1 0 1	. . .
(4)	1 0 0	1 0 0	. . .
(3)	0 1 1	0 1 1	. . .
(2)	0 1 0	0 1 0	etc.
(1)	0 0 1	0 0 1	
(0)	0 0 0	0 0 0	etc.

A, *B*, and *C* represent the FF output states as the counter goes through its sequence. It can be seen that the A FF (LSB) changes states (toggles) at each step in the sequence just as it does in the up counter. The B FF changes states each time *A* goes from LOW to HIGH; *C* changes states each time *B* goes from LOW to HIGH. Thus, in a down counter each FF, except the first, must toggle when the preceding FF goes from LOW to HIGH. If the FFs have *CLK* inputs that respond to negative transitions (HIGH to LOW), then an inverter can be placed in front of each *CLK* input; however, the same effect can be accomplished by driving each FF *CLK* input from the *inverted* output of the preceding FF. This is illustrated in Figure 7.13 for a MOD-8 down counter.

The input pulses are applied to the *A* FF; the $\overline{A}$ output serves as the *CLK* input for the *B* FF; the $\overline{B}$ output serves as the *CLK* input for the *C* FF. The waveforms at *A*, *B*, and *C* show that *B* toggles whenever *A* goes LOW to HIGH (so $\overline{A}$ goes HIGH to LOW) and *C* toggles whenever *B* goes LOW to HIGH. This results in the desired down-counting sequence at the *C*, *B*, and *A* outputs.

Down counters are not as widely used as up counters. Their major application is in situations where it must be known when a desired number of input pulses has occurred. In these situations the down counter is *preset* to the desired number and then allowed to count down as the pulses are applied. When the counter reaches the *zero* state it is detected by a logic gate whose output then indicates that the preset number of pulses has occurred. We shall discuss presettable counters in a later section.

All J, K inputs are HIGH.

Input

FIGURE 7.13   MOD-8 down counter.

1. What is the difference between the counting sequence of an up counter and a down counter?
2. Describe how an asynchronous down-counter circuit differs from an up-counter circuit.

7.5 PROPAGATION DELAY IN RIPPLE COUNTERS

Ripple counters are the simplest type of binary counters, since they require the fewest components to produce a given counting operation. They do, however, have one major drawback, which is caused by their basic principle of operation. Each FF is triggered by the transition at the output of the preceding FF. Because of the inherent propagation delay time (t_{pd}) of each FF, this means that the second FF will not respond until a time t_{pd} after the first FF receives an input pulse; the third FF will not respond until a time equal to $2 \times t_{pd}$ after the clock pulse occurs; and so on. In other words, the propagation delays of the FFs accumulate so that the Nth FF cannot change states until a time equal to $N \times t_{pd}$ after the clock pulse occurs. This is illustrated in Figure 7.14, where the waveforms for a 3-bit ripple counter are shown.

The first set of waveforms in Figure 7.14(a) shows a situation where an input pulse occurs every 1000 ns (the clock period $T = 1000$ ns) and it is assumed that each FF has a propagation delay of 50 ns ($t_{pd} = 50$ ns). Notice that the A FF output toggles 50 ns after the falling edge of each input pulse. Similarly, B toggles 50 ns after A goes from 1 to 0 and C toggles 50 ns after B goes from 1 to 0. As a result, when the fourth

(a)

(b)

FIGURE 7.14 Waveforms of 3-bit ripple counter illustrating effects of FF propagation delays for different input pulse frequencies.

input pulse occurs, the C output goes HIGH after a delay of 150 ns. In this situation the counter does operate properly in the sense that the FFs do eventually get to their correct states, representing the binary count. However, the situation worsens if the input pulses are applied at a much higher frequency.

The waveforms in Figure 7.14(b) show what happens if the input pulses occur once every 100 ns. Again, each FF output responds 50 ns after the 1-to-0 transition at its *CLK* input (note the change in the relative time scale). Of particular interest is the situation after the falling edge of the fourth input pulse where the C output does not go HIGH until 150 ns later, which is the same time that the A output goes HIGH in response to the fifth input pulse. In other words, the condition $C = 1, B = A = 0$ (count of 100), never appears because the input frequency is too high. This could cause a serious problem if this condition were supposed to be used to control some other operation in a digital system. Problems such as this can be avoided if the period between input pulses is made longer than the total propagation delay of the counter. That is,

$$T_{clock} \geq N \times t_{pd} \tag{7.2}$$

where N = number of FFs. Stated in terms of input-clock frequency, the maximum frequency that can be used is given by

$$f_{max} = \frac{1}{N \times t_{pd}} \tag{7.3}$$

For example, suppose a 4-bit ripple counter is constructed using the 74LS112 J-K FF. Table 5.1 or the data sheet in Appendix II shows that the 74LS112 has $t_{PLH} = 16$ ns and $t_{PHL} = 24$ ns as the propagation delays from *CLK* to Q. To calculate f_{max}, we will assume the "worst case"; that is, we will use $t_{pd} = t_{PHL} = 24$ ns, so that

$$f_{max} = \frac{1}{4 \times 24 \text{ ns}} = 10.4 \text{ MHz}$$

Clearly, as the number of bits in the counter increases, the total propagation delay increases and f_{max} decreases. For example, a ripple counter that uses six 74LS112 FFs will have

$$f_{max} = \frac{1}{6 \times 24 \text{ ns}} = 6.9 \text{ MHz}$$

Thus, asynchronous counters are not useful at very high frequencies, especially for large numbers of bits. Another problem caused by propagation delays in asynchronous counters occurs when the counter outputs are *decoded*. This problem is discussed in a later section. Despite these problems, the simplicity of asynchronous counters makes them useful for applications where their frequency limitation is not critical.

REVIEW QUESTION

1. Explain why a ripple counter's frequency limitation changes as more FFs are added to the counter.

7.6 SYNCHRONOUS (PARALLEL) COUNTERS

The problems encountered with ripple counters are caused by the accumulated FF propagation delays; stated another way, the FFs do not all change states simultaneously in synchronism with the input pulses. These limitations can be overcome with the use of *synchronous* or *parallel* counters in which all the FFs are triggered simultaneously (in parallel) by the clock input pulses. Since the input pulses are applied to all the FFs, some means must be used to control when each FF is to toggle or remain unaffected by a clock pulse. This is accomplished by using the J and K inputs and is illustrated in Figure 7.15(a) for a 4-bit, MOD-16 parallel counter.

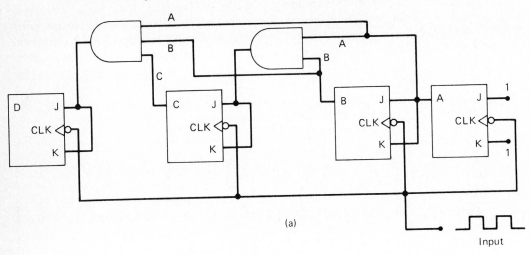

(a)

Count	D	C	B	A
0	0	0	0	0
1	0	0	0	1
2	0	0	1	0
3	0	0	1	1
4	0	1	0	0
5	0	1	0	1
6	0	1	1	0
7	0	1	1	1
8	1	0	0	0
9	1	0	0	1
10	1	0	1	0
11	1	0	1	1
12	1	1	0	0
13	1	1	0	1
14	1	1	1	0
15	1	1	1	1

(b)

FIGURE 7.15   Parallel MOD-16 counter.

Each of the FFs has the input pulses applied to its *CLK* input. The A FF has $J = K = 1$, so it will toggle on the NGT of *each* input pulse. The B FF has its J and K inputs connected to the A output, so the value of A determines whether B will toggle on a particular input pulse. If $A = 0$ *before* the input pulse occurs, then $J_B = K_B = 0$ for the B FF, and it *will not* toggle when the pulse does occur (e.g., going from count of 2 to 3). If $A = 1$ *before* the input pulse occurs, then $J_B = K_B = 1$ and the B FF *will* toggle (e.g., going from count of 3 to 4).

Similarly, the C FF will toggle only if $A = B = 1$ before the input pulse occurs because of AND gate 1 (e.g., going from count of 3 to 4). The D FF will toggle only if $A = B = C = 1$ before the pulse occurs due to AND gate 2. Thus, the counter will progress through the binary counting sequence as shown in the table in Figure 7.15(b).

The most important characteristic of this parallel counter is the fact that all FFs will change states simultaneously in synchronism with the input pulses. The total propagation delay that must be allowed for is simply the t_{pd} of a *single* FF plus the delay needed for the levels to propagate through a *single* AND gate. This means that a much higher input-pulse frequency can be used than would be possible for a corresponding ripple counter. Of course, the parallel counter requires more logic circuitry and a greater number of connections.

EXAMPLE 7.11

(a) Determine f_{max} for the counter of Figure 7.15(a) if t_{pd} for each FF is 50 ns and t_{pd} for each AND gate is 20 ns. Compare this to f_{max} for a MOD-16 ripple counter.
(b) What has to be done to convert this counter to MOD-32?
(c) Determine f_{max} for the MOD-32 parallel counter.

Solution:
(a) The total delay that must be allowed between input clock pulses is equal to FF t_{pd} + AND gate t_{pd}. Thus, $T_{clock} \geq 50 + 20 = 70$ ns, so the parallel counter has

$$f_{max} = \frac{1}{70 \text{ ns}} = 14.3 \text{ MHz} \quad \text{(parallel counter)}$$

A MOD-16 ripple counter uses four FFs with $t_{pd} = 50$ ns. Thus, f_{max} for the ripple counter is

$$f_{max} = \frac{1}{4 \times 50 \text{ ns}} = 5 \text{ MHz} \quad \text{(ripple counter)}$$

(b) A fifth FF must be added, since $2^5 = 32$. The *CLK* input of this FF is also fed by the input pulses. Its J and K inputs are fed by the output of a four-input AND gate whose inputs are A, B, C, and D.
(c) f_{max} is still determined as in (a) regardless of the number of FFs in the parallel counter. Thus, f_{max} is still 14.3 MHz.

REVIEW QUESTION

1. What is the advantage of a synchronous counter over an asynchronous counter? What is the disadvantage?

7.7 PARALLEL DOWN AND UP/DOWN COUNTERS

In Section 7.4 we saw that a ripple counter could be made to count down by using the inverted outputs of each FF to drive the next FF in the counter. A parallel down counter can be constructed in a similar manner—that is, by using the inverted FF outputs to feed the various logic gates. For example, the parallel up counter of Figure 7.15 can be converted to a down counter by connecting the $\overline{A}$, $\overline{B}$, and $\overline{C}$ outputs to the AND gates in place of A, B, and C, respectively. The counter will then proceed through the following sequence as input pulses are applied.

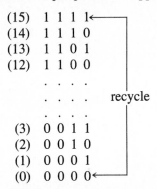

To form a parallel up/down counter (see Figure 7.16) the control inputs (COUNT-UP and COUNT-DOWN) are used to control whether the normal FF outputs or the inverted FF outputs are fed to the J and K inputs of the following FFs. The counter in Figure 7.16 is a MOD-8 up/down counter that will count from 000 up to 111 when the COUNT-UP control input is 1 and from 111 down to 000 when the COUNT-DOWN control input is 1.

A logical 1 on the COUNT-UP line while COUNT-DOWN = 0 enables AND gates 1 and 2 and disables AND gates 3 and 4. This allows the A and B outputs through to the J and K inputs of the following FFs so that the counter will count up as pulses are applied. The opposite action takes place when COUNT-UP = 0 and COUNT-DOWN = 1.

EXAMPLE 7.12

Describe how the circuit of Figure 7.16 will operate when COUNT-UP and COUNT-DOWN are both LOW, and when they are both HIGH.

Solution: When both these inputs are LOW, each AND-gate and each OR-gate output will be LOW. Thus, the J-K inputs of FFs B and C will be LOW, and neither of these FFs can change states. The A FF will still toggle on each input pulse.

When both inputs are HIGH, each OR-gate output will be HIGH because one of the AND outputs feeding each OR gate will be HIGH, depending on the state of the corresponding FF. Thus, all three FFs will have $J = K = 1$ and will toggle on each input pulse.

FIGURE 7.16 Parallel up/down counter (MOD-8).

Input

Count-up

Count-down

All FFs
are master/slave

7.8 PRESETTABLE COUNTERS

Most integrated-circuit parallel counters are designed to be *presettable;* in other words, they can be preset to any desired starting count either asynchronously (independent of the clock signal) or synchronously (on the active transition of the clock signal).

Figure 7.17 shows a 3-bit presettable parallel up counter. It uses asynchronous presetting through the asynchronous PRESET and CLEAR inputs. We can preset the counter to any desired starting count at any time by doing the following:

1. Apply the desired starting count to the preset inputs P_C, P_B, and P_A.
2. Momentarily pulse the PARALLEL LOAD input ($\overline{PL}$) to the LOW state. This input is active-LOW.

This procedure will perform an asynchronous parallel transfer of the P_C, P_B, and P_A levels into C, B, and A, respectively. This *jam transfer* (described in Section 5.15) will take place independent from the J, K, and CLK inputs. Once $\overline{PL}$ returns HIGH, the counter can resume counting the input pulses starting from the count that was loaded into the counter.

The asynchronous presetting method is used in many IC counters such as the TTL 74192 and 74193, and the CMOS 40193 and 74C193. Actually, the CMOS 40193 and 74C193 are logically equivalent to and have the same pin layout as the TTL 74193. Another way to say this is that they are logically compatible and pin-compatible.

Synchronous Presetting There are several parallel-counter ICs that use *synchronous presetting,* whereby the counter is preset on the active transition of the same clock signal that is used for counting. The logic level applied to the PARALLEL LOAD input determines whether the active clock transition is used to preset the counter, or whether it will be counted as in normal counter operation.

Examples of this type of counter are the TTL 74161 and 74163 and their CMOS equivalents, the 40161 and 40163.

REVIEW QUESTION

1. Explain the difference between asynchronous and synchronous presetting.

7.9 THE 74193 COUNTER

Figure 7.18 shows the logic symbol and the input/output description for the 74193 counter. This counter can be described as a MOD-16, presettable up/down counter with synchronous counting, asynchronous preset, and asynchronous master reset. Let us look at the function of each input and output.

Clock Inputs CP_U and CP_D The counter will respond to the positive-going transitions at one of two clock inputs. CP_U is the *count-up clock* input. When pulses are applied to this input, the counter will increment (count up) on each PGT to a maximum count

FIGURE 7.17 Presettable parallel counter with asynchronous preset.

Pin Names	Description
CP_U	Count-up clock input (active rising edge)
CP_D	Count-down clock input (active rising edge)
MR	Asynchronous master reset input (active HIGH)
$\overline{PL}$	Asynchronous parallel load input (active LOW)
P_0–P_3	Parallel data inputs
Q_0–Q_3	Flip-flop outputs
$\overline{TC}_D$	Terminal count-down (borrow) output (active LOW)
$\overline{TC}_U$	Terminal count-up (carry) output (active LOW)

(a) (b)

FIGURE 7.18 74193 presettable up/down counter: (a) logic symbol; (b) input/output description. (Courtesy of Fairchild, a Schlumberger company)

of 1111; then it recycles to 0000 and starts over. CP_D is the *count-down clock* input. When pulses are applied to this input, the counter will decrement (count down) on each PGT to a minimum count of 0000; then it recycles to 1111 and starts over. Thus, one clock input or the other will be used for counting while the other clock input is inactive (kept HIGH).

Master Reset (MR)   This is an active-HIGH asynchronous input that resets the counter to the 0000 state. *MR* is a dc reset, so it will hold the counter at 0000 as long as *MR* = 1. It also overrides *all* other inputs.

Preset Inputs   The counter FFs can be preset to the logic levels present on the parallel data inputs P_3–P_0 by momentarily pulsing the parallel load input $\overline{PL}$ from HIGH to LOW. This is an asynchronous preset that overrides the counting operation. $\overline{PL}$ will have no effect, however, if the *MR* input is in its active HIGH state.

Count Outputs   The current count is always present at the FF outputs Q_3–Q_0, where Q_0 is the LSB, and Q_3 is the MSB.

Terminal Count Outputs   These outputs are used when two or more 74193s are connected as a multistage counter to produce a larger MOD number. In the count-up mode, the $\overline{TC}_U$ output of the lower-order counter is connected to the CP_U input of the next higher-order counter. In the count-down mode, the $\overline{TC}_D$ output of the lower-order counter is connected to the CP_D input of the next higher-order counter.

$\overline{TC}_U$ is the *terminal count-up* (also called the *carry*) output. It is generated on the 74193 chip using the logic shown in Figure 7.19(a). Clearly, $\overline{TC}_U$ will be LOW only when the counter is in the 1111 state and CP_U is LOW. Thus, $\overline{TC}_U$ will remain HIGH as the counter counts up from 0000 to 1110. On the next PGT of CP_U, the count goes to 1111, but $\overline{TC}_U$ does not go LOW until CP_U returns LOW. The next PGT at CP_U recycles the count to 0000 and also causes $\overline{TC}_U$ to return HIGH. This PGT at TC_U occurs when the counter recycles from 1111 to 0000, and can be used to clock a second 74193 up counter to its next higher count.

$\overline{TC}_D$ is the *terminal count-down* (also called the *borrow*) output. It is generated as shown in Figure 7.19(b). It is normally HIGH and does not go LOW until the counter has counted down to the 0000 state and CP_D is LOW. When the next PGT at CP_D recycles the counter to 1111, it causes $\overline{TC}_D$ to return HIGH. This PGT at $\overline{TC}_D$ can be used to clock a second 74193 down counter to its next lower count.

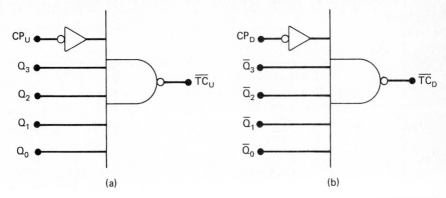

FIGURE 7.19 (a) Logic on 74193 for generating $\overline{TC}_U$; (b) logic for generating $\overline{TC}_D$.

EXAMPLE 7.13

Refer to Figure 7.20(a), where a 74193 is wired as an up counter. The parallel data inputs are permanently connected as 1011, and the CP_U, $\overline{PL}$, and MR input waveforms are shown in Figure 7.20(b). Assume that the counter is initially in the 0000 state, and determine the counter output waveforms.

Solution: Initially (at t_0) the counter FFs are all LOW. This causes $\overline{TC}_U$ to be HIGH. Just prior to time t_1 the $\overline{PL}$ input is pulsed LOW. This immediately loads the counter with 1011 to produce $Q_3 = 1$, $Q_2 = 0$, $Q_1 = 1$, and $Q_0 = 1$. At t_1 the CP_U input makes a PGT, but the counter cannot respond to this because $\overline{PL}$ is still active at that time. At times t_2, t_3, t_4, and t_5 the counter counts up on each PGT at CP_U. After the PGT at t_5 the counter is in the 1111 state, but $\overline{TC}_U$ does not go LOW until CP_U goes LOW at t_6. When the next PGT occurs at t_7, the counter recycles to 0000, and $\overline{TC}_U$ returns HIGH.

The counter will count up in response to the PGTs at t_8 and t_9. The PGT at t_{10} will have no effect because the MR goes HIGH prior to t_{10} and remains active at t_{10}. This will reset all FFs to 0 and overrides the CP_U signal.

FIGURE 7.20   Example 7.13

EXAMPLE 7.14

Figure 7.21(a) shows the 74193 wired as a down counter. The parallel data inputs are permanently wired as 0111, and the CP_D and $\overline{PL}$ waveforms are shown in Figure 7.21(b). Assume that the counter is initially in the 0000 state, and determine the output waveforms.

Solution: At t_0 all the FF outputs are LOW and CP_D is LOW. These are the conditions that produce $\overline{TC}_D = 0$. Prior to t_1 the $\overline{PL}$ input is pulsed LOW. This immediately presets the counter to 0111 and therefore causes $\overline{TC}_D$ to go HIGH. The PGT of CP_D at t_1 will have no effect, since $\overline{PL}$ is still active. The counter will respond to the PGTs at t_2–t_8 and counts down to 0000 at t_8. $\overline{TC}_D$ does not go LOW until t_9, when CP_D goes LOW. At t_{10} the PGT of CP_D causes the counter to recycle to 1111 and also drives $\overline{TC}_D$ back HIGH.

FIGURE 7.21   Example 7.14

Variable MOD Number Using the 74193   Presettable counters can be easily wired for different MOD-numbers without the need for additional logic circuitry. We will demonstrate this for the 74193 using the circuit of Figure 7.22(a). Here, the 74193 is used as a down counter with its parallel load inputs permanently connected as 0101 (5_{10}). Note that the $\overline{TC}_D$ output is connected back to the $\overline{PL}$ input.

We will begin our analysis by assuming that the counter has been counting down and is in the 0101 state at time t_0. Refer to Figure 7.22(b) for the counter waveforms.

The counter will decrement (count down) on the PGTs of CP_D at times t_1–t_5. At t_5 the counter is in the 0000 state. When CP_D goes LOW at t_6, it drives $\overline{TC}_D$ LOW. This immediately activates the $\overline{PL}$ input and presets the counter back to the 0101 state. Note that $\overline{TC}_D$ stays LOW for only a short interval because once the counter outputs go to 0101 in response to $\overline{PL} = 0$, the condition needed to keep $\overline{TC}_D = 0$ is removed. Thus, there is only a narrow glitch at $\overline{TC}_D$.

This same sequence is repeated at times t_7–t_{12} and at equal intervals thereafter. If we examine the Q_2 waveform, we can see that it goes through one complete cycle for every *five* cycles of CP_D. For example, there are *five* clock cycles between the PGT

FIGURE 7.22 (a) 74193 wired as a MOD-5; (b) waveforms.

of Q_2 at t_6 and the PGT of Q_2 at t_{11}. Thus, the frequency of the Q_2 waveform is $\frac{1}{5}$ of the clock frequency.

It is no coincidence that the frequency-division ratio (5) is the same as the number applied to the parallel data inputs (0101 = 5). In fact, we can vary the frequency division by changing the logic levels applied to the parallel data inputs.

A *variable frequency-divider* circuit can be easily implemented by connecting switches to the parallel data inputs of the circuit in Figure 7.22. The switches can be set to a value equal to the desired MOD number.

1. Describe the function of the $\overline{PL}$ and P_0–P_3 inputs.
2. Describe the function of the MR input.
3. True or false: The 74193 cannot be preset while MR is active.
4. What logic levels must be present at CP_D, $\overline{PL}$, and MR in order for the 74193 to count pulses that appear at CP_U? (*Ans.* 1, 1, 0, respectively)

7.10 DECODING A COUNTER

Digital counters are often used in applications where the count represented by the states of the FFs must somehow be determined or displayed. One of the simplest means for displaying the contents of a counter involves just connecting the output of each FF to a small indicator lamp. In this way the states of the FFs are visibly represented by the lamps (bright = 1, dark = 0) and the count can be mentally determined by decoding the binary states of the lamps. For instance, suppose that this method is used for a BCD counter and the states of the lamps are respectively dark–bright–bright–dark. This would represent 0110, which we would mentally decode as decimal 6. Other combinations of lamp states would represent the other possible counts.

The indicator lamp method becomes inconvenient as the size (number of bits) of the counter increases, because it is much harder to mentally decode the displayed results. For this reason it would be preferable to develop a means for *electronically* decoding the contents of a counter and displaying the results in a form that would be immediately recognizable and would require no mental operations.

An even more important reason for electronic decoding of a counter occurs because of the many applications in which counters are used to control the timing or sequencing of operations *automatically* without human intervention. For example, a certain system operation might have to be initiated when a counter reaches the 101100 state (count of 44_{10}). A logic circuit can be used to decode for or detect when this particular count is present and then initiate the operation. Many operations may have to be controlled in this manner in a digital system. Clearly, human intervention in this process would be undesirable except in extremely slow systems.

Active-HIGH Decoding A MOD-X counter has X different states; each state is a particular pattern of 0s and 1s stored in the counter FFs. A decoding network is a logic circuit that generates X different outputs, each of which detects (decodes) the presence of one particular state of the counter. The decoder outputs can be designed to produce either a HIGH or a LOW level when the detection occurs. An active-HIGH decoder produces HIGH outputs to indicate detection. Figure 7.23 shows the complete active-HIGH decoding logic for a MOD-8 counter. The decoder consists of eight three-input AND gates. Each AND gate produces a HIGH output for one particular state of the counter.

For example, AND gate 0 has as its inputs the FF outputs $\overline{C}$, $\overline{B}$, and $\overline{A}$. Thus, its output will be LOW at all times *except* when $A = B = C = 0$—that is, on the count of 000 (zero). Similarly, AND gate 5 has as its inputs the FF outputs C, $\overline{B}$, and A, so its output will go HIGH only when $C = 1$, $B = 0$, and $A = 1$—that is, on the

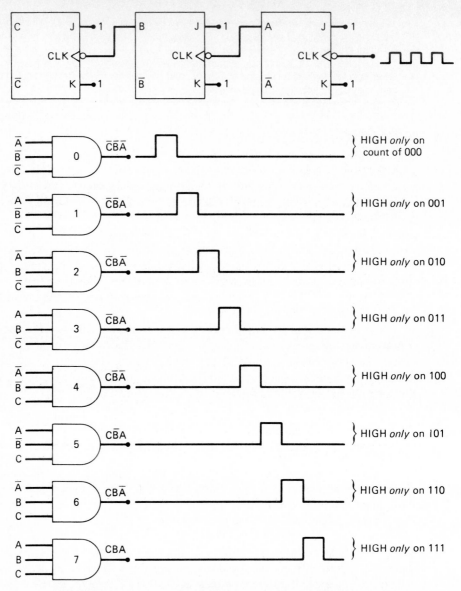

FIGURE 7.23 Using AND gates to decode a MOD-8 counter.

count of 101 (decimal 5). The rest of the AND gates perform in the same manner for the other possible counts. At any one time only one AND-gate output is HIGH, the one which is decoding for the particular count that is present in the counter. The waveforms in Figure 7.23 show this clearly.

The eight AND outputs can be used to control eight separate indicator lamps, which represent the decimal numbers 0–7. Only one lamp will be on at a given time, indicating the proper count.

The AND-gate decoder can be extended to counters with any number of states. The following example illustrates.

EXAMPLE 7.15

How many AND gates are required to completely decode all the states of a MOD-32 binary counter? What are the inputs to the gate that decodes for the count of 21?

Solution: A MOD-32 counter has 32 possible states. One AND gate is needed to decode for each state; therefore, the decoder requires 32 AND gates. Since $32 = 2^5$, the counter contains five FFs. Thus, each gate will have five inputs, one from each FF. To decode for the count of 21 that is 10101_2 requires AND-gate inputs of E, $\overline{D}$, C, $\overline{B}$, and A, where E is the MSB flip-flop.

Active-LOW Decoding　If NAND gates are used in place of AND gates, the decoder outputs will produce a normally HIGH signal, which goes LOW only when the number being decoded occurs. Both types of decoders are used, depending on the type of circuits being driven by the decoder outputs.

EXAMPLE 7.16

Figure 7.24 shows a common situation in which a counter is used to help generate a control waveform which could be applied to devices such as a motor, solenoid valve, or heater. The MOD-16 counter cycles and recycles through its counting sequence. Each time it goes to the count of 8 (1000), the upper NAND gate will produce a LOW output, which sets FF X to the 1 state. FF X stays HIGH until the counter reaches the count of 14 (1110), at which time the lower NAND

FIGURE 7.24

gate decodes it and produces a LOW output to clear X to the 0 state. Thus, the X output is HIGH between the counts of 8 and 14 for each cycle of the counter.

BCD Counter Decoding A BCD counter has 10 states, which can be decoded using the techniques previously described. BCD decoders provide 10 outputs corresponding to the decimal digits 0 through 9 represented by the states of the counter FFs. These 10 outputs can be used to control 10 indicator lamps for a visual display. More often, instead of using 10 separate lamps, a single display device is used to display the decimal numbers 0 through 9. One such device, called a *nixie tube,* contains 10 very thin numerically shaped filaments stacked on top of each other. The BCD decoder outputs control which filament is illuminated. Another class of decimal displays contains seven small segments made of a material (usually incandescent filaments or light-emitting diodes) which emits light when an electric current is passed through it. The BCD decoder outputs control which segments are illuminated in order to produce a pattern representing one of the decimal digits.

We will go into more detail concerning these types of decoders and displays in Chapter 9. However, since BCD counters and their associated decoders and displays are very commonplace, we will use the decoder/display unit (see Figure 7.25) to represent the complete circuitry used to visually display the contents of a BCD counter as a decimal digit.

FIGURE 7.25 BCD counters usually have their count displayed on a single display device.

REVIEW QUESTION

1. Describe the decoding gate needed to produce a LOW output when a MOD-64 counter is at the count of 23. (*Ans.* A six-input NAND gate with inputs A, B, C, $\overline{D}$, E, and $\overline{F}$)

7.11 *DECODING GLITCHES*

Section 7.5 discussed the effects of FF propagation delays in ripple counters. As we saw then, the accumulated propagation delays serve to essentially limit the frequency response of ripple counters. The delays between FF transitions can also cause problems when decoding a ripple counter. The problem occurs in the form of glitches or spikes at the outputs of some of the decoding gates. This is illustrated in Figure 7.26 for a MOD-4 ripple counter.

The waveforms at the outputs of each FF and decoding gate are shown in the figure. Notice the propagation delay between the clock waveform and the A output

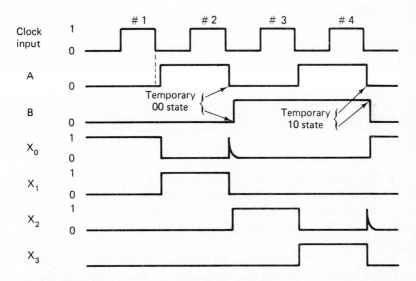

FIGURE 7.26 FF and decoding waveforms for a MOD-4 ripple counter showing glitches at X_0 and X_2 outputs.

waveform and between the A waveform and the B waveform. The glitches in the X_0 and X_2 decoding waveforms are caused by the delay between the A and B waveforms. X_0 is the output of the AND gate decoding for the normal 00 count. The 00 condition also occurs momentarily as the counter goes from the 01 to the 10 count, as shown by the waveforms. This is because B cannot change states until A goes LOW. This momentary 00 condition only lasts for several nanoseconds (depending on t_{pd} of FF

B) but can be detected by the decoding gate if the gate's response is fast enough. Hence, the spike at the X_0 output.

A similar situation produces a glitch at the X_2 output. X_2 is decoding for the 10 condition, and this condition occurs momentarily as the counter goes from 11 to 00 in response to the fourth clock pulse, as shown in the waveforms. Again, this is due to the delay of FF B's response after A has gone LOW.

Although the situation is illustrated for a MOD-4 counter, the same type of situation can occur for *any* ripple counter. This is because ripple counters work on the "chain-reaction" principle, whereby each FF triggers the next one and so on. The spikes at the decoder outputs may or may not present a problem, depending on how the counter is being used. When the counter is being used only to count pulses and display the results, the decoding spikes are of no consequence because they are very short in duration and will not even show up on the display. However, when the counter is used to control other logic circuits, such as was done in Figure 7.24, the spikes can cause improper operation. For example, in Figure 7.24, a spike at the output of either decoding NAND gate would cause FF X to be set or cleared at the wrong time.

In situations where the decoding spikes cannot be tolerated, there are two basic solutions to the problem. The first possibility is to use a parallel counter instead of a ripple counter. Recall that in a parallel counter the FFs are all triggered at the same time by the clock pulses so that it appears that the conditions which produced the decoder spikes cannot occur. However, even in a parallel counter the spikes may occur because the FFs will not all necessarily have the same t_{pd}, especially when some FFs may be loaded more heavily than others.

Strobing A more reliable method for eliminating the decoder spikes is to use a technique called *strobing*. This technique uses a signal called a *strobe-signal* to keep the decoding AND gates disabled (outputs at 0) until all the FFs have reached a stable state in response to the negative clock transition. This is illustrated in Figure 7.27(a), where the strobe signal is connected as an input to each decoding gate. The accompanying waveforms show that the strobe signal goes LOW when the clock pulse goes HIGH. During the time that the strobe is LOW, the decoding gates are kept LOW. The strobe signal goes HIGH some time t_D *after* the clock pulse goes LOW to enable the decoding gates. t_D is chosen to be greater than the total time it takes the counter to reach a stable count and depends, of course, on the FF delays and the number of FFs in the counter. In this way the decoding gate outputs will not contain any spikes because they are disabled during the time the FFs are in transition.

Figure 7.27(b) shows a simple circuit used to generate the strobe signal for a typical situation. The OS is triggered by the *positive* transition of the 1-MHz clock and its $\overline{Q}$ output goes LOW for $t_p = 600$ ns. Thus, the strobe output will stay LOW until 100 ns after the clock goes LOW. While the strobe signal is HIGH, the decoder outputs will be activated.

The strobe method is not used if a counter is only used for display purposes, since the decoding spikes are too narrow to affect the display. The strobe signal is used when the counter is used in control applications like that of Figure 7.24, where the spikes could cause erroneous operation.

FIGURE 7.27 (a) Use of strobe signal to eliminate decoding spikes; (b) one possible way to generate strobe waveform.

REVIEW QUESTIONS

1. Explain why the decoding gates for an asynchronous counter may have glitches on their outputs.
2. How does strobing eliminate decoding glitches?

7.12 CASCADING BCD COUNTERS

BCD counters are used whenever pulses are to be counted and the results displayed in decimal. A single BCD counter can count from 0 through 9 and then recycles to 0. To count to larger decimal values, we can cascade BCD counters as illustrated in Figure 7.28. This arrangement operates as follows:

1. Initially all counters are cleared to the zero state. Thus, the decimal display is 000.
2. As input pulses arrive, the units BCD counter advances one count per pulse. After nine pulses have occurred, the hundreds and tens BCD counters are still at zero and the units counter is at 9 (binary 1001). Thus, the decimal display reads 009.
3. On the tenth input pulse the units counter recycles to zero, causing its D FF output to go from 1 to 0. This 1-to-0 transition acts as the clock input for the tens counter

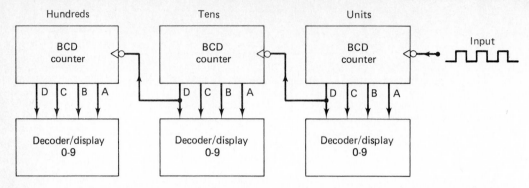

FIGURE 7.28 **Cascading BCD counters to count and display numbers from 000 to 999.**

and causes it to advance one count. Thus, after 10 input pulses, the decimal readout is 010.

4. As additional pulses occur, the units counter advances one count per pulse, and each time the units counter recycles to zero, it advances the tens counter one count. Thus, after 99 input pulses have occurred, the tens counter is at 9, as is the units counter. The decimal readout is thus 099.

5. On the hundredth input pulse, the units counter recycles to zero, which in turn causes the tens counter to recycle to zero. The D FF output of the tens counter thus makes a 1-to-0 transition which acts as the clock input for the hundreds counter and causes it to advance one count. Thus, after 100 pulses the decimal readout is 100.

6. This process continues up until 999 pulses. On the 1000th pulse, all the counters recycle back to zero.

It should be obvious that this arrangement can be expanded to any desired number of decimal digits simply by adding on more stages. For example, to count up to 999,999 will require six BCD counters and associated decoders and displays. In general, then, we need one BCD counter per decimal digit.

Incidentally, the BCD counters used in Figure 7.28 could be 7493s wired as MOD-10 counters, or they could be IC counters such as the 7490 or 74192 that are internally wired as BCD counters.

7.13 SHIFT-REGISTER COUNTERS

Shift registers can be arranged to form several types of counters. All *shift-register counters* use *feedback*, whereby the output of the last FF in the shift register is in some way connected to the first FF. The most widely used shift-register counters are the ring counter and the Johnson counter.

Ring Counter The simplest shift-register counter is essentially a *circulating* shift register connected so that the last FF shifts its value into the first FF. This arrangement is shown in Figure 7.29 using D-type FFs (J-K FFs can also be used). The FFs are connected so that information shifts from left to right and back around from Q_0 to Q_3. In most instances only a single 1 is in the register and it is made to circulate around

(a)

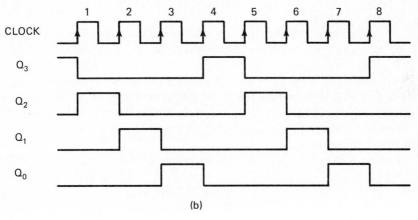

(b)

Q_3	Q_2	Q_1	Q_0	CLOCK pulse
1	0	0	0	0
0	1	0	0	1
0	0	1	0	2
0	0	0	1	3
1	0	0	0	4
0	1	0	0	5
0	0	1	0	6
0	0	0	1	7
⋮	⋮	⋮	⋮	⋮

(c)

FIGURE 7.29   (a) A 4-bit ring counter; (b) waveforms; (c) sequence table.

the register as long as clock pulses are applied. For this reason it is called a *ring counter*.

The waveforms and sequence table in Figure 7.29 show the various states of the FFs as pulses are applied, assuming a starting state of $Q_3 = 1$ and $Q_2 = Q_1 = Q_0 = 0$. After the first pulse, the 1 has shifted from Q_3 to Q_2 so that the counter is in the 0100 state. The second pulse produces the 0010 state, and the third pulse produces the 0001 state. On the *fourth* clock pulse the 1 from Q_0 is transferred to Q_3, resulting in the 1000 state, which is, of course, the initial state. Subsequent pulses cause the sequence to repeat.

This counter functions as a MOD-4 counter, since it has *four* distinct states before the sequence repeats. Although this circuit does not progress through the normal binary counting sequence, it is still a counter because each count corresponds to a particular state of the FFs. Note that each FF output waveform has a frequency equal to $\frac{1}{4}$ of the clock frequency, since this is a MOD-4 ring counter.

Ring counters can be constructed for any desired MOD number; a MOD-N ring counter uses N FFs connected in the arrangement of Figure 7.29. In general, a ring counter will require more FFs than a binary counter for the same MOD number; for example, a MOD-8 ring counter requires eight FFs while a MOD-8 binary counter requires only three.

Despite the fact that it is less efficient in the use of FFs, a ring counter is still useful because it can be decoded without the use of decoding gates. The decoding signal for each state is obtained at the output of its corresponding FF. Compare the FF waveforms of the ring counter with the decoding waveforms in Figure 7.23. In some cases a ring counter might be a better choice than a binary counter with its associated decoding gates. This is especially true in applications where the counter is being used to control the sequencing of operations in a system.

Starting a Ring Counter   To operate properly a ring counter must start off with only one FF in the 1 state and all the other FFs at 0. When power is first applied to the circuit, there is only a remote possibility that the FFs will come up in such a state. Thus, it is necessary to preset the counter to the required starting state before clock pulses are applied. A simple way to accomplish this is to apply a momentary pulse to the DC SET input of one of the FFs and to the DC CLEAR inputs of all the others. This will place a single 1 into the ring counter.

Johnson Counter   The basic ring counter can be modified slightly to produce another type of shift register counter, which will have somewhat different properties. The *Johnson* or *twisted-ring counter* is constructed exactly like a normal ring counter except that the *inverted* output of the last FF is connected to the input of the first FF. A 3-bit Johnson counter is shown in Figure 7.30. Note that the $\overline{Q}_0$ output is connected back to the D input of Q_2. This means that the *inverse* of the level stored in Q_0 will be transferred to Q_3 on the clock pulse.

The Johnson-counter operation is easy to analyze if we realize that on each positive clock-pulse transition the level at Q_2 shifts into Q_1, the level at Q_1 shifts into Q_0, and the *inverse* of the level at Q_0 shifts into Q_2. Using these ideas and assuming that all FFs are initially 0, the waveforms and sequence table of Figure 7.30 can be generated.

Examination of the waveforms and sequence table reveals the following important points:

1. This counter has six distinct states: 000, 100, 110, 111, 011, and 001 before it repeats the sequence. Thus, it is a MOD-6 Johnson counter. Note that it does not count in a normal binary sequence.

FIGURE 7.30 (a) MOD-6 Johnson counter; (b) waveforms; (c) sequence table.

2. The waveform of each FF is a squarewave (50 percent duty cycle) at $\frac{1}{6}$ the frequency of the clock. In addition, the FF waveforms are shifted by one clock period with respect to each other.

The MOD number of a Johnson counter will always be equal to *twice* the number of FFs. For example, if we connect five FFs in the arrangement of Figure 7.30, the result is a MOD-10 Johnson counter, where each FF output waveform is a squarewave at $\frac{1}{10}$ the clock frequency. Thus it is possible to construct a MOD-N counter (where N is an even number) by connecting $N/2$ FFs in a Johnson-counter arrangement.

Decoding a Johnson Counter For a given MOD number, a Johnson counter requires only half the number of FFs that a ring counter requires. However, a Johnson counter requires decoding gates whereas a ring counter does not. As in the binary counter, the Johnson counter uses one logic gate to decode for each count, but each gate requires only two inputs, regardless of the number of FFs in the counter. Figure 7.31 shows the decoding gates for the six states of the Johnson counter of Figure 7.30.

Notice that each decoding gate only has two inputs, even though there are three FFs in the counter. This is because for each count, two of the three FFs are in a unique combination of states. For example, the combination $Q_2 = Q_0 = 0$ occurs only once in the counting sequence, at the count of 0. Thus, AND gate 0 with inputs $\overline{Q}_2$ and $\overline{Q}_0$ can be used to decode for this count. This same characteristic is shared by all the other states in the sequence, as the reader can verify. In fact, for *any* Johnson counter, only two-input decoding gates are required.

Johnson counters represent a middle ground between ring counters and binary counters. A Johnson counter requires less FFs than a ring counter but generally more than a binary counter; it has more decoding circuitry than a ring counter but less than a binary counter. Thus, they sometimes represent a logical choice for certain applications.

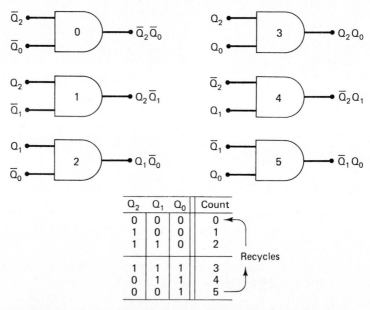

Q_2	Q_1	Q_0	Count
0	0	0	0
1	0	0	1
1	1	0	2
1	1	1	3
0	1	1	4
0	0	1	5

Recycles

FIGURE 7.31 Decoding logic for MOD-6 Johnson counter.

IC Shift-Register Counters There are not too many ring counters or Johnson counters available as ICs. The reason is that it is relatively simple to take a shift-register IC and to wire it as either a ring or Johnson counter. A couple of CMOS Johnson-counter ICs also include the complete decoding circuitry on the same chip.

1. Which shift-register counter requires the most FFs for a given MOD number?
2. Which shift-register counter requires the most decoding circuitry?
3. How can a ring counter be converted to a Johnson counter?
4. Answer true or false:
    (a) The outputs of a ring counter are always squarewaves.
    (b) The decoding circuitry for a Johnson counter is simpler than for a binary counter.
    (c) Ring and Johnson counters are synchronous counters.

7.14 COUNTER APPLICATIONS: FREQUENCY COUNTER

There are numerous applications for the many types of counters we have been discussing. In this section and the next we will look at two representative applications
that illustrate the uses of counters in digital systems.

A *frequency counter* is a circuit that can measure and display the frequency of
a pulse signal. One of the most straightforward methods for constructing a frequency
counter is shown in Figure 7.32(a) in simplified form. It contains a counter with its
associated decoder/display circuitry and an AND gate. The AND-gate inputs include
the pulses with unknown frequency, f_x, and a SAMPLE pulse that controls how long
these pulses are allowed to pass through the AND gate into the counter. The counter
is usually made up of cascaded BCD counters (Figure 7.28), and the decoder/display
unit converts the BCD outputs into a decimal display for easy monitoring.

The waveforms in Figure 7.32(b) show that a CLEAR pulse is applied to the
counter at t_0 to start the counter at zero. Prior to t_1 the SAMPLE pulse waveform is
LOW, so the AND output, Z, will be LOW and the counter will not be counting. The
SAMPLE pulse goes HIGH from t_1 to t_2; this is called the *sampling interval*. During
this sampling interval the unknown frequency pulses will pass through the AND gate
and will be counted by the counter. After t_2 the AND output returns LOW and the
counter stops counting. Thus, the counter will have counted the number of pulses that
occurred during the sampling interval, and the resulting contents of the counter is a
direct measure of the frequency of the pulse waveform.

EXAMPLE 7.17

The unknown frequency is 3792 pps.* The counter is cleared to the zero state
prior to t_1. Determine the counter reading after a sampling interval of (a) 1 s,
(b) 0.1 s, and (c) 10 ms.

Solution:
(a) Within a sampling interval of 1 s there will be 3792 pulses entering the
counter, so after t_2 the contents of the counter will read 3792.
(b) With a 0.1-s sampling interval the number of pulses passing through the
AND gate into the counter will be 3792 pulses/second $\times$ 0.1 s = 379.2. This
means that either 379 or 380 pulses will be counted, depending on what part of
a pulse cycle that t_1 occurs.

*Recall, pps stands for *pulses per second*.

(c) With a 10-ms = 0.01-s sampling interval, the counter will read either 37 or 38.

(a)

(b)

FIGURE 7.32 Basic frequency-counter method.

The accuracy of this method depends almost entirely on the duration of the sampling interval, which must be very accurately controlled. A commonly used method for obtaining very accurate sample pulses is shown in Figure 7.33. A crystal-controlled oscillator is used to generate a very accurate 100-kHz waveform, which is shaped into square pulses and fed to a series of decade counters which are being used

FIGURE 7.33 Method for obtaining accuate sampling intervals for frequency counter.

to successively divide this 100-kHz frequency by 10. The frequencies at the outputs of each decade counter are as accurate (percentagewise) as the crystal frequency.

The switch is used to select one of the decade-counter output frequencies to be fed to a single FF to be divided by 2. For example, in switch position 1 the 1-Hz pulses are fed to FF Q, which is acting as a toggle FF so that its output will be a squarewave with a period of $T = 2$ s and a pulse duration of $t_p = T/2 = 1$ s. This pulse duration is the desired 1-s sampling interval. In position 2 the sampling interval would be 0.1 s, and so on for the other positions.

EXAMPLE 7.18

Assume that the counter in Figure 7.32 is made up of three cascaded BCD counters and their associated displays. If the unknown input frequency is between 1 and 9.99 kpps, what is the best setting for the switch position in Figure 7.33?

Solution: With 3 BCD counters the total capacity of the counter is 999. A 9.99 kpps frequency would produce a count of 999 if a 0.1-s sample interval were used. Thus, in order to use the full capacity of the counter, the switch should be set to position 2. If a 1-s sampling interval were used, the counter capacity would always be exceeded for frequencies in the specified range. If a lower interval were used, the counter would count only between 0 and 99; this would give a reading to only two significant figures and would be a waste of the counter's capacity.

We will now look at a more complete frequency-counter circuit in Figure 7.34(a). The circuit now contains a one-shot and a J-K FF operating in the toggle mode, and the AND gate has three inputs, one of which is the FF output X. The SAMPLE pulses are connected to the AND gate and also to the *CLK* input of the FF. These SAMPLE pulses would be generated from a circuit such as that in Figure 7.33. The following step-by-step description refers to the waveforms in Figure 7.34(b).

1. Assume that FF X is in the 0 state (it has toggled to 0 on the falling edge of the previous sample pulse).
2. This LOW from X is fed to the AND gate, disabling its output, so no pulses are fed to the counter even when the first sample pulse occurs between t_1 and t_2.
3. At t_2 the NGT of the first sample pulse toggles FF X to the 1 state (note that $J = K = 1$). This positive transition at X triggers the OS, which generates a 100-ns pulse to *clear* the counter. The counter now displays *zero*.
4. At t_3 the second sample pulse enables the AND gate (since X is now 1) and allows the unknown frequency into the counter to be counted until t_4.
5. At t_4 the sample pulse returns LOW and toggles X LOW, disabling the AND gate. The counter stops counting.
6. Between t_4 and t_6 the counter holds and displays the count that it had reached at t_4. Note that the third sample pulse does not enable the AND gate because FF X is LOW.
7. At t_6 the NGT of the sample pulse toggles X HIGH and the operation follows the same sequence that began at t_2.

(a)

(b)

FIGURE 7.34   Frequency counter.

This frequency counter, then, goes through a repetitive sequence of counting, holding for display, clearing to zero, counting, and so on. Since the display would show the clearing and counting action of the counter, it would be very difficult to read the display to determine the unknown frequency except at very low frequencies. This problem can be overcome by inserting a *buffer register* between the counter and the decoder/display unit. We will consider this in Problem 7.31 at the end of the chapter.

7.15 COUNTER APPLICATIONS: DIGITAL CLOCK

One of the most popular applications of counters is the digital clock—a time clock which displays the time of day in hours, minutes, and sometimes seconds. In order to construct an accurate digital clock, a very closely controlled basic clock frequency is required. For battery-operated digital clocks (or watches) the basic frequency is normally obtained from a quartz-crystal oscillator. Digital clocks operated from the ac power line can use the 60-Hz power frequency as the basic clock frequency. In either case, the basic frequency has to be divided down to a frequency of 1 Hz or 1 pulse per second (pps). Figure 7.35 shows the basic block diagram for a digital clock operating from 60 Hz.

The 60-Hz signal is sent through a shaping circuit to produce square pulses at the rate of 60 pps. This 60-pps waveform is fed into a MOD-60 counter which is used to divide the 60 pps down to 1 pps. The 1-pps signal is fed into the SECONDS section, which is used to count and display seconds from 0 through 59. The BCD counter advances one count per second. After 9 seconds the BCD counter recycles to 0, which triggers the MOD-6 counter and causes it to advance one count. This continues for 59 s when the MOD-6 counter is in the 101 (5) count and the BCD counter is at 1001 (9), so the display reads 59 s. The next pulse recycles the BCD counter to 0, which in turn recycles the MOD-6 counter to 0 (remember: the MOD-6 counts from 0 through 5).

The output of the MOD-6 counter in the SECONDS section has a frequency of 1 pulse per minute (the MOD-6 recycles every 60 s). This signal is fed to the MINUTES section, which counts and displays minutes from 0 through 59. The MINUTES section is identical to the SECONDS section and operates in exactly the same manner.

The output of the MOD-6 counter in the MINUTES section has a frequency of 1 pulse per hour (the MOD-6 recycles every 60 min). This signal is fed to the HOURS section, which counts and displays hours from 1 through 12. This HOURS section is different from the SECONDS and MINUTES sections in that it never goes to the zero state. The circuitry in this section is sufficiently unusual to warrant a closer investigation.

Figure 7.36 shows the detailed circuitry contained in the HOURS section. It includes a BCD counter to count units of hours, and a single FF (MOD-2) to count tens of hours. The BCD counter is a 74192 which operates exactly like the 74193 that we studied earlier except that it counts only between 0000 and 1001. In other words, the 74192 can either count up in BCD fashion (i.e., 0 to 9 and then back to 0) or count down in BCD fashion (i.e., 9 to 0 and then back to 9). Here it is used to count up in response to the 1-pulse/hour signal coming from the MINUTES section. The IN-

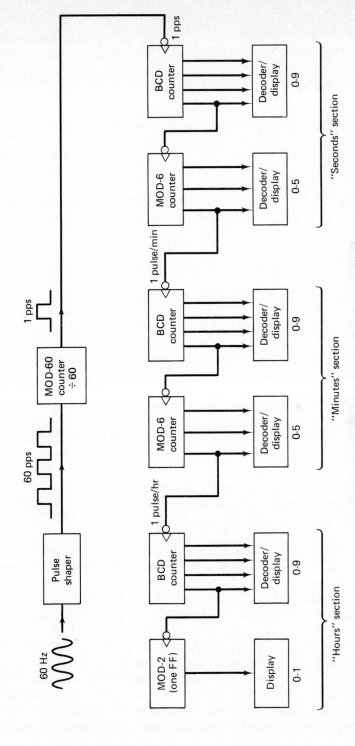

FIGURE 7.35 Block diagram for digital clock.

FIGURE 7.36 Detailed circuitry for the hours section.

VERTER on the CP_U input is needed because the 74192 responds to PGTs, and we want it to respond to the NGT that occurs when the MINUTES section recycles back to zero.

The incoming pulses advance the BCD counter once per hour. For example, at 7 o'clock this counter will be at 0111, and its decoder/display circuitry will display the numeral 7. At the same time, X will be LOW and its display will show a zero. Thus, the two displays will show "07". When the BCD counter is in the 1001 (9) state and the next input pulse occurs, it will recycle back to 0000. The NGT at Q_3 will toggle FF X from 0 to 1. This produces a numeral 1 on the X display and a numeral 0 on the BCD display so that the combined displays show "10" for 10 o'clock.

The next two pulses advance the BCD counter so that "11" and "12" are displayed at 11 o'clock and 12 o'clock, respectively. The next pulse advances the BCD counter to 0011 (3). In this state, the counter's Q_1 and Q_0 outputs are both HIGH, and X is still HIGH. Thus, the NAND-gate output goes LOW and activates the DC CLEAR of FF X and the $\overline{PL}$ input of the 74192. This clears X to 0 and presets the BCD counter to 0001. The result is a display of "01" for 1 o'clock. Several of the end-of-chapter problems will provide more details on the digital clock circuit.

--- **REVIEW QUESTIONS**

1. Name the basic blocks that make up a digital clock circuit.
2. Why is an INVERTER needed in Figure 7.36?

7.16 INTEGRATED-CIRCUIT REGISTERS

The various types of registers can be classified according to the manner in which data can be entered into the register for storage and the manner in which data are output from the register. The various classifications are listed below.

1. Parallel in/parallel out.
2. Serial in/serial out.
3. Parallel in/serial out.
4. Serial in/parallel out.

Each of these types is available in IC form so that a logic designer can usually find exactly what is required for a given application. In the following sections we will examine a representative IC from each of the above categories.

7.17 PARALLEL IN/PARALLEL OUT—THE 74174 AND 74178

There are actually two types of registers in this category; one is strictly parallel, the other is actually a shift register that can be loaded with parallel data and has parallel outputs available.

The 74174 Figure 7.37(a) shows the logic diagram for the 74174, a 6-bit register that has parallel inputs (D_5–D_0) and parallel outputs Q_5–Q_0. Parallel data are loaded into the register on the PGT of the clock input CP. A master reset input $\overline{MR}$ can be used to asynchronously reset all the register FFs to 0.

The logic symbol for the 74174 is shown in Figure 7.37(b). This symbol is used in circuit diagrams to represent the circuitry of Figure 7.37(a).

The 74178 Figure 7.38(a) shows the logic diagram for the 74178, a 4-bit shift register that has parallel data entry (P_0–P_3) and parallel outputs (Q_0–Q_3). It has a serial data input, D_S, and two enable inputs: PE (parallel enable) and SE (serial enable).

Figure 7.38(b) is the mode-select table that describes the various modes of operation for this IC. The first entry gives the input conditions necessary for the shift-right operation. With $SE = 1$, the data will shift from left to right on the NGT of clock input $\overline{CP}$ regardless of the logic level at the PE input (recall that X represents the "don't care" condition). You can verify this by tracing through the logic diagram and noting that when $SE = 1$, the D_S input passes through the logic gates and appears at the D input of the Q_0 FF. Likewise, Q_0 will appear at the D input of Q_1; Q_1 will appear at the D input of Q_2; and Q_2 will appear at the D input of Q_3.

FIGURE 7.37 (a) Circuit diagram of 74174; (b) logic symbol. (Courtesy of Fairchild, a Schlumberger company)

Mode select table

Inputs			Response
SE	PE	$\overline{CP}$	
H	X	⌐⌐	Right shift $D_S \rightarrow Q_0$; $Q_0 \rightarrow Q_1$, etc.
L	H	⌐⌐	Parallel load $P_0 \rightarrow Q_0$; $P_1 \rightarrow Q_1$, etc.
L	L	X	Hold

H = HIGH voltage level
L = LOW voltage level
X = immaterial

(b)

FIGURE 7.38 (a) Logic diagram for the 74178; (b) mode-select table. (courtesy of Fairchild Semiconductor)

269

The second entry in the table gives the conditions needed to produce parallel transfer from the parallel data inputs (P_0–P_3) to the outputs (Q_0–Q_3). When $SE = 0$ and $PE = 1$, this parallel transfer occurs on the NGT of $\overline{CP}$. Note that this is a *synchronous* transfer. Again, by tracing through the logic diagram, you can see that when $SE = 0$ and $PE = 1$ the parallel data inputs will get through the logic gates and will appear at the D inputs of their respective FFs.

The final entry in the table indicates that the $SE = 0$, $PE = 0$ condition will cause the register FFs to hold their current levels regardless of what happens at the clock input. For this input condition, each FF output is allowed to pass through its respective logic gates and appear at the D input of the same FF. Thus, a NGT on $\overline{CP}$ will not change the state of the FF.

Figure 7.39 shows the logic symbol for the 74178.

FIGURE 7.39   Logic symbol for the 74178.

EXAMPLE 7.19

Describe how the 74178 can be connected as a ring counter.

FIGURE 7.40   74178 wired as a ring counter.

Solution: The correct connections are shown in Figure 7.40. The Q_3 output is connected back around to the D_S input. The parallel data inputs are permanently set at 0001 and the PE input is connected permanently HIGH. The SE input is initially connected LOW. This allows each NGT at $\overline{CP}$ to parallel-load the shift register with 0001. Then, SE is switched to the HIGH state (shift mode), where it remains as the NGTs at $\overline{CP}$ cause the single 1 to shift and recirculate through the register.

7.18   SERIAL IN/SERIAL OUT—THE 4731B

The 4731B is a CMOS *quad* 64-bit shift register. It contains four identical 64-bit shift registers on one chip. Figure 7.41 shows the logic diagram for one of the 64-bit registers. It has a serial input, D_S, a clock input $\overline{CP}$ that responds to NGTs, and a serial output from the last FF, Q_{63}. This is the only output that is externally accessible. Note that this output goes through a *buffer* circuit (triangle symbol with no inversion bubble). A buffer does not change the signal's logic level; it is used to provide a greater output-current capability than normal. Also note that there is no means for parallel data entry into the register FFs.

FIGURE 7.41   Logic diagram for one of four 64-bit shift registers on a 7431B. (Courtesy of Fairchild, a Schlumberger company)

EXAMPLE 7.20

How can the 4731B chip be wired as a single 256-bit shift register?

FIGURE 7.42 The four 64-bit shift registers on the 4731B chip can be connected together as a 256-bit register.

Solution: The four shift registers can be strung together as one long register by connecting the Q_{63} output of one to the D_S input of the next, and by connecting all $\overline{CP}$ inputs together as a common clock input. This is shown in Figure 7.42, where the logic symbol for each shift register is used for convenience.

7.19 PARALLEL IN/SERIAL OUT—THE 74165

The 74165, an 8-bit register, is shown in Figure 7.43(a). It actually has both serial data entry via D_S and parallel data entry via P_0–P_7. The only accessible FF outputs are Q_7 and $\overline{Q}_7$.

Note that the FFs are the clocked *SR* type (same as clocked *SC*) that respond to NGTs at their clock inputs. Each FF has asynchronous inputs, PRESET and *CL* (same as CLEAR), that are used for parallel data entry.

Also note that there are two clock inputs, CP_1 and CP_2, either of which can be used to produce the shift operation on PGTs.

EXAMPLE 7.21

Determine the necessary input conditions if pulses at CP_1 are to produce the shift operation.

Solution: In order for the CP_1 pulses to get through the AND gate, the $\overline{PL}$ input has to be HIGH. The CP_1 pulses out of the AND gate will be inverted by the NOR gate only if the other NOR input is LOW. Thus, CP_2 must be held LOW. The inverted CP_1 pulses will trigger the FFs. The FFs trigger on the NGT of the inverted CP_1 pulses, which means that they trigger on the PGT of the pulses applied to the CP_1 input.

EXAMPLE 7.22

How can the 74165 be placed in the hold mode so that the register contents does not change as pulses are applied to CP_1?

Solution: If CP_2 is kept HIGH while $\overline{PL}$ is HIGH, the bottom AND-gate output

FIGURE 7.43 **(a) Logic diagram of the 74165; (b) logic symbol. (Courtesy of Fairchild Semiconductor)**

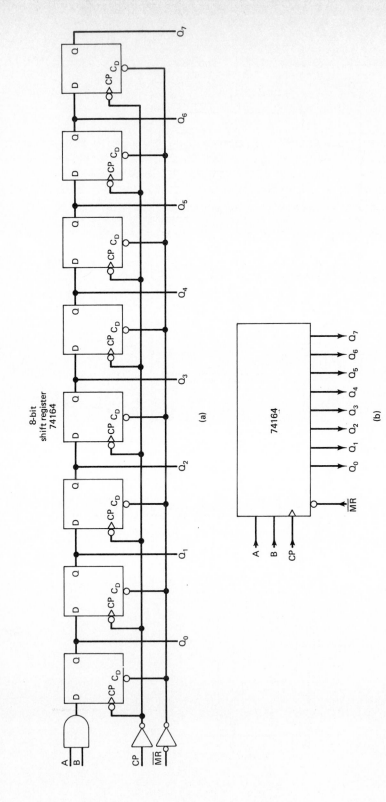

FIGURE 7.44 **(a)** Logic diagram for the 74164; **(b)** logic symbol. (Courtesy of Fairchild, a Schlumberger company)

will be HIGH and will keep the NOR output at a constant LOW regardless of the CP_1 input.

7.20 SERIAL IN/PARALLEL OUT—THE 74164

The logic diagram for the 74164 is shown in Figure 7.44(a). It is an 8-bit shift register with each FF output externally accessible. Instead of a single serial input, an AND gate combines inputs A and B to produce the serial input to FF Q_0.

The shift operation occurs on the PGTs of the clock input CP. The $\overline{MR}$ input provides asynchronous resetting of all FFs on a LOW level.

EXAMPLE 7.23

Assume that the initial contents of the 74164 register in Figure 7.45(a) is 00000000. Determine the sequence of states as clock pulses are applied.

(a)

Input pulse number	Q_0	Q_1	Q_2	Q_3	Q_4	Q_5	Q_6	Q_7	
0	0	0	0	0	0	0	0	0	
1	1	0	0	0	0	0	0	0	
2	1	1	0	0	0	0	0	0	
3	1	1	1	0	0	0	0	0	
4	1	1	1	1	0	0	0	0	Recycles
5	1	1	1	1	1	0	0	0	
6	1	1	1	1	1	1	0	0	
7	1	1	1	1	1	1	1	0	
8	1	1	1	1	1	1	1	1	

Temporary
state
(b)

FIGURE 7.45 Example 7.23

Solution: The correct sequence is given in Figure 7.45(b). With $A = B = 1$, the serial input is 1, so that 1s will shift into the register on each PGT of CP. Since Q_7 is initially at 0, the $\overline{MR}$ input is inactive.

On the eighth pulse, the register tries to go to the 11111111 state as the 1 from Q_6 shifts into Q_7. This state occurs only momentarily because $Q_7 = 1$ produces a LOW at $\overline{MR}$ that immediately resets the register back to 00000000. The sequence is then repeated on the next eight clock pulses.

The IC registers that have been presented here are representative of the various types that are commercially available. Although there are many variations on these basic registers, most of them should now be relatively easy to understand from the manufacturers' data sheets.

We will present several register applications in the end-of-chapter problems and in the material covered in subsequent chapters.

PROBLEMS

7.1 An 8-MHz squarewave clocks a 5-bit ripple counter. What is the frequency at the output of the last FF? What is the duty cycle of this output waveform?

7.2 Repeat Problem 7.1 if the input has a 20 percent duty cycle.

7.3 Assume that the 5-bit binary counter starts in the 00000 state. What will be the state of the counter after 144 input pulses?

7.4 Use J-K FFs and any other necessary logic to construct a MOD-24 asynchronous counter.

7.5 Draw the waveforms for all the FFs in the decade counter of Figure 7.5(b) in response to a 1-kHz clock frequency. Show any glitches that might appear on any of the FF outputs. Determine the frequency at the D output.

7.6 When a NAND gate is used to clear a counter back to zero as in the counters of Figures 7.4–7.6, there may be a possible problem because the propagation delays from CLEAR input to FF output may vary from FF to FF. For example, suppose that one FF clears in 10 ns and another clears in 50 ns. When the faster FF clears to 0 it will cause the NAND gate output to go back HIGH, thereby terminating the CLEAR pulse and the slower FF may not be cleared. This problem is especially prevalent when the FF outputs are unevenly loaded, since, in general, the propagation delay of a FF increases as it is loaded.

A possible means for eliminating this problem is shown in Figure 7.46 for a MOD-6 counter. The SET-CLEAR FF is the cross-coupled NAND-gate type studied in Chapter 5. Analyze the operation of this circuit and explain how the clearing function is performed without the problem mentioned above.

7.7 Refer to the counter shown in Figure 7.47. How can you tell that it is a down counter? It has been modified so that it does *not* count through the entire binary sequence 111 to 000. Determine the actual sequence it counts through.

7.8 A counter or group of counters is often used to divide a high-frequency clock signal down to a lower-frequency output. When these counters are binary counters (i.e., they count in the binary sequence), the output will not be a symmetrical squarewave if the binary sequence has been shortened in order to produce the desired MOD number. For example, refer to the C waveform of the MOD-6 counter in Figure 7.4.

FIGURE 7.46

FIGURE 7.47

When a counter is being used only for frequency division, it is not necessary that it count in a binary sequence as long as it has the desired MOD number. A symmetrical squarewave output can be obtained for any *even* MOD number by breaking the MOD number into the product of two MOD numbers, one of which is a power of 2. For example, a MOD-6 counter can be formed from a MOD-3 counter and a MOD-2 counter as shown in Figure 7.48.

Here FFs A, B, and the NAND gate make up the MOD-3 counter whose B output has $\frac{1}{3}$ the frequency of the input pulses. This B output is connected to the input of FF

FIGURE 7.48

C, which is acting as a MOD-2 to divide the frequency down to $\frac{1}{6}$ the frequency of the input pulses.

(a) Assume all FFs are initially LOW, and sketch the waveforms at each FF output for 12 cycles of the input.

(b) List the sequence of FFs states in a table and show that it is not a normal binary sequence.

7.9 Show how a 7493 counter can be used to produce a 1.2-kpps output from an 18-kpps input.

7.10 Show how two 7493s can be connected to divide an input frequency by 60 while producing a symmetrical squarewave output.

7.11 Determine the frequency at output X in Figure 7.49.

FIGURE 7.49

7.12 A 4-bit ripple counter is driven by a 20-MHz clock signal. Draw the waveforms at the output of each FF if each FF has t_{pd} = 20 ns. Determine which counter states, if any, will not occur because of the propagation delays.

7.13 What is the maximum clock frequency that can be used with the counter of Problem 7.12? What would be f_{max} if the counter were expanded to 6 bits?

7.14 (a) Draw the circuit diagram for a MOD-64 parallel counter.

(b) Determine f_{max} for this counter if each FF has t_{pd} = 20 ns and each gate has t_{pd} = 10 ns.

(c) A more efficient gating arrangement can be used for this counter by deriving the inputs for each AND gate from the output of the preceding gate (see Figure 7.50). For example, the output of the first AND gate (equal to $A \cdot B$) can be fed to the input of the second AND gate, together with C, to produce $A \cdot B \cdot C$. Similarly, the output of the second AND gate (equal to $A \cdot B \cdot C$) is fed to the third AND gate along with D to produce $A \cdot B \cdot C \cdot D$, and so on. This arrangement requires only two-input AND gates. However, it increases the total propagation delay, since the FF output signals must now propagate through more than one gate. Calculate the total delay for this arrangement and then calculate f_{max} and compare it to the value calculated in (b). Assume the same delays as given in (b).

7.15 Figure 7.51 shows a 4-bit parallel counter which is designed so that it does not sequence through the entire 16 binary states. Analyze its operation by drawing the waveforms at each FF output. Then determine the sequence which the counter goes through. Assume that all FFs are initially 0.

7.16 Refer to Appendix II for the 7490 data sheets.

(a) Show how to connect the 7490 as a BCD counter.

(b) Show how to connect the 7490 so that it divides the input frequency by 10 and produces a symmetrical squarewave output.

7.17 Figure 7.52 shows how a presettable down counter can be used in a *programmable timer* circuit. The input clock frequency is an accurate 1 Hz derived from the 60-Hz line frequency after division by 60. Switches S1–S4 are used to preset the counter to a desired starting count when a momentary pulse is applied to $\overline{PL}$. The timer operation is initiated by depressing the START pushbutton switch. FF Z is used to eliminate effects of bounce in the START switch. The OS is used to provide a very narrow pulse to the $\overline{PL}$ input. The output of FF X will be a waveform that goes HIGH for a number of seconds equal to the number set on the switches.

(a) Assume that all FFs and the counter are in the 0 state and analyze and explain the circuit operation, showing waveforms when necessary, for the case where S1 and S4 are LOW and S2 and S3 are HIGH. Be sure to explain the function of FF X.

(b) Why can't the timer output be taken at the $\overline{TC}_D$ output?

(c) Why can't the START switch be used to trigger the OS directly?

(d) What will happen if the START switch is held down too long? Add the necessary logic needed to ensure that holding the START switch down will not affect the timer operation.

7.18 Modify the circuit of Figure 7.22 so that it functions as a MOD-10 counter. The frequency at the Q_3 output should be $\frac{1}{10}$ the frequency of the CP_D input. Draw the waveforms at Q_3, Q_2, Q_1, Q_0, and $\overline{TC}_D$.

7.19 Change the parallel data inputs in Figure 7.22 to 1001. Draw the waveforms at Q_3, Q_2, Q_1, Q_0, and $\overline{TC}_D$. What is the MOD number?

7.20 Draw the gates necessary to decode all the states of a MOD-16 counter using active-LOW outputs.

FIGURE 7.50

FIGURE 7.51

FIGURE 7.52

7.21 Draw the AND gates necessary to decode the 10 states of the BCD counter of Figure 7.5(b).

7.22 Figure 7.53 shows a counter being used to help generate control waveforms. Control waveforms 1 and 2 could be used for many purposes, including control of motors, solenoids, valves, and heaters. Determine the control waveforms, assuming that all FFs are initially LOW. Ignore decoding glitches. Assume clock frequency = 1 kpps.

FIGURE 7.53

7.23 Draw the complete waveforms at the output of the decoding gates of a MOD-16 *ripple* counter, including any glitches or spikes that can occur due to the FF delays. Why are the gates which are decoding for *even* numbers the only ones that have glitches?

7.24 The circuit of Figure 7.53 might malfunction because of glitches at the outputs of the decoding NAND gates.
(a) Determine at what point(s) the glitches can cause erroneous operation.
(b) What are two ways that can be used to eliminate the possibility of erroneous operation?

7.25 How many FFs are used in Figure 7.28? Indicate the states of each of these FFs after 795 pulses have occurred.

7.26 How many cascaded BCD counters are needed to be able to count up to 8000? How many FFs does this require? Compare this to the number of FFs required for a normal binary counter to count up to 8000. Since it uses more FFs, why is the cascaded BCD method used?

7.27 Draw the diagram for a 5-bit ring counter using J-K flip flops.

7.28 Combine the ring counter of Problem 7.27 with a *single* J-K FF to produce a MOD-10 counter. Determine the sequence of states for this counter. This is an example of a decade counter that is not a BCD counter.

7.29 Draw the diagram for a MOD-10 Johnson counter using J-K flip-flops and determine its counting sequence. Draw the decoding circuit needed to decode each of the 10 states. This is another example of a decade counter that is not a BCD counter.

7.30 Determine the frequency of the pulses at points w, x, y, and z in the circuit of Figure 7.54.

FIGURE 7.54

7.31 The frequency counter of Figure 7.34 has the disadvantage of a blinking display. This can be remedied through the use of D-type FFs, which are used to *store* the contents of the counter at the end of the counting interval (t_3-t_4 in Figure 7.34) and hold it for display until the end of the next counting interval (t_7-t_8). Figure 7.55 shows this modified

FIGURE 7.55

frequency counter, which is the same as Figure 7.34 except that D FFs have been added between the counter FFs and the decoder/displays. Each counter FF output is fed to a D FF input. The D FF outputs drive the decoder/display. Analyze this circuit and determine how its operation differs from Figure 7.34, especially concerning the readings of the displays.

7.32 The frequency counter of Figure 7.55 uses three BCD counters and a sampling interval of 100 μs. Determine the readings on the three frequency-counter displays for each of the following input frequencies: (a) 220 kpps; (b) 4.5 Mpps; (c) 750 pps.

7.33 Design the complete circuit for the SECONDS section of the digital clock circuit of Figure 7.35. Use a 7493 for the MOD-6 and a 7490 for the BCD.

7.34 The digital clock of Figure 7.35 has to have some means for manually setting the HOURS and MINUTES section to the correct starting time. For example, this can be done by switching the 1-pps signal into the input of the MINUTES section when a SET MINUTES pushbutton is activated. A similar operation can be done with a SET HOURS pushbutton. Design the necessary logic to provide this capability using two pushbutton switches.

7.35 Modify the HOURS section of the digital clock (Figure 7.36) so that it counts and displays military time (i.e., 00 to 23 hours).

7.36 Modify the frequency counter of Figure 7.55 so that it uses 74174 ICs. Assume that the counter contains three BCD counters and a 3-digit display.

7.37 Show how to connect two 74178s as an 8-bit ring counter.

7.38 Figure 7.56 shows how a 74178 can be used as a *parallel-to-serial converter*. The

FIGURE 7.56

parallel data that are entered at P_0–P_3 are shifted out serially so that a serial waveform appears at Q_3. The shifting out of the parallel data is controlled by the occurrence of the START pulse. This START pulse occurs asynchronously to the clock pulses, so the D FFs are used to synchronize the parallel loading and shifting of the 74178.

Assume that the START pulse has been LOW and that clock pulses have been continually applied to the circuit for a long time before time t_0 (see waveforms). Draw the waveforms that appear at Q_X, $\overline{Q}_X$, Q_Y and Q_3 in response to the START pulse shown in the figure.

7.39 A shift register is often used to delay a logic signal by an integral number of clock cycles. If the waveforms in Figure 7.57 are applied to the inputs of one of the 4731B shift registers, how long will it take the D_S waveform to appear at the Q_{63} output? (Assume that the t_H requirement of the register FFs is met by the D waveform.)

FIGURE 7.57

7.40 Refer to the circuit of Figure 7.58.
 (a) Draw the waveforms at each FF output in response to the input waveforms shown.
 (b) Add the necessary logic to produce a timing-signal output that goes HIGH only during time intervals t_1–t_2 and t_8–t_9.
 (c) Add the necessary logic to produce a timing signal that goes LOW only during the t_4–t_7 interval.

FIGURE 7.58

8

INTEGRATED-CIRCUIT LOGIC FAMILIES

O U T L I N E

Almost all modern digital systems utilize digital integrated circuits (ICs) because they result in an increase in reliability and a reduction in weight and size. Digital IC technology has advanced rapidly from small-scale integration (SSI), with less than 13 equivalent logic gates per chip, through medium-scale integration (MSI), with between 13 and 99 equivalent logic gates per chip, to large-scale integration (LSI) and very-large-scale integration (VLSI) which can have thousands of gates per chip. With the widespread use of ICs comes the necessity to know and understand the characteristics of the most commonly used logic families. (A logic family refers to a specific class of logic circuits that are manufactured using the same manufacturing techniques.) In this chapter we will examine the TTL, ECL, NMOS, PMOS, and CMOS logic families in sufficient detail to understand the relative advantages and disadvantages of each.

The various logic families fall into two broad categories based on the major device used in the circuit fabrication. The *bipolar* families, TTL and ECL, use the bipolar transistor (NPN, PNP) as their principal circuit element. The *metal-oxide-semiconductor* (*MOS*) families use MOS field-effect transistors (MOSFETs) as their principal circuit element. PMOS, NMOS, and CMOS are all MOS families.

At present, the TTL and CMOS logic families are dominating those application areas that require SSI chips (i.e., gates, FFs) and MSI chips (i.e., counters, registers). In general, all the MOS families are better suited for LSI and VLSI applications (i.e., microprocessors, memory chips) than the bipolar families, because MOS circuits require less chip area and consume less power than their bipolar counterparts. On the other hand, MOS ICs generally operate at slower speeds than bipolar and require special handling and storage precautions.

8.1 DIGITAL IC TERMINOLOGY

Although there are many digital IC manufacturers, much of the nomenclature and terminology is fairly standardized. The most useful terms are defined and discussed below.

Current and Voltage Parameters

V_{IH} **(min)—high-level input voltage:**   The voltage level required for a logical 1 at an *input*. Any voltage below this level will not be accepted as a HIGH by the logic circuit.

V_{IL} **(max)—low-level input voltage:**   The voltage level required for a logical 0 at an *input*. Any voltage above this level will not be accepted as a LOW by the logic circuit.

V_{OH} **(min)—high-level output voltage:**   The voltage level at a logic circuit *output* in the logical 1 state. The minimum value of V_{OH} is usually specified.

V_{OL} **(max)—low-level output voltage:**   The voltage level at a logic circuit *output* in the logical 0 state. The maximum value of V_{OL} is usually specified.

I_{IH}**—high-level input current:**   The current that flows into an input when a specified high-level voltage is applied to that input.

I_{IL}**—low-level input current:**   The current that flows into an input when a specified low-level voltage is applied to that input.

I_{OH}—**high-level output current:**   The current that flows from an output in the logical 1 state under specified load conditions.

I_{OL}—**low-level output current:**   The current that flows from an output in the logical 0 state under specified load conditions.

Fan-Out   In general, a logic-circuit output is required to drive several logic inputs. The *fan-out* (also called *loading factor*) is defined as the *maximum* number of standard logic inputs that an output can drive reliably. For example, a logic gate that is specified to have a fan-out of 10 can drive 10 standard logic inputs. If this number is exceeded, the output logic-level voltages cannot be guaranteed.

Propagation Delays   A logic signal always experiences a delay in going through a circuit. The two propagation delay times are defined as

t_{PLH}:   delay time in going from logical 0 to logical 1 state (LOW to HIGH).

t_{PHL}:   delay time in going from logical 1 to logical 0 state (HIGH to LOW).

Figure 8.1 illustrates these propagation delays for an INVERTER. Note that t_{PHL} is the delay in the output's response as it goes from HIGH to LOW. It is measured between the 50 percent points on the input and output transitions. The t_{PLH} value is the delay in the output's response as it goes from LOW to HIGH.

In general, t_{PHL} and t_{PLH} are not the same value, and both will vary depending on loading conditions. The values of propagation times are used as a measure of the relative speed of logic circuits. For example, a logic circuit with values of 10 ns is a faster logic circuit than one with values of 20 ns.

Power Requirements   Every IC requires a certain amount of electrical power to operate. This power is supplied by one or more power-supply voltages connected to the power pin(s) on the chip. Usually there is only one power supply terminal on the chip, and it is labeled V_{CC} (for TTL) or V_{DD} (for MOS devices).

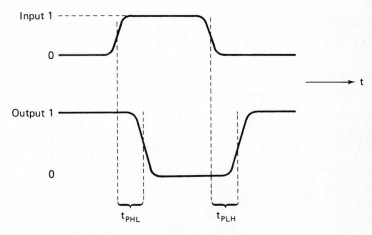

FIGURE 8.1   Propagation delays.

The amount of power that an IC requires is usually specified in terms of the current, I_{CC}, that it draws from the V_{CC} supply, and the actual power is the product $I_{CC} \times V_{CC}$. For many ICs the current drain on the supply will vary depending on the logic states of the circuits on the chip. For example, Figure 8.2(a) shows a NAND chip where *all* the gate *outputs* are HIGH. The current drain on the V_{CC} supply for this case is called I_{CCH}. Likewise, Figure 8.2(b) shows the current drain when *all* the gate *outputs* are LOW. This current is called I_{CCL}.

In general, I_{CCH} and I_{CCL} will be different values. The average current is

$$I_{CC}(\text{avg}) = \frac{I_{CCH} + I_{CCL}}{2}$$

and it can be used to calculate average power drain as

$$P_D(\text{avg}) = I_{CC}(\text{avg}) \times V_{CC}$$

Noise Immunity Stray electrical and magnetic fields can induce voltages on the connecting wires between logic circuits. These unwanted, spurious signals are called *noise* and can sometimes cause the voltage at the input to a logic circuit to drop below $V_{IH}(\text{min})$ or rise above $V_{IL}(\text{max})$, which could produce unreliable operation. The *noise immunity* of a logic circuit refers to the circuit's ability to tolerate noise voltages on

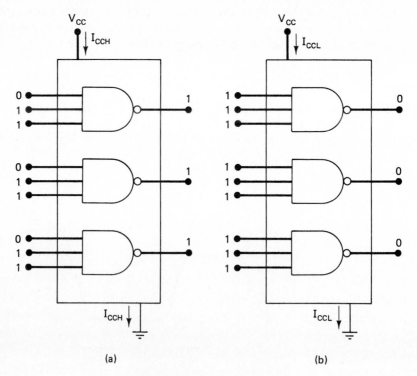

(a)   (b)

FIGURE 8.2 I_{CCH} and I_{CCL}.

its inputs. A quantitative measure of noise immunity is called *noise margin* and is illustrated in Figure 8.3.

Figure 8.3(a) is a diagram showing the range of voltages that can occur at a logic circuit output. Any voltages greater than $V_{OH}(min)$ are considered a logic 1, and any voltages lower than $V_{OL}(max)$ are considered a logic 0. Voltages in the indeterminate range should not appear at a logic circuit output under normal conditions. Figure 8.3(b) shows the voltage requirements at a logic circuit input. The logic circuit will respond to any input greater than $V_{IH}(min)$ as a logic 1, and will respond to voltages lower than $V_{IL}(max)$ as a logic 0. Voltages in the indeterminate range will produce an unpredictable response and should not be used.

The *high-state noise margin* V_{NH} is defined as

$$V_{NH} = V_{OH}(min) - V_{IH}(min) \tag{8.1}$$

as illustrated in Figure 8.3. V_{NH} is the difference between the lowest possible HIGH output and the minimum input voltage required for a HIGH. When a HIGH logic output is driving a logic circuit input, any negative noise spikes greater than V_{NH} appearing on the signal line can cause the voltage to drop into the indeterminate range, where unpredictable operation can occur.

The *low-state noise margin* V_{NL} is defined as

$$V_{NL} = V_{IL}(max) - V_{OL}(max) \tag{8.2}$$

and it is the difference between the largest possible LOW output and the maximum input voltage required for a LOW. When a LOW logic output is driving a logic input, any positive noise spikes greater than V_{NL} can cause the voltage to rise into the indeterminate range.

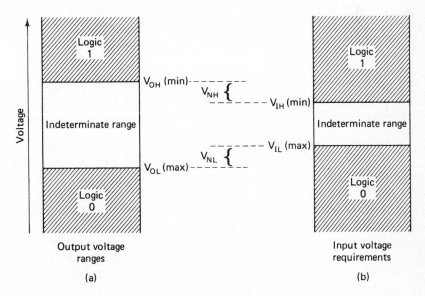

FIGURE 8.3 DC noise margins.

EXAMPLE 8.1

The input/output voltage specifications for the standard TTL family are listed below. Use these values to determine

(a) The maximum-amplitude noise spike that can be tolerated when a HIGH output is driving an input.

(b) The maximum-amplitude noise spike that can be tolerated when a LOW output is driving an input.

Parameter	Min (V)	Typical (V)	Max (V)
V_{OH}	2.4	3.6	
V_{OL}		0.2	0.4
V_{IH}	2.0*		
V_{IL}			0.8*

*Normally only the minimum V_{IH} and maximum V_{IL} values are given.

Solution:

(a) When an output is HIGH, it may be as low as V_{OH} (min) = 2.4 V. The minimum voltage that an input will respond to as a HIGH is V_{IH} (min) = 2.0 V. A negative noise spike can drive the voltage below 2.0 V if its amplitude is greater than

$$V_{NH} = V_{OH}(\text{min}) - V_{IH}(\text{min})$$

$$= 2.4 \text{ V} - 2.0 \text{ V} = 0.4 \text{ V}$$

(b) When an output is LOW, it may be as high as $V_{OL}(\text{max})$ = 0.4 V. The maximum voltage that an input will respond to as a LOW is $V_{IL}(\text{max})$ = 0.8 V. A positive noise spike can drive the voltage above the 0.8-V level if its amplitude is greater than

$$V_{NL} = V_{IL}(\text{max}) - V_{OL}(\text{max})$$

$$= 0.8 \text{ V} - 0.4 \text{ V} = 0.4 \text{ V}$$

AC Noise Margin   Strictly speaking, the noise margins predicted by expressions (8.1) and (8.2) are termed dc noise margins. The term "dc noise margin" might seem somewhat inappropriate when dealing with noise, which is generally thought of as an ac signal of the transient variety. However, in today's high-speed integrated circuits, a pulse width of 1 μs is extremely long and may be treated as dc as far as the response of a logic circuit is concerned. As pulse widths decrease to the low-nanosecond region, a limit is reached where the *pulse duration* is too short for the circuit to respond. At this point, the pulse amplitude would have to be increased appreciably to produce a change in the circuit output. What this means is that a logic circuit can tolerate a large noise amplitude if the noise pulse is of a very short duration compared to the circuit's response time (i.e., propagation delays). In other words, a logic circuit's *ac noise margins* are generally substantially greater than its dc noise margins given by (8.1) and (8.2).

Current-Sourcing and Current-Sinking Logic Logic families can be categorized according to how current flows between the output of one logic circuit and the input of another. Figure 8.4(a) illustrates *current-sourcing* logic. When the output of gate 1 is in the HIGH state, it supplies a current I_{IH} to the input of gate 2, which acts essentially as a resistance to ground. Thus, the output of gate 1 is acting as a *source* of current for the gate 2 input.

(a)

(b)

FIGURE 8.4 **Comparison of current-sourcing and current-sinking actions.**

Current-sinking logic is illustrated in Figure 8.4(b). Here the input circuitry of gate 2 is represented as a resistance tied to $+V_{CC}$, the positive terminal of a power supply. When the gate 1 output goes to its LOW state, current will flow in the direction shown from the input circuit of gate 2 back through the output resistance of gate 1 to ground. In other words, in the LOW state the circuit driving an input of gate 2 must be able to *sink* a current, I_{IL}, coming from that input.

The distinction between current-sourcing and current-sinking logic circuits is an important one which will become more apparent as we examine the various logic families.

REVIEW QUESTIONS

1. Define each of the following: V_{OH}, V_{IL}, I_{OL}, I_{IH}, t_{PLH}, t_{PHL}, I_{CCL}, I_{CCH}.
2. True or false: If a logic circuit has a fan-out of 5, the circuit has five outputs.
3. Define V_{NL} and V_{NH}.

8.2 THE TTL LOGIC FAMILY

At this writing, the transistor-transistor logic (TTL) family still enjoys the most widespread use in applications that require SSI and MSI devices. The basic TTL logic circuit is the NAND gate. Its detailed circuit diagram, shown in Figure 8.5(a), has several distinctive characteristics. First, note that transistor Q_1 has two emitters; thus, it has two emitter-base (E-B) junctions that can be used to turn Q_1 ON. This *multiple-emitter* input transistor can have up to eight emitters for an eight-input NAND gate.

(a)   (b)

FIGURE 8.5 (a) Basic TTL NAND gate; (b) diode equivalent for Q_1.

Also note that on the output side of the circuit, transistors Q_3 and Q_4 are in a *totem-pole* arrangement. As we will see shortly, in normal operation either Q_3 or Q_4 will be conducting, depending on the logic state of the output.

Circuit Operation—LOW State Although this circuit looks extremely complex, we can simplify its analysis somewhat by using the diode equivalent of the multi-emitter transistor Q_1 as shown in Figure 8.5(b). Diodes D_2 and D_3 represent the two E-B junctions of Q_1, and D_4 is the collector-base (C-B) junction. In the following analysis we will use this representation for Q_1.

First, let's consider the case where the output is LOW. Figure 8.6(a) shows this situation with inputs A and B both at +5 V. The +5 V at the cathodes of D_2 and D_3 will turn these diodes OFF and they will conduct almost no current. The +5-V supply will push current through R_1 and D_4 into the base of Q_2, which turns ON. Current from Q_2's emitter will flow into the base of Q_4 and turn Q_4 ON. At the same time, the flow of Q_2 collector current produces a voltage drop across R_2 that reduces Q_2's collector voltage to a low value that is insufficient to turn ON Q_3.

Input conditions	Output conditions
A and B are both HIGH (≥ 2 V)	Q_3 OFF
Input currents are very low. I_{IH} = 10 μA	Q_4 ON so that V_X is LOW (≤ 0.4 V)

(a) LOW output

Input conditions	Output conditions
A or B or both are LOW (≤ 0.8 V)	Q_4 OFF
Current flows back to ground through LOW input terminal. I_{IL} = 1.1 mA	Q_3 acts as emitter-follower and V_{OH} ≥ 2.4 V, typically 3.6 V

(b) HIGH output

FIGURE 8.6   TTL NAND gate in its two output states.

The voltage at Q_2's collector is shown as approximately 0.8 V. This is because Q_2's emitter is at 0.7 V relative to ground due to Q_4's E-B forward voltage, and Q_2's collector is at 0.1 V relative to its emitter due to V_{CE}(sat). This 0.8 V at Q_3's base is not enough to forward-bias both Q_3's E-B junction and diode D_1. In fact, D_1 is needed to keep Q_3 OFF in this situation.

With Q_4 ON, the output terminal, X, will be at a very low voltage, since Q_4's ON-state resistance will be low (1–25 Ω). Actually, the output voltage, V_{OL}, will depend on how much collector current Q_4 conducts. With Q_3 OFF, there is no current coming from the +5-V terminal through R_4. As we shall see, Q_4's collector current will come from the TTL inputs that terminal X is connected to.

It is important to note that the HIGH inputs at A and B will have to supply only a very small diode leakage current. Typically, this current I_{IH} is only around 10 μA at room temperature.

Circuit Operation—HIGH State   Figure 8.6(b) shows the situation where the circuit output is HIGH. This situation can be produced by connecting either or both inputs LOW. Here, input B is connected to ground. This will forward-bias D_3 so that current will flow from the +5-V source terminal, through R_1 and D_3, and through terminal B to ground. The forward voltage across D_3 will hold point Y at approximately 0.7 V. This voltage is not enough to forward-bias D_4 and the E-B junction of Q_2 sufficiently for conduction.

With Q_2 OFF, there is no base current for Q_4, and it turns OFF. Since there is no Q_2 collector current, the voltage at Q_3's base will be large enough to forward-bias Q_3 and D_1, so Q_3 will conduct. Actually, Q_3 acts as an emitter-follower, because output terminal X is essentially at its emitter. With no load connected from point X to ground, V_{OH} will be around 3.6–3.8 V, because two 0.7-V diodes drops (E-B of Q_3, and D_1) subtract from the 5 V applied to Q_3's base. This voltage will decrease under load because the load will draw emitter current from Q_3 which draws base current through R_2, thereby increasing the voltage drop across R_2.

It's important to note that there is a substantial current flowing back through input terminal B to ground. This current, I_{IL}, is typically around 1.1 mA. The LOW B input acts as a *sink* to ground for this current.

Current-Sinking Action   TTL logic circuits are current-sinking circuits in which outputs *receive* current from the inputs they are driving in the LOW state. Figure 8.7 shows the output of one TTL gate driving the input of another. When the driving gate output is LOW, transistor Q_4 is saturated and Q_3 is OFF. The LOW voltage at X forward-biases the emitter of Q_1 and current flows, as shown, back through Q_4. The saturation collector current of Q_4 is provided by the gate being driven. Q_4 is acting as a current sink.

Totem-Pole Output Circuit   Several points should be mentioned concerning the totem-pole arrangement of the TTL output circuit, since it is not readily apparent why it is used. The same logic could be accomplished by eliminating Q_3 and D_1 and connecting the bottom of R_4 to the collector of Q_4. But this would mean that Q_4 would conduct a fairly heavy current in its saturation state (5 V/130 $\Omega \approx$ 40 mA). With Q_3

| Output circuit | Input circuit | **FIGURE 8.7   TTL current-** |
| of driving gate | of driven gate | **sinking action.** |

in the circuit, there will be no current through R_4 in the output LOW state. This is important because it keeps the circuit power dissipation down.

Another advantage of this arrangement occurs in the output HIGH state. Here Q_3 is acting as an emitter-follower with its associated low output impedance (typically 10 Ω). This low output impedance provides a short time constant for charging up any capacitive load on the output. This action (commonly called *active pull-up*) provides very fast rise-time waveforms at TTL outputs.

A disadvantage of the totem-pole output arrangement occurs during the transition from LOW to HIGH. Unfortunately, Q_4 turns OFF more slowly than Q_3 turns ON, so there is a period of a few nanoseconds during which both transistors are conducting and a relatively large current (30–40 mA) will be drawn from the 5-V supply. This can present a problem that will be examined later.

SUMMARY

All TTL circuits have a structure similar to this basic NAND gate. The input of any TTL circuit will be the emitter of an NPN transistor, or the cathode of a diode, so that a HIGH input voltage will turn OFF the device and only a small leakage current (I_{IH}) will flow. Conversely, a LOW input voltage turns ON the device and a relatively large current (I_{IL}) will flow. Most, but not all, TTL circuits will have some type of totem-pole output configuration. There are some exceptions that will be discussed later.

REVIEW QUESTIONS

1. True or false: A TTL output acts as a current sink in the LOW state.
2. In which TTL input state does the largest amount of input current flow?
3. State the advantages and disadvantages of a totem-pole output.

8.3 STANDARD TTL SERIES CHARACTERISTICS

In 1964 Texas Instruments introduced the first standard product line of TTL circuits. The 5400/7400 series, as it is called, has been one of the most widely used families of IC logic. We will simply refer to it as the 7400 series, since the only difference between the 5400 and 7400 versions is that the 5400 series is meant for military use and can operate over a wider temperature and power-supply range. Many IC manufacturers now produce the 7400 line of ICs, although some use their own identification numbers. For example, Fairchild has a series of TTL ICs, which uses numbers such as 9N00, 9300, 9600, and so on. However, on the Fairchild specification sheets the equivalent 7400 series number is usually indicated.

The 7400 series operates reliably over the temperature range 0–70°C and with a supply voltage (V_{CC}) of from 4.75 to 5.25 V. The 5400 series is somewhat more flexible, since it can tolerate a −55 to +125°C temperature range and a supply variation of 4.5–5.5 V. Both series typically have a fan-out of 10, indicating that they can reliably drive 10 other inputs.

7400 Voltage Levels   Table 8.1 lists the required input and output voltage levels for the standard 7400 series. The minimum and maximum values shown are for worst-case conditions of power supply, temperature, and loading conditions. Inspection of the table reveals a guaranteed maximum logical 0 output $V_{OL} = 0.4$ V, which is 400 mV less than the logical 0 voltage needed at the input $V_{IL} = 0.8$ V. This means that the guaranteed LOW-state dc noise margin is 400 mv. That is,

TABLE 8.1   STANDARD 7400 SERIES VOLTAGE LEVELS

	Minimum	Typical	Maximum
V_{OL}	—	0.2	0.4
V_{OH}	2.4	3.6	—
V_{IL}	—	—	0.8
V_{IH}	2.0	—	—

$$V_{NL} = V_{IL}(\text{max}) - V_{OL}(\text{max}) = 0.8 \text{ V} - 0.4 \text{ V} = 0.4 \text{ V} = 400 \text{ mV}$$

Similarly, the logical 1 output V_{OH} is a guaranteed minimum of 2.4 V, which is 400 mV greater than the logical 1 voltage needed at the input $V_{IH} = 2.0$ V. Thus, the HIGH-state dc noise margin is 400 mV.

$$V_{NH} = V_{OH}(\text{min}) - V_{IH}(\text{min}) = 2.4 \text{ V} - 2.0 \text{ V} = 0.4 \text{ V} = 400 \text{ mV}$$

Thus, the *guaranteed worst-case* dc noise margins for the 7400 series are both 400 mV. In actual operation the *typical* dc noise margins are somewhat higher ($V_{NL} = 1$ V and $V_{NH} = 1.6$ V).

Maximum Voltage Ratings   The voltage values in Table 8.1 do not include the absolute maximum ratings beyond which the useful life of the IC may be impaired. The voltages applied to any input of a standard 7400 series IC must never exceed

+5.5 V. A voltage greater than +5.5 V applied to an input emitter can cause reverse breakdown of the E-B junction of Q_1.

There is also a limit on the maximum *negative* voltage that can be applied to a TTL input. This limit, -0.5 V, is caused by the fact that most TTL circuits employ protective shunt diodes on each input, as illustrated in Figure 8.8. These diodes were purposely left out of our earlier analysis, since they do not enter into the normal circuit operation. They are connected from each input to ground to limit the negative input voltage excursions that often occur when logic signals have excessive ringing. With these diodes, we should not apply more than -0.5 V to an input, because the protective diodes would begin to conduct and draw substantial current.

FIGURE 8.8   Protective diodes on TTL inputs.

Power Dissipation   The basic TTL logic circuit is the NAND gate of Figure 8.5. Typically, it draws an average supply current I_{CC} of 2 mA, resulting in a power dissipation of 2 mA $\times$ 5 V = 10 mW.

Propagation Delay   The basic TTL NAND gate has typical propagation delays of $t_{PLH} = 11$ ns and $t_{PHL} = 7$ ns, which is an *average* propagation delay of 9 ns.

Table 8.2 summarizes these characteristics of the standard 7400 series NAND gate.

TABLE 8.2   STANDARD 7400 SERIES NAND-GATE CHARACTERISTICS

Noise margins (worst-case)	$V_{NL} = V_{NH} = 400$ mV
Average power dissipation (basic gate)	$P_D = 10$ mW
Average propagation delay	9 ns
Typical fan-out	10

EXAMPLE 8.2

Refer to the data sheet for the 7400 quad two-input NAND IC in Appendix II. Determine the maximum average power dissipation and maximum average propagation delay of a *single* gate.

Solution: Look under DC AND AC CHARACTERISTICS for the maximum I_{CCH} and I_{CCL} values for the 54/74 series. The values are 8 mA and 22 mA, respectively. The average I_{CC} is therefore $(8 + 22)/2 = 15$ mA. The average power is obtained by multiplying by V_{CC}. The data sheet indicates that these I_{CC} values were obtained when V_{CC} was at its maximum value (5.25 V for the 7400 series). Thus we have

$$P_D(\text{avg}) = 15 \text{ mA} \times 5.25 \text{ V} = 78.75 \text{ mW}$$

as the power drawn by the *complete* IC. We can determine the power drain of one NAND gate by dividing this by four:

$$P_D(\text{avg}) = \textbf{19.7 mW per gate}$$

Since this average power drain was calculated using the maximum current and voltage values, it is the maximum average power that a 7400 NAND gate will draw under "worst-case" conditions. Designers often use worst-case values to ensure that their circuits will work under all conditions.

The maximum propagation delays for a 7400 NAND gate are listed as

$$t_{PLH} = 22 \text{ ns}, \qquad t_{PHL} = 15 \text{ ns}$$

so that the average propagation delay is

$$t_{pd}(\text{avg}) = \frac{22 + 15}{2} = \textbf{18.5 ns}$$

Again, this is a worst-case maximum possible average propagation delay.

8.4 OTHER TTL SERIES

The standard 7400 series ICs offer a combination of speed and power dissipation suited for many applications. ICs offered in this series include a wide variety of gates, flip-flops, and one-shots in the small-scale integration (SSI) line and shift registers, counters, decoders, memories, and arithmetic circuits in the medium-scale integration (MSI) line.

Several other TTL series have been developed since the introduction of the standard 7400 series. These other series provide a wide choice of speed and power characteristics. We will describe these series in the following paragraphs. Notice that whenever we use the term "TTL," we are usually referring to the standard 7400 series.

Low-Power TTL, 74L00 Series (L-TTL) Low-power TTL circuits designated as the 74L00 series have essentially the same basic circuit as the standard 7400 series except that all the resistor values are *increased*. The larger resistors reduce the power requirements but at the expense of longer propagation delays. A typical NAND gate in this series has an average power dissipation of 1 mW and an average propagation delay of 33 ns.

The 74L00 series is ideal for applications in which power dissipation is more critical than speed. Low-frequency, battery-operated circuits such as calculators are well suited for this TTL series. The term "L-TTL" will be used to refer to the low-power TTL series.

High-Speed TTL, 74H00 Series (H-TTL) The 74H00 series is a high-speed TTL series. The basic circuitry for this series is essentially the same as the standard 7400 series except that *smaller* resistor values are used and the emitter-follower transistor Q_4 is replaced by a Darlington pair. These differences result in a much faster switching speed with an average propagation delay of 6 ns. However, the increased speed is accomplished at the expense of increased power dissipation. The basic NAND gate in this series has an average P_D of 23 mW. This H-TTL series has essentially become obsolete since the development of the Schottky TTL series described in the following paragraphs.

Schottky TTL, 74S00 Series (S-TTL) The TTL, L-TTL, and H-TTL series all operate using saturated switching in which many of the transistors, when conducting, will be in the saturated condition. This operation causes a storage-time delay, t_S, when the transistors switch from ON to OFF, and limits the circuit's switching speed.

The 74S00 series reduces this storage-time delay by not allowing the transistor to go as deeply into saturation. It accomplishes this by using a Schottky barrier diode (SBD) connected between the base and collector of each transistor as shown in Figure 8.9(a). The SBD has a forward voltage of only 0.25 V. Thus, when the C-B junction becomes forward-biased at the onset of saturation, the SBD will conduct and divert some of the input current away from the base. This reduces the excess base current and decreases the storage-time delay at turn-OFF.

As shown in Figure 8.9(a), the transistor/SBD combination is given a special symbol. This symbol is used for all the transistors in the circuit diagram for the 74S00 NAND gate shown in Figure 8.9(b). This 74S00 NAND gate has an average propagation delay of only 3 ns, which is twice as fast as the 74H00. Note the presence of shunt diodes D_1 and D_2 to limit negative input voltages.

Circuits in the S-TTL series also use smaller resistor values to help improve switching times. This increases the circuit average power dissipation to about 23 mW, the same as for H-TTL. The S-TTL circuits also use a Darlington pair (Q_3 and Q_4) to provide a more rapid output rise time when switching from ON to OFF.

Thus, S-TTL has twice the speed of H-TTL at about the same power requirement. This is why H-TTL is becoming obsolete.

Low-Power Schottky TTL, 74LS00 Series (LS-TTL) This series is a lower-powered, slower-speed version of the 74S00 series. It uses the Schottky-clamped transistor, but with larger resistor values than the S-TTL series (see Figure 8.10). The larger resistor values reduce the circuit power requirement, but at the expense of an increase in switching times.

A NAND gate in the LS-TTL series will typically have an average propagation delay of 9.5 ns and an average power dissipation of 2 mW. Since it has about the same switching speed as the standard TTL series at a much lower power requirement, the LS-TTL series has gradually replaced the 7400 series in those applications where relatively high-speed operation is required at minimum power consumption. In other

FIGURE 8.9 (a) Schottky-clamped transistor; (b) basic NAND gate in S-TTL series. (Courtesy of Fairchild, a Schlumberger company)

words, LS-TTL is becoming the "mainstay" of the TTL family, and it can be found in almost all new TTL designs that do not require the maximum speed (where S-TTL is used) or the minimum power drain (where L-TTL is used).

Note that the 74LS00 NAND gate in Figure 8.10 does not use the multiple-emitter input transistor. It uses input diodes (D_1 and D_2) instead, but the basic circuit operation is the same as for the multiple-emitter inputs.

Comparison of TTL Series Some of the more important characteristics of the different TTL series are shown in Table 8.3. This table gives typical values that may vary slightly from one manufacturer to another. The manufacturer's data should always be consulted for specific characteristics of a given IC. Note that all the TTL

FIGURE 8.10   74LS00 NAND gate. (Courtesy of Fairchild, a Schlumberger company)

series use the same nominal +5 V for V_{CC}, but there is some variation in the output voltage limits, V_{OH}(min) and V_{OL}(max).

The last table entry represents the maximum clock frequency at which an average J-K FF can reliably operate.

TABLE 8.3   TYPICAL TTL SERIES CHARACTERISTICS

	Standard	L-TTL	H-TTL	S-TTL	LS-TTL
V_{CC}	←		5 V		→
V_{OH}(min)	2.4 V	2.4 V	2.4 V	2.7 V	2.7 V
V_{IH}(min)	2 V	2 V	2 V	2 V	2 V
V_{OL}(max)	0.4 V	0.3 V	0.4 V	0.5 V	0.5 V
V_{IL}(max)	0.8 V	0.8 V	0.8 V	0.8 V	0.8 V
Propagation delay	9 ns	33 ns	6 ns	3 ns	9.5 ns
Power required	10 mW	1 mW	23 mW	23 mW	2 mW
Max. clock rate	25 MHz	3 MHz	40 MHz	80 MHz	30 MHz

Sec. 8.4   Other TTL Series

EXAMPLE 8.3

Use Table 8.3 to calculate the dc noise margins for a typical LS-TTL IC. How does this compare to the standard TTL noise margins obtained in Section 8.3?

Solution:

$$V_{NH} = V_{OH}(\text{min}) - V_{IH}(\text{min})$$

$$= 2.7 \text{ V} - 2.0 \text{ V}$$

$$= \mathbf{0.7 \text{ V}}$$

as compared to $V_{NH} = 0.4$ V for standard TTL.

$$V_{NL} = V_{OL}(\text{max}) - V_{IL}(\text{max})$$

$$= 0.8 \text{ V} - 0.5 \text{ V}$$

$$= \mathbf{0.3 \text{ V}}$$

as compared to $V_{NL} = 0.4$ V for TTL.

REVIEW QUESTIONS

1. Which TTL series is best for high-speed operation?
2. Which series is best for low-speed battery-powered operation?
3. Which series uses a special type of diode to reduce switching time?
4. Which series has become obsolete?
5. Which series does not use multiple-emitter transistors?
6. Which series has the largest LOW-state noise margin? (*Ans.* L-TTL)
7. Which series has essentially replaced the standard 7400 series in all new designs?

8.5 TTL LOADING RULES

In designing digital systems using TTL devices it is important to know how to determine and use the fan-out or drive capability of each circuit. Figure 8.11(a) shows a single TTL output in the LOW state connected to several TTL inputs. Transistor Q_4 is ON and is acting as a current sink for all the currents (I_{IL}) coming back from each input. Although Q_4 is saturated, its ON-state resistance is some value other than zero, so the current I_{OL} produces an output voltage drop V_{OL}. The value of V_{OL} must not exceed $V_{OL}(\text{max})$ (Table 8.3), and this limits the value of I_{OL} and thus the number of loads that can be driven.

The HIGH-state situation is shown in Figure 8.11(b). Here Q_3 is acting as an emitter-follower and is sourcing (supplying) current to each TTL input. These currents (I_{IH}) are just reverse-bias leakage currents, since the TTL input emitter-base junctions or diodes are reverse-biased. If too many loads are driven, however, the total output current I_{OH} can become too large, causing larger drops across R_2, Q_3, and D_1, thereby lowering V_{OH} below $V_{OH}(\text{min})$.

FIGURE 8.11   TTL output-drive capabilities.

Unit Loads   In order to simplify designing with TTL circuits, the manufacturers have established standardized input and output loading factors in terms of *current*. These currents are called unit loads (UL) and are defined as follows:

$$1 \text{ unit load (UL)} = \begin{cases} 40 \ \mu\text{A in the HIGH state} \\ 1.6 \text{ mA in the LOW state} \end{cases}$$

These unit load factors actually represent the *maximum* input currents for the standard TTL series. In other words, the maximum current that flows into a standard TTL input in the HIGH state is I_{IH}(max) = 40 μA, and the maximum current that flows out of a standard TTL input in the LOW state is I_{OL}(max) = 1.6 mA. Even though the unit load factors are based on standard TTL, they are used to express the input requirements and output drive capabilities of all the TTL series.

Table 8.4 gives the typical input and output factors for the five TTL series. These values are typical, so there may be some variations, depending on the particular device or manufacturer. The device data sheet should be consulted to determine exact values.

TABLE 8.4

TTL Series	Input Loading		Fan-Out (UL)		
	High	Low	High	Low	
7400	1 UL	1 UL	10 UL	10 UL	
74H00	1.25 UL	1.25 UL	12.5 UL	12.5 UL	$1 \text{ UL} = \begin{cases} 40 \ \mu\text{A (HIGH)} \\ 1.6 \text{ mA (LOW)} \end{cases}$
74L00	0.5 UL	$\approx$0.1 UL*	10 UL	2.5 UL	
74S00	1.25 UL	1.24 UL	25 UL	12.5 UL	
74LS00	0.5 UL	$\approx$0.25 UL*	10 UL	5 UL	

*Actual I_{IL} for 74L00 series is 0.18 mA, slightly more than 0.1 UL, and actual I_{IL} for 74LS00 series is 0.36 mA, slightly less than 0.25 UL.

Note that the higher-speed series (74H00, 74S00) have larger input loading factors and fan-outs than the standard 7400 series. On the other hand, the lower power series (74L00, 74LS00) have lower input loading requirements and lower fan-outs.

EXAMPLE 8.4

Determine the input and output loading factors for the 7404 INVERTER chip (Appendix II).

Solution: This information can be found on the 7404 data sheet under the heading "INPUT LOADING/FAN-OUT." Look under "54/74(UL)" and you will find the first entry, 1.0/1.0. This means that the input requirements are 1 UL in each state. In other words, the 7404 input draws a maximum of 40 μA of current from the input signal source in the HIGH state, and sources a maximum of 1.6 mA back through a LOW input. These are the maximum worst-case currents that flow under extreme conditions. In practice, you might measure I_{IH} = 10 μA and I_{IL} = 1.1 mA at the input. Most designers assume the worst-case values that the manufacturer provides on the data sheets.

The second entry under "54/74(UL)" is 20/10. This means that the 7404 output is rated at 20 UL in the HIGH state and 10 UL in the LOW state. In other words, the 7404 output can *supply* up to

$$20 \times 40 \ \mu\text{A} = 800 \ \mu\text{A}$$

to load devices in the HIGH state without its V_{OH} dropping below V_{OH}(min) = 2.4 V, and it can *sink* up to

$$10 \times 1.6 \text{ mA} = 16 \text{ mA}$$

in the LOW state without its V_{OL} rising above V_{OL}(max) = 0.4 V.

EXAMPLE 8.5

Repeat for the 74LS04 IC.

Solution: Again, look under the heading "INPUT LOADING/FAN-OUT" for the 54/74LS entry, and we see that the inputs are rated at 0.5/0.25. Thus, the 74LS04 has an input loading factor of 0.5 UL in the HIGH state, and 0.25 UL in the LOW state.

The outputs are shown rated at 10/5.0, which means that a 74LS04 output can drive 10 UL in the HIGH state, and 5 UL in the LOW state.

EXAMPLE 8.6

The output of a 7404 INVERTER is providing the clock signal to a parallel register made up of 74107 J-K FFs. What is the maximum number of FFs that this clock signal can drive?

Solution: This type of problem is solved by first determining the unit load capabilities of a 7404 output and then determining the unit load input requirements of a 74107's *CLK* input. We already know the 7404's fan-out from Example 8.4 as 20 UL (HIGH) and 10 UL (LOW). The 74107 data sheet shows that its $\overline{CP}$ input requirements are 2 UL in both states. Thus, the number of 74107 $\overline{CP}$ loads that the 7404 output can drive is

$$\# \text{ of loads} = \frac{\text{output rating}}{\text{input rating}} = \frac{10 \text{ UL}}{2 \text{ UL}} = \mathbf{5}$$

Note that we used the 7404's LOW-state output rating because it is more restrictive than its HIGH-state output rating.

EXAMPLE 8.7

Repeat Example 8.6 using a 74LS04 and a 74LS107.

Solution: The 74LS04 has output ratings of 10 UL (HIGH) and 5 UL (LOW), while the 74LS107 clock inputs require 2 UL (HIGH) and 0.5 UL (LOW). Here it will be necessary to calculate the number of 74LS107 loads that a 74LS04 can drive in each state, and then take the smaller of the two numbers.

$$\# \text{ of loads (HIGH)} = \frac{10 \text{ UL}}{2 \text{ UL}} = 5$$

$$\# \text{ of loads (LOW)} = \frac{5 \text{ UL}}{0.5 \text{ UL}} = 10$$

Thus, a 74LS04 can drive **5** 74LS107 $\overline{CP}$ inputs.

REVIEW QUESTIONS

1. Describe what is meant by a unit load.
2. A certain TTL output has a fan-out of 5 UL in both states. How much current can it supply to loads in the HIGH state? (*Ans.* 200 µA)

3. What can happen if a TTL output is connected to more unit loads than its output rating specification?
4. How many 74107 $\overline{CP}$ inputs can be driven by a 74LS04 output? (*Ans. 2*)

8.6 OTHER TTL CHARACTERISTICS

Several other characteristics of TTL logic must be understood if one is to intelligently use TTL in a digital-system application.

Unconnected Inputs (Floating)   Any input to a TTL circuit that is left disconnected (open) acts exactly like a logical 1 applied to that input, because in either case the emitter-base junction or diode at the input will not be forward-biased. This means that on *any* TTL IC, *all* the inputs are 1s if they are not connected to some logic signal or to ground. When an input is left unconnected, it is said to be "floating."

Unused Inputs   Frequently, all the inputs on a TTL IC are not being used in a particular application. A common example is when all the inputs to a logic gate are not needed for the required logic function. For example, suppose that we needed the logic operation $\overline{AB}$ and we were using a chip that had a three-input NAND gate. The possible ways of accomplishing this are shown in Figure 8.12.

   In Figure 8.12(a) the unused input is left disconnected, which means that it acts as a logical 1. The NAND-gate output is therefore $x = \overline{A \cdot B \cdot 1} = \overline{A \cdot B}$, which is the desired result. Although the logic is correct, it is usually undesirable to leave an input disconnected because it will act like an antenna, which is liable to pick up stray radiated signals that could cause the gate to operate improperly. A better technique is shown in Figure 8.12(b). Here the unused input is connected to +5 V through a 1-kΩ resistor, so the logic level is a 1. The 1-kΩ resistor is simply for current protection of the emitter-base junctions of the gate inputs in case of spikes on the power-supply line. This same technique can be used for AND gates, since a 1 on an unused input will not affect the output. As many as 30 unused inputs can share the same 1-kΩ resistor tied to V_{CC}.

   A third possibility is shown in Figure 8.12(c), where the unused input is tied to a used input. This is satisfactory provided that the circuit driving input B is not going to have its fan-out exceeded. This technique can be used for *any* type of gate.

   For OR gates and NOR gates, the unused inputs cannot be left disconnected or

FIGURE 8.12   Three ways to handle unused logic inputs.

tied to +5 V since this would produce a constant-output logic level (1 for OR, 0 for NOR) regardless of the other inputs. Instead, for these gates the unused inputs must either be connected to ground (0 V) for a logic 0 or they can be tied to a used input as in Figure 8.12(c).

Tied-Together Inputs   When two or more TTL-gate inputs are connected together to form a common input as in Figure 8.12(c), the common input will generally have an input loading factor that is the sum of the input loading factors for each input. The only exception is for NAND and AND gates. For these gates, the LOW-state input loading factor will be the *same as a single input* no matter how many inputs are tied together.

To illustrate, assume that each input of the three-input NAND gate in Figure 8.12(c) is rated at 1 UL in each state. The common B input will therefore present an input loading factor of 2 UL in the HIGH state but only 1 UL in the LOW state. The same would be true if this were an AND gate. If it were an OR or NOR gate, the common B input would present an input loading factor of 2 UL in both states.

The reason for this characteristic can be found by looking back at the circuit diagram of the TTL NAND gate in Figure 8.6(b). The current I_{IL} is limited by the resistance R_1. Even if inputs A and B were tied together and grounded, this current would not change; it would merely divide up and flow through the parallel paths provided by diodes D_2 and D_3. The situation is different for OR and NOR gates, since they do not use multiple-emitter transistors, but rather have a separate input transistor for each input.

EXAMPLE 8.8

Determine the number of ULs that the X output is driving in Figure 8.13. Assume each gate input is rated at 1 UL in each state.

Loading on gate 1 output	
High state	Low state
2 UL – gate 2	1 UL – gate 2
1 UL – gate 3	1 UL – gate 3
3 UL – gate 4	3 UL – gate 4
6 UL – total	5 UL – total

FIGURE 8.13   Example 8.8

Solution: The loading on X will be different for its two different logic states as shown in the figure. Note that the tied-together inputs of NAND gate 2 are counted as only 1 UL in the LOW state. The NOR-gate inputs are treated as separate loads for either state.

Biasing TTL Inputs to 0   Occasionally, the situation arises where a TTL input must be held normally LOW and then caused to go HIGH by the actuation of a mechanical switch. This is illustrated in Figure 8.14 for the input to a one-shot. This OS triggers on a positive transition that occurs when the switch is momentarily closed. The resistor R serves to keep the T input LOW while the switch is open. Care must be taken to keep the value of R low enough so that the voltage developed across it by the current I_{IL} that flows out of the OS input to ground will not exceed $V_{IL}(\text{max})$. Thus, the largest value of R is given by

$$I_{IL} \times R_{\text{max}} = V_{IL}(\text{max})$$

$$R_{\text{max}} = \frac{V_{IL}(\text{max})}{I_{IL}}$$

(8.3)

R must be kept below this value to ensure that the OS input will be at an acceptable LOW level while the switch is open. The minimum value of R is determined by the current drain on the 5-V supply when the switch is closed. In practice, this current drain should be minimized by keeping R slightly below R_{max}.

FIGURE 8.14

EXAMPLE 8.9

Determine an acceptable value for R if the OS is a standard TTL IC with an input rating of 1 UL.

Solution: At 1 UL, the value of I_{IL} will be a maximum of 1.6 mA. This maximum value should be used to calculate R_{max}. For standard TTL, $V_{IL}(\text{max}) = 0.8$ V. Thus, we have

$$R_{\text{max}} = \frac{0.8 \text{ V}}{1.6 \text{ mA}} = 500 \ \Omega$$

A good choice here would be $R = 470 \ \Omega$, a standard resistor value.

EXAMPLE 8.10

Repeat Example 8.8 if the OS belongs to the LS-TTL series with an input rating of 0.25 UL.

Solution: At 0.25 UL, we have $I_{IL}(\text{max}) = 0.4$ mA. For LS-TTL, $V_{IL}(\text{max}) = 0.8$ V. Thus,

$$R_{max} = \frac{0.8 \text{ V}}{0.4 \text{ ma}} = 2 \text{ k}\Omega$$

A good choice here would be 1.8 kΩ, a standard value.

Transition Times The input signals that drive TTL circuits must have relatively fast transitions for reliable operation. Rise and fall times that exceed 50 ns can cause oscillations at the output of logic gates and INVERTERs, as shown in Figure 8.15(a), and erratic triggering of FFs and one-shots.

A slow-changing signal can be sharpened up by passing it through a *Schmitt-trigger* circuit. A Schmitt-trigger circuit produces very fast output transitions (typically 10 ns) independent of the input transition times. Several TTL ICs include Schmitt-trigger circuitry. One is the 7414, which is a hex INVERTER chip that contains six INVERTERs. Its operation is illustrated in Figure 8.15(b). Note the special Schmitt-trigger symbol. Another one is the 7413 dual four-input NAND chip.

If an IC is classified as having Schmitt-trigger operation, it will respond reliably to slow-changing signals; otherwise, it will not. Designers use a Schmitt-trigger IC to convert a slow-changing signal to a fast-changing signal that can reliably drive any IC.

Current Transients TTL logic circuits suffer from internally generated current transients or spikes because of the totem-pole output structure. When the output is switching from the LOW state to the HIGH state (see Figure 8.16), the two output transistors are changing states; Q_3 OFF to ON and Q_4 ON to OFF. Since Q_4 is changing from the saturated condition, it will take longer than Q_3 to switch states. Thus, there is a short interval of time (about 2 ns) during the switching transition where both transistors are conducting and a relatively large surge of current (30–50 mA) is drawn from the +5-V supply. The duration of this current transient is extended by the effects of any load capacitance on the circuit output. This capacitance consists of stray wiring

t_R and $t_F > 50$ ns

(a)

7414

$t_R = t_F \approx 10$ ns

(b)

FIGURE 8.15 (a) Slow t_R and t_F cause TTL outputs to oscillate; (b) Schmitt-trigger IC used to reduce t_R and t_F.

Sec. 8.6   **Other TTL Characteristics**

311

FIGURE 8.16 A large current spike is drawn from V_{CC} when a totem-pole output switches from LOW to HIGH.

capacitance and the input capacitance of any load circuits and must be charged up to the HIGH-state output voltage. This overall effect can be summarized as follows: *Whenever a totem-pole TTL output goes from LOW to HIGH, a high-amplitude current spike is drawn from the V_{CC} supply.*

In a complex digital circuit or system there may be many TTL outputs switching states at the same time, each one drawing a narrow spike of current from the power supply. The accumulative effect of all these current spikes will be to produce a voltage spike on the common V_{CC} line, mostly due to the distributed inductance on the supply line [remember: $V = L(di/dt)$ for inductance and di/dt is very large for a 2-ns current spike]. This voltage spike can cause serious malfunctions during switching transitions unless some type of filtering is used. The most common technique uses small *RF* capacitors connected from V_{CC} to GROUND to essentially "short out" these high-frequency spikes. This is called *power-supply decoupling*.

It is standard practice to connect a 0.01-μF or 0.1-μF low-inductance, ceramic disk capacitor between V_{CC} and ground near each TTL IC on a circuit board. The capacitor leads are kept very short to minimize inductance.

In addition, it is standard practice to connect a single large capacitor (2 to 20 μF) between V_{CC} and ground on each board to filter out possible variations in V_{CC} caused by the large changes in I_{CC} levels as outputs switch states.

REVIEW QUESTIONS

1. What will be the output of a TTL NAND gate that has all its inputs unconnected?
2. What are two acceptable ways to handle unused inputs to an AND gate?
3. Repeat for a NOR gate.
4. True or false: When NAND-gate inputs are tied together, they are always treated as a single load on the signal source.

5. True or false: When NOR-gate inputs are tied together, the input loading factors are added.
6. Explain why there is an upper limit on the resistor value that is used to keep a TTL input LOW.
7. How does a Schmitt-trigger IC differ from a conventional IC?
8. Describe how TTL circuits can cause spikes on the V_{CC} line, and what can be done to minimize the spikes.

8.7 TTL OPEN-COLLECTOR OUTPUTS

Consider the logic circuit of Figure 8.17(a). NAND gates 4 and 5 provide the AND function, which is ANDing the outputs of NAND gates 1, 2, and 3, so the final output X has the expression

$$X = \overline{AB} \cdot \overline{CD} \cdot \overline{EF}$$

The circuit of Figure 8.17(b) shows the same logic operation obtained by simply tying together the outputs of NAND gates 1, 2, and 3. In other words, the AND operation is performed by tying the outputs together. This can be reasoned as follows: With all outputs tied together, when any one of the gate outputs goes to the LOW state, the common output point must go LOW as a result of the "shorting-to-ground" action of

FIGURE 8.17 Wired-AND operation.

the Q_4 transistor in that gate. The common output point will be HIGH only when all gate outputs are in the HIGH state. Clearly, this is the AND operation.

The arrangement in Figure 8.17(b) has an advantage over the conventional arrangement of Figure 8.17(a). It requires fewer gates to produce the desired output. This configuration is called the *wired-AND* operation because it produces the AND operation by connecting output wires together. It is sometimes misleadingly called the *wired-OR* operation.

Totem-Pole Outputs Cannot Be Wired-ANDed In order to take advantage of the wired-AND configuration the outputs of two or more gates must be tied together without harmful effects. Unfortunately, the totem-pole output circuitry of conventional TTL circuits prohibits tying outputs together. This is illustrated in Figure 8.18, where the totem-pole outputs of two separate gates are connected together at point X. Suppose that the gate-A output is in the HIGH state (Q_{3A} ON, Q_{4A} OFF) and the gate-B output is in the LOW state (Q_{3B} OFF, Q_{4B} ON). In this situation Q_{4B} is a very low resistance load on Q_{3A} and will draw a current which can go as high as 55 mA. This current can easily damage Q_{4B}, which is usually guaranteed to sink only 16 mA (I_{OL}). The situation is even worse when more than two TTL outputs are tied together.

Sometimes totem-pole outputs are unintentionally tied together because of wiring errors or accidental shorting on a printed-circuit board. When this happens, the signal at the common point will usually be the AND combination of the two output

FIGURE 8.18 Totem-pole outputs tied together can produce damaging current through Q_4.

signals that have been shorted together. This is true only of TTL circuits; MOS and CMOS behave differently when outputs are tied together.

Open-Collector Outputs   To permit wire-ANDing, some TTL circuits are designed with *open-collector* outputs. As shown in Figure 8.19(a) the open-collector-type circuit eliminates Q_3, D_1, and R_4. The output is taken at Q_4's collector, which is unconnected. In the output LOW state, Q_4 is ON (has base current), and in the HIGH state it is OFF (essentially an open circuit). For proper operation an external *pull-up* collector resistor R_c should be connected as shown in Figure 8.19(b), so a high voltage level will appear at the output in the HIGH state.

With open-collector outputs, the wired-AND operation can be accomplished safely. Figure 8.20 shows three two-input *open-collector* NAND gates (7401s) that are wired-ANDed together. Notice that the open-collector NAND gates have no special symbol. Also notice the presence of the external pull-up resistor (R_c) and the sometimes-used wired-AND symbol.

Value of R_c The value of R_c must be chosen so that when one gate output goes LOW while the others are HIGH, the sink current through the LOW output does not exceed its I_{OL} limit (usually 16 mA). A value of $R_c = 1$ kΩ will produce a sink current of 5 mA through the LOW gate's output transistor. Of course, the output is usually driving other TTL loads that will add to this sink current, whose total must not exceed I_{OL}(max). It might appear that a large value of R_c would therefore be advisable.

(a)   (b)

FIGURE 8.19   (a) Open-collector TTL circuit; (b) with external pull-up resistor.

FIGURE 8.20   Wired-AND operation using open-collector gates.

However, it must be realized that any load capacitance will be charged up through R_c, so that R_c should be made as small as possible to enhance switching speed. Even with R_c minimized, this open-collector arrangement is much slower than totem-pole TTL outputs, where Q_3 acts as a low-impedance emitter-follower to charge up load capacitance. For this reason, open-collector circuits should not be used in applications where switching speed is a principal consideration.

EXAMPLE 8.11

The 7405 IC (see Appendix II) contains six inverters with open-collector outputs. These six inverters are connected in a wired-AND arrangement in Figure 8.21(a).
(a) Determine the logic expression for output X.
(b) Determine a value for R_c assuming that output X is to drive other circuits with a total loading factor of 4 UL.

Solution:
(a) Each inverter output is the inverse of its input. The wired-AND connection simply ANDs each inverter output. Thus,

$$X = \bar{A} \cdot \bar{B} \cdot \bar{C} \cdot \bar{D} \cdot \bar{E} \cdot \bar{F}$$

Using DeMorgan's theorem, this is equivalent to

$$A = \overline{A + B + C + D + E + F}$$

which is the NOR operation.
(b) If we assume that only one of the INVERTERs has a HIGH input, then the

FIGURE 8.21   Example 8.11

output transistor of that INVERTER has to be able to sink the currents I_{RC} and I_{IL} as shown in Figure 8.21(b). Referring to the data sheet for the 7405, we find that it has a fan-out of 10 in the LOW state. Thus, it can sink a total current $I_{OL}(\text{max}) = 16$ mA. Summing the currents at X, we have

$$I_{OL}(\text{max}) = I_{RC} + I_{IL}$$

$$16 \text{ mA} = I_{RC} + 6.4 \text{ mA}$$

$$\therefore I_{RC} = 9.6 \text{ mA}$$

The voltage across R_c will be $V_{CC} - V_{OL}(\text{max})$; we use $V_{OL}(\text{max})$ since the transistor is sinking its maximum current. For standard TTL, we know that $V_{OL}(\text{max}) = 0.4$ V. Thus, we have

$$R_c(\text{min}) = \frac{V_{CC} - V_{OL}(\text{max})}{I_{RC}} = \frac{5 \text{ V} - 0.4 \text{ V}}{9.6 \text{ mA}} = 480 \ \Omega$$

In practice, we would use a value slightly larger than this to be on the safe side. Remember, we want to keep R_c as small as possible to avoid increasing the switching time. A good choice would be the standard value of 560 Ω.

EXAMPLE 8.12

A technician breadboards the circuit of Figure 8.21 with $R_c = 10$ kΩ. He figures that since the circuit is operating at a very low frequency, there is no need to worry about the decrease in switching speed produced by increasing R_c. Besides, he wants to minimize the current drawn from V_{CC}.

He tests the circuit and finds that the voltage at point X is slightly less than 2 V in the HIGH state. He replaces the 7405 chip and observes the same result. What is causing V_{CC} to be so low?

Solution: In order for X to go HIGH, each INVERTER input has to be LOW. This will turn OFF the output transistor of each open-collector output. Unfortunately, these transistors are not ideal, and they do have some leakage current in the OFF state. This leakage current, I_{OH}, is specified as a *maximum* of 250 μA for the standard TTL series, and 100 μA for LS-TTL. Under *normal* conditions it will usually be about 50 μA for TTL and 20 μA for LS-TTL. These leakage currents, however, are additive, so the total leakage current of the six output transistors can be substantial. This current has to be supplied by V_{CC} through R_c. In addition V_{CC} has to supply the I_{IH} for each of the four ULs. As a result, there may typically be anywhere from 300 to 400 μA flowing from V_{CC} through R_c. With $R_c = 10$ kΩ, it is not surprising that around 3 V is dropped across R_c, leaving only about 2 V at point X.

Because the open-collector leakage currents and the I_{IH} of the loads has to flow through R_c, it is best to keep R_c as close as possible to the minimum value that is calculated based on the open-collector output's maximum sink current, I_{OL}(max).

Open-Collector Buffer/Drivers   Any logic circuit that is called a *buffer*, a *driver*, or a *buffer/driver* is designed to have a greater output current and/or voltage capability than an ordinary logic circuit. Buffer/driver ICs are available with totem-pole outputs and with open-collector outputs.

The 7406 is a popular open-collector buffer/driver IC that contains six INVERTERs with open-collector outputs that can sink up to 40 mA in the LOW state. In addition, the 7406 can handle output voltages up to 30 V. This means that the output transistor can be connected to a voltage greater than 5 V.

This is illustrated in Figure 8.22, where a 7406 is used as a buffer between a 74107 FF and an incandescent indicator lamp that is rated at 24 V, 25 mA. The 7406 controls the lamp's ON/OFF status to indicate the state of FF output Q. Note that the lamp is powered from +24 V, and it acts as the pull-up resistor for the open-collector output.

When $Q = 1$, the 7406 output goes LOW and its output transistor sinks the 25 mA of lamp current supplied by the 24-V source, and the lamp is ON. When $Q = 0$, the 7406 output transistor turns OFF; there is no path for current and the lamp turns OFF. In this state, the full 24 V will appear across the OFF output transistor so that $V_{OH} = 24$ V, which is lower than the 7406 maximum V_{OH} rating.

FIGURE 8.22 An open-collector buffer/driver drives a high-current, high-voltage load.

REVIEW QUESTIONS

1. Why can't the outputs of ordinary (totem-pole) TTL ICs be connected together in the wired-AND configuration?
2. What is an advantage of using open-collector outputs?
3. What is a disadvantage?
4. Why is it best to keep R_c as small as possible?

8.8 TRISTATE TTL

This is a third type of TTL output configuration. It utilizes the high-speed operation of the totem-pole arrangement while permitting outputs to be paralleled (connected together). It is called tristate TTL because it allows *three* possible output states: HIGH, LOW, and high-impedance (Hi-Z). The Hi-Z state is a condition where both transistors in the totem-pole arrangement are turned OFF so that the output terminal is a high impedance to ground and to V_{CC}. In other words, the output is an open or floating terminal that is neither a LOW nor a HIGH. In practice, the output terminal is not an exact open circuit, but has a resistance of several megohms or more relative to ground and V_{CC}.

The tristate operation is obtained by modifying the basic totem-pole circuit. Figure 8.23(a) shows the circuit for a tristate INVERTER where the portion enclosed in dotted lines has been added to the basic circuit. The circuit has two inputs; A is the normal logic input, E is an ENABLE input that can produce the Hi-Z state. We will examine the operation for both states of E.

The Enabled State   With $E = 1$ the circuit operates as a normal INVERTER because the HIGH voltage at E has no effect on Q_1 or D_2. In this enabled condition, the output is simply the inverse of logic input A.

The Disabled State (Hi-Z)   When $E = 0$ the circuit goes into its Hi-Z state regardless of the state of logic input A. The LOW at E forward-biases the emitter-base

(a)

(b)

Operating conditions	
ENABLE (E)	**OUTPUT**
HIGH	<u>Enabled:</u> operates as an INVERTER A ▷○ X
LOW	<u>Disabled:</u> output is Hi-Z state. Input A has no effect A ▷○- X

(c)

FIGURE 8.23 Tristate TTL INVERTER.

junction of Q_1 and shunts the R_1 current away from Q_2 so that Q_2 turns OFF, which turns Q_4 OFF. The LOW at E also forward-biases diode D_2 to shunt current away from the base of Q_3, so that Q_3 also turns OFF.

    With both totem-pole transistors in the nonconducting state, the output terminal is essentially an open circuit. This is shown symbolically in the table of Figure 8.23(c).

    The logic symbol for the tristate INVERTER is shown in Figure 8.23(b). Note where the ENABLE input is placed on the INVERTER symbol. Also note that E is active-HIGH; that is, the INVERTER is enabled when $E = 1$.

Advantage of Tristate The outputs of tristate ICs can be connected together (paralleled) without sacrificing switching speed. This is because a tristate output, when enabled, operates as a totem-pole output with its associated low-impedance, high-speed characteristic. It is important to realize, however, that when tristate outputs are paralleled, only one of them should be enabled at one time. Otherwise, two active totem-pole outputs would be connected together and damaging currents could flow. We will elaborate on this in our discussion of tristate buffers.

Tristate Buffers A *tristate buffer* is a circuit that is used to control the passage of a logic signal from input to output. Some tristate buffers also invert the signal as it goes through. The circuit in Figure 8.23 can be called an *inverting tristate buffer*.

   Two of the most commonly used tristate buffer ICs are the 74125 and 74126. Both contain four noninverting tristate buffers like those shown in Figure 8.24. The 74125 and 74126 differ only in the active state of their ENABLE inputs. The 74125 allows the input signal A to reach the output when $E = 0$, while the 74126 passes the input when $E = 1$.

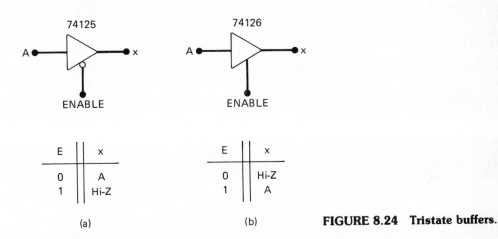

FIGURE 8.24   Tristate buffers.

   Tristate buffers have many applications in circuits where several signals are connected to common lines (buses). We will examine some of these applications in Chapter 9, but we can get the basic idea from Figure 8.25. Here we have three logic signals A, B, and C connected to a common bus line through 74126 tristate buffers. This arrangement permits us to transmit any one of the signals over the bus line to other circuits by enabling the appropriate buffer.

   For example, if we make $E_B = 1$ and $E_A = E_C = 0$, the B signal will pass to the output of its tristate buffer and will appear on the bus. The other buffers will be in their Hi-Z state so their outputs will be essentially disconnected from the bus. As stated earlier, we should never enable two tristate outputs at the same time because this would connect two totem-pole outputs together and could damage the ICs. Even if damage did not occur, this situation would produce on the bus a signal that is a combination of the two signals. This is commonly called *bus contention*.

Sec. 8.8   Tristate TTL   **321**

FIGURE 8.25 Tristate buffers used to connect several signals to a common bus.

A — 74126 — E_A

B — 74126 — E_B — Common bus

C — 74126 — E_C — To other circuits

MOS Tristate Tristate ICs are also available in the MOS and CMOS logic families. In fact, the development of MOS tristate circuits was a prime reason for the rapid development of microcomputer systems. We will not discuss tristate operation when we discuss the MOS and CMOS family characteristics because it is basically the same as described for tristate TTL.

REVIEW QUESTIONS

1. What are the three states of a tristate logic circuit?
2. What happens when a tristate INVERTER is disabled?
3. Why shouldn't more than one tristate output be enabled at one time?
4. What conditions are necessary to transmit input *C* to the bus in Figure 8.25?

8.9 THE ECL DIGITAL IC FAMILY

The TTL family uses transistors operating in the saturated mode. As a result, their switching speed is limited by the storage delay time associated with a transistor that is driven into saturation. Another *bipolar* logic family has been developed that prevents transistor saturation, thereby increasing overall switching speed. This logic family is called *emitter-coupled logic* (ECL), and it operates on the principle of current switching whereby a fixed bias current less than $I_{C(sat)}$ is switched from one transistor's collector to another. Because of this current-mode operation, this logic form is also referred to as *current-mode logic* (CML).

Basic ECL Circuit The basic circuit for emitter-coupled logic is essentially the differential amplifier configuration of Figure 8.26(a). The V_{EE} supply produces an essentially fixed current I_E which remains around 3 mA during normal operation. This

V_{IN}	Outputs	
−1.7 V (logic 0)	$V_{C1} = 0$ V $V_{C2} = -0.9$ V	Q_2 conducts
−0.8 V (logic 1)	$V_{C1} = -0.9$ V $V_{C2} = 0$ V	Q_1 conducts

(a)

Input/output
logic levels
"0" = −1.7 V
"1" = −0.8 V

(b)

FIGURE 8.26   (a) Basic ECL circuit; (b) with addition of emitter-followers.

Sec. 8.9   **The ECL Digital IC Family**   **323** current is allowed to flow through either Q_1 or Q_2, depending on the voltage level at V_{IN}. In other words, this current will switch between Q_1's collector and Q_2's collector as V_{IN} switches between its two logic levels of -1.7 V (logical 0 for ECL) and -0.8 V (logical 1 for ECL). The table in Figure 8.26(a) shows the resulting output voltages for these two conditions at V_{IN}. Two important points should be noted: (1) V_{C1} and V_{C2} are the *complements* of each other, and (2) the output voltage levels are not the same as the input logic levels.

The second point noted above is easily taken care of by connecting V_{C1} and V_{C2} to emitter-follower stages (Q_3 and Q_4), as shown in Figure 8.26(b). The emitter-followers perform two functions: (1) they subtract approximately 0.8 V from V_{C1} and V_{C2} to shift the output levels to the correct ECL logic levels, and (2) they provide a very low output impedance (typically 7 Ω), which provides for large fan-out and fast charging of load capacitance. This circuit produces two complementary outputs: V_{OUT1}, which equals $\overline{V_{IN}}$, and V_{OUT2}, which is equal to V_{IN}.

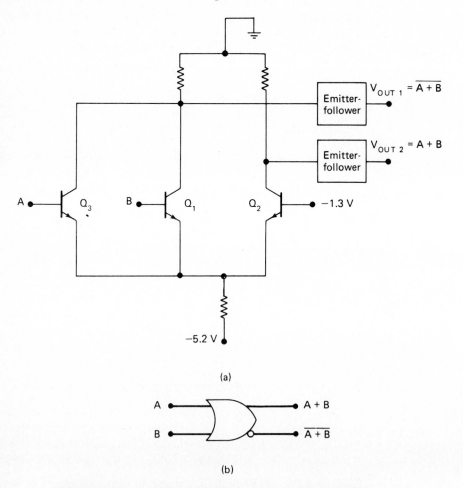

(a)

(b)

FIGURE 8.27 (a) ECL NOR/OR circuit; (b) logic symbol.

ECL OR/NOR Gate   The basic ECL circuit of Figure 8.26(b) can be used as an inverter if the output is taken at V_{OUT1}. This basic circuit can be expanded to more than one input by paralleling transistor Q_1 with other transistors for the other inputs, as in Figure 8.27(a). Here either Q_1 or Q_3 can cause the current to be switched out of Q_2, resulting in the two outputs V_{OUT1} and V_{OUT2} being the logical NOR and OR operations, respectively. This OR/NOR gate is symbolized in Figure 8.27(b) and is the fundamental ECL gate.

ECL Characteristics   The following are the most important characteristics of the ECL family of logic circuits:

1. The transistors never saturate, so switching speed is very high. Typical propagation delay time is 2 ns, which makes ECL a little faster than Schottky TTL (74S00 series).
2. The logic levels are nominally -0.8 V and -1.70 V for the logical 1 and 0, respectively.
3. Worst-case ECL noise margins are approximately 250 mV. These low noise margins make ECL somewhat unreliable for use in heavy industrial environments.
4. An ECL logic block usually produces an output and its complement. This eliminates the need for inverters.
5. Fan-outs are typically around 25, owing to the low-impedance emitter-follower outputs.
6. Typical power dissipation for a basic ECL gate is 25 mW, just slightly higher than Schottky TTL.
7. The total current flow in an ECL circuit remains relatively constant regardless of its logic state. This helps to maintain an unvarying current drain on the circuit power supply even during switching transitions. Thus, no noise spikes will be internally generated like those produced by TTL totem-pole circuits.

Table 8.5 shows how ECL compares to the TTL logic families.

TABLE 8.5

Logic family	t_{pd} (ns)	P_D (mW)	Worst-case noise margin (mV)	Maximum clock rate (MHz)
7400	9	10	400	25
74L00	33	1	400	3
74H00	6	23	400	40
74S00	3	23	300 (V_{NL})	80
74LS00	9.5	2	300 (V_{NL})	30
ECL	2	25	250	120

The ECL family is not as widely used as the TTL and MOS families except in very high frequency applications where its speed is superior. Its relatively low noise margins and high power drain are disadvantages compared to other logic families. Another drawback is its negative supply voltage and logic levels, which are not compatible with the other logic families; this makes it difficult to use ECL in conjunction with TTL and MOS circuits.

1. True or false
   (a) ECL obtains high-speed operation by preventing transistor saturation.
   (b) ECL circuits have two outputs.
   (c) The noise margins for ECL circuits are larger than TTL noise margins.
   (d) ECL circuits do not generate noise spikes during state transitions.
   (e) ECL devices require less power than standard TTL.

8.10 MOS DIGITAL INTEGRATED CIRCUITS

MOS (metal-oxide-semiconductor) technology derives its name from the basic MOS structure of a metal electrode over an oxide insulator over a semiconductor substrate. The transistors of MOS technology are field-effect transistors called MOSFETs. Most of the MOS digital ICs are constructed entirely of MOSFETs and no other components.

   The chief advantages of the MOSFET are that it is relatively simple and inexpensive to fabricate, it is small, and it consumes very little power. The fabrication of MOS ICs is approximately one-third as complex as the fabrication of bipolar ICs (TTL, ECL, etc.). In addition, MOS devices occupy much less space on a chip than bipolar transistors; typically, a MOSFET requires 1 square mil of chip area while a bipolar transistor requires about 50 square mils. More importantly, MOS digital ICs normally do not use the IC resistor elements, which take up so much of the chip area of bipolar ICs.

   All of this means that MOS ICs can accommodate a much larger number of circuit elements on a single chip than bipolar ICs. This advantage is evidenced by the fact that MOS ICs are surpassing bipolar ICs in the area of large-scale integration (LSI). The high packing density of MOS ICs results in a greater system reliability because of the reduction in the number of necessary external connections.

   The principal disadvantage of MOS ICs is their relatively slow operating speed when compared to the bipolar IC families. In many applications this is not a prime consideration, so MOS logic offers an often superior alternative to bipolar logic. We will examine the MOS logic families after a brief discussion of MOSFETs.

8.11 THE MOSFET

There are presently two general types of MOSFETs: *depletion* and *enhancement*. MOS digital ICs use enhancement MOSFETs exclusively, so only this type will be considered in the following discussion. Furthermore, we will concern ourselves only with the operation of these MOSFETs as ON/OFF switches.

   Figure 8.28 shows the schematic symbols for the *N*-channel and *P*-channel enhancement MOSFETs, where the direction of the arrow indicates either *P*- or *N*-channel. The symbols show a broken line between the *source* and *drain* to indicate that there is *normally* no conducting channel between these electrodes. The symbol also shows a separation between the *gate* and the other terminals to indicate the very high resistance (typically greater than 10,000 MΩ) between the gate and channel.

FIGURE 8.28　Schematic symbols for enhancement MOSFETs.

Basic MOSFET Switch　Figure 8.29 shows the switching operation of an *N*-channel MOSFET. For the *N*-channel device the drain is always biased positive relative to the source. Note also that the substrate is connected to the source, which is most often the case. The gate-to-source voltage V_{GS} is the input voltage, which is used to control the resistance between drain and source (i.e., the channel resistance) and therefore determines whether the device is ON or OFF.

　　When $V_{GS} = 0$ V, there is no conductive channel between source and drain, and the device is OFF. Typically the channel resistance in this OFF state is 10^{10} Ω, which for most purposes is an *open circuit*. The MOSFET will remain OFF as long as V_{GS} is zero or negative. As V_{GS} is made positive (gate positive relative to source), a threshold voltage (V_T) is reached, at which point a conductive channel begins to form between source and drain. Typically $V_T = +1.5$ V for N-MOSFET,* so any $V_{GS} \geq 1.5$ V will cause the MOSFET to conduct. Generally, a value of V_{GS} much larger than V_T is used to turn ON the MOSFET more completely. As shown in Figure 8.29(b), when $V_{GS} = +5$ V, the channel resistance between source and drain has dropped to a value of $R_{ON} = 1000$ Ω.

FIGURE 8.29　*N*-channel MOSFET switching states.

*The newer N-MOSFETs use a special silicon gate to produce lower V_T values. These devices are called low-threshold MOSFETs.

In essence, then, the N-MOSFET will switch from a very high resistance to a low resistance as the gate voltage switches from a LOW voltage to a HIGH voltage. It is helpful simply to think of the MOSFET as a switch that is either opened or closed between source and drain.

The *P*-channel MOSFET operates in exactly the same manner as the *N*-channel except that it uses voltages of the opposite polarity. For P-MOSFETs the drain is connected to a voltage source V_{DD}, which is negative relative to the source. To turn the P-MOSFET ON, a negative voltage that exceeds V_T must be applied to the gate. Table 8.6 summarizes the *P*- and *N*-channel switching characteristics.

TABLE 8.6

	Drain-to-source bias	Gate-to-source voltage (V_{GS}) needed for conduction	R_{ON}	R_{OFF}
P-channel	Negative	Negative and greater than V_T	1000 Ω (typical)	10^{10} Ω
N-channel	Positive	Positive and greater than V_T	1000 Ω (typical)	10^{10} Ω

8.12 DIGITAL MOSFET CIRCUITS

Digital circuits employing MOSFETs are broken down into three categories: (1) P-MOS, which uses *only P*-channel enhancement MOSFETs; (2) N-MOS, which uses *only N*-channel enhancement MOSFETs; and (3) CMOS (complementary MOS), which uses both *P*- and *N*-channel devices.

P-MOS and N-MOS digital ICs have a greater packing density (more transistors per chip) and are therefore more economical than CMOS. N-MOS has about twice the packing density of P-MOS. In addition, N-MOS is also about twice as fast as P-MOS, owing to the fact that free electrons are the current carrier in N-MOS while holes (slower-moving positive charges) are the current carriers for P-MOS. CMOS has the greatest complexity and lowest packing density of the MOS families, but it possesses the important advantages of higher speed and much lower power dissipation.

In this section we will look at some of the basic N-MOS logic circuits, keeping in mind that the P-MOS circuits would be the same except for the voltage polarities. Since P-MOS and N-MOS find their widest applications in LSI (microprocessors, memories, ROMs, etc.), we will defer any applications of these families until later. CMOS, which, like TTL, is widely used in MSI applications, will be covered in more detail beginning with Section 8.14.

N-MOS INVERTER   Figure 8.30 shows the basic N-MOS INVERTER circuit. It contains two *N*-channel MOSFETs: Q_1 is called the *load* MOSFET and Q_2 the *switching* MOSFET. Q_1 has its gate *permanently* connected to +5 V, so it is *always* in the ON state and essentially acts as a load resistor of value R_{ON}. Q_2 will switch from ON to OFF in response to V_{IN}. The Q_1 MOSFET is designed to have a narrower channel

V_{IN}	Q_1	Q_2	$V_{OUT} = \overline{V_{IN}}$
0 V (logic 0)	$R_{ON} =$ 100 kΩ	$R_{OFF} =$ 10^{10} Ω	+5 V (logic 1)
+5 V (logic 1)	$R_{ON} =$ 100 kΩ	$R_{ON} =$ 1 kΩ	+0.05 V (logic 0)

(a)   (b)

FIGURE 8.30 N-MOS INVERTER.

than Q_2, so Q_1's R_{ON} is much greater than Q_2's. Typically, R_{ON} for Q_1 is 100 kΩ and R_{ON} for Q_2 is 1 kΩ. R_{OFF} for Q_2 is usually around 10^{10} Ω.

The two states of the INVERTER are summarized in Figure 8.30(b). The best way to analyze this circuit is to consider each MOSFET channel as a resistance so that the output voltage is taken from a voltage divider formed by the two resistances. With V_{IN} = 0 V, transistor Q_2 is OFF, with a very large resistance of 10^{10} Ω. Since Q_1 has R_{ON} = 100 kΩ, the voltage-divider output will be essentially +5 V. With V_{IN} = +5 V, Q_2 is ON, with R_{ON} = 1 kΩ. The voltage divider is now 100 kΩ and 1 kΩ, so V_{OUT} = 1/101 × (+5 V) ≈ 0.05 V.

The circuit functions as an INVERTER since a LOW input produces a HIGH output and vice versa. This basic INVERTER can be modified to form NAND and NOR logic gates.

N-MOS NAND Gate   The NAND operation is performed by the circuit of Figure 8.31(a), where Q_1 is again acting as a load resistance while Q_2 and Q_3 are switches controlled by input levels A and B. If either A or B is at 0 V (logical 0), the corresponding FET is OFF, thereby presenting a high resistance from the output terminal to ground so that output X is HIGH (+5 V). When both A and B are +5 V (logical 1), both Q_2 and Q_3 are ON, so output X is LOW. Clearly, the output equals the NAND of the inputs ($x = \overline{AB}$).

N-MOS NOR Gate   The NOR gate of Figure 8.31(b) uses Q_2 and Q_3 as parallel switches with Q_1 again acting as a load resistance. When either input A or B is at +5 V, the corresponding MOSFET is ON, forcing the output to be LOW. When both inputs are at 0 V, both Q_2 and Q_3 are OFF, so the output goes HIGH. Clearly, this is the NOR operation with $X = \overline{A + B}$.

FIGURE 8.31 (a) N-MOS NAND gate; (b) NOR gate.

N-MOS OR gates and AND gates are easily formed by combining the NOR or NAND with inverters.

N-MOS Flip-Flops Flip-flops are formed by using two cross-coupled NOR gates or NAND gates. The MOS FF is very important in MOS memories, which will be discussed later.

8.13 CHARACTERISTICS OF MOS LOGIC

Compared to the bipolar logic families the MOS logic families are slower in operating speed, require much less power, have a better noise margin, a greater supply voltage range, and a higher fan-out, and, as was mentioned earlier, require much less "real estate" (chip area).

Operating Speed A typical N-MOS NAND gate has a propagation delay time of 50 ns. This is due to *two* factors: the relatively high output resistance (100 kΩ) in the HIGH state and the capacitive loading presented by the inputs of the logic circuits being driven. MOS logic inputs have very high input resistance ($> 10^{12}$ Ω), and they have a reasonably high gate capacitance (MOS capacitor), typically 2 to 5 picofarads. This combination of large R_{OUT} and large C_{LOAD} serves to increase switching time.

Noise Margin Typically N-MOS noise margins are around 1.5 V when operated from $V_{DD} = 5$ V and will be proportionally higher for larger values of $V_{DD.}$

Fan-Out Because of the extremely high input resistance at each MOSFET input, one would expect that the fan-out capabilities of MOS logic would be virtually unlimited. This is essentially true for dc or low-frequency operation. However, for frequencies greater than around 100 kHz, the gate input capacitances cause a deterioration in switching time which increases in proportion to the number of loads being driven. Even so, MOS logic can easily operate at a fan-out of 50, which is somewhat better than the bipolar families.

Power Drain MOS logic circuits draw small amounts of power because of the relatively large resistances being used. To illustrate, we can calculate the power dissipation of the INVERTER of Figure 8.30 for its two operating states.

  1. $V_{IN} = 0$ V: $R_{ON(Q_1)} = 100$ kΩ; $R_{OFF(Q_2)} = 10^{10}$ Ω. Therefore, I_D, current from V_{DD} supply, ≈ 0.05 nA, and $P_D = 5$ V × 0.05 nA = 0.25 nW.
  2. $V_{IN} = +5$ V: $R_{ON(Q_1)} = 100$ kΩ; $R_{ON(Q_2)} = 1$ kΩ. Therefore, $I_D = 5$ V/101 kΩ $\approx$ 50 μA and $P_D = 5$ V × 50 μA = 0.25 mW.

This gives an *average* P_D of a little over 0.1 mW for the INVERTER. The low power drain of MOS logic makes it suitable for LSI, where many gates, FFs, etc., can be on one chip without causing overheating that can damage the chip.

Process Complexity MOS logic is the simplest logic family to fabricate since it uses only one basic element, an N-MOS (or P-MOS) transistor. It requires no other elements, such as resistors, diodes, etc. This characteristic, together with its lower P_D, makes it ideally suited for LSI (large memories, calculator chips, etc.) and this is where MOS logic has made its greatest impact in the digital field. The operating speed of P-MOS and N-MOS is not comparable with TTL, so very little has been done with them in SSI and MSI applications. In fact, there are very few MOS logic circuits in the SSI or MSI categories (gates, FFs, counters, etc.) CMOS, however, is competitive in the MSI area, which was heretofore dominated by TTL.

REVIEW QUESTIONS

1. Describe the relative advantages and disadvantages of N-MOS circuits as compared to TTL.
2. What determines the fan-out limitations of MOS logic?
3. What factors limit the switching speed of N-MOS circuits?
4. What makes MOS circuitry so well suited for LSI?

8.14 COMPLEMENTARY MOS LOGIC

The *complementary MOS* (CMOS) logic family uses *both* P- and N-channel MOSFETs in the same circuit to realize several advantages over the P-MOS and N-MOS families. Generally speaking, CMOS is faster and consumes even less power than the other MOS families. These advantages are offset somewhat by the increased complexity of the IC fabrication process and a lower packing density. Thus, CMOS cannot yet hope to compete with MOS in applications requiring the utmost in LSI.

However, CMOS logic has undergone a constant growth in the MSI area, mostly at the expense of TTL, with which it is directly competitive. The CMOS fabrication process is simpler than TTL and has a greater packing density, therefore permitting more circuitry in a given area and reducing the cost per function. CMOS uses only a fraction of the power needed for even the low-power TTL series (74L00) and is thus ideally suited for applications using battery power or battery back-up power. CMOS operating speed, however, is typically two to four times slower than the standard and LS-TTL series.

CMOS Inverter   The circuitry for the basic CMOS INVERTER is shown in Figure 8.32. For this diagram and those that follow, the standard symbols for the MOSFETs have been replaced by blocks labeled P and N to denote a P-MOSFET and N-MOSFET, respectively. This is done simply for convenience in analyzing the circuits. The CMOS INVERTER has two MOSFETs in series such that the P-channel device has its source connected to $+V_{DD}$ (a positive voltage) and the N-channel device has its source connected to ground.* The gates of the two devices are connected together as

V_{IN}	Q_1	Q_2	V_{OUT}
$+V_{DD}$ (logic 1)	OFF $R_{OFF} = 10^{10}\ \Omega$	ON $R_{ON} = 1\ k\Omega$	$\simeq 0\ V$
$0\ V$ (logic 0)	ON $R_{ON} = 1\ K\Omega$	OFF $R_{OFF} = 10^{10}\ \Omega$	$\simeq +V_{DD}$

$$V_{OUT} = \overline{V_{IN}}$$

FIGURE 8.32   Basic CMOS INVERTER.

*Some manufacturers label the ground terminal as V_{SS}.

a common input. The drains of the two devices are connected together as the common output.

The logic levels for CMOS are essentially $+V_{DD}$ for logical 1 and 0 V for logical 0. Consider, first, the case where $V_{IN} = +V_{DD}$. In this situation the gate of Q_1 (P-channel) is at 0 V relative to the source of Q_1. Thus, Q_1 will be in the OFF state with $R_{OFF} \approx 10^{10}$ Ω. The gate of Q_2 (N-channel) will be at $+V_{DD}$ relative to its source. Thus, Q_2 will be ON with typically $R_{ON} = 1$ kΩ. The voltage divider between Q_1's R_{OFF} and Q_2's R_{ON} will produce $V_{OUT} \approx 0$ V.

Next, consider the case where $V_{IN} = 0$ V. Q_1 now has its gate at a negative potential relative to its source while Q_2 has $V_{GS} = 0$ V. Thus, Q_1 will be ON with $R_{ON} = 1$ kΩ and Q_2 OFF with $R_{OFF} = 10^{10}$ Ω, producing a V_{OUT} of approximately $+V_{DD}$. These two operating states are summarized in Figure 8.32(b), showing that the circuit does act as a logic INVERTER.

CMOS NAND Gate Other logic functions can be constructed by modifying the basic INVERTER. Figure 8.33 shows a NAND gate formed by adding a parallel P-channel MOSFET and a series N-channel MOSFET to the basic INVERTER. To analyze this circuit it helps to realize that a 0-V input turns ON its corresponding P-MOSFET and turns OFF its corresponding N-MOSFET, and vice versa for a $+V_{DD}$ input. Thus, it can be seen that the only time a LOW output will occur is when inputs

A	B	X
LOW	LOW	HIGH
LOW	HIGH	HIGH
HIGH	LOW	HIGH
HIGH	HIGH	LOW

FIGURE 8.33 CMOS NAND gate.

A and B are both HIGH ($+V_{DD}$) to turn ON both N-MOSFETs, thereby providing a low resistance from the output terminal to ground. For all other input conditions, at least one P-MOSFET will be ON while at least one N-MOSFET will be OFF. This produces a HIGH output.

CMOS NOR Gate A CMOS NOR gate is formed by adding a series P-MOSFET and a parallel N-MOSFET to the basic inverter as shown in Figure 8.34. Once again this circuit can be analyzed by realizing that a LOW at any input turns ON its corresponding P-MOSFET and turns OFF its corresponding N-MOSFET, and vice versa for a HIGH input. It is left to the reader to verify that this circuit operates as a NOR gate.

CMOS AND and OR gates can be formed by combining NANDs and NORs with inverters.

CMOS SET-CLEAR FF Two CMOS NOR gates or NAND gates can be cross-coupled to form a simple SET-CLEAR FF. Additional gating circuitry is used to convert the basic SET-CLEAR FF to clocked D and J-K FFs.

A	B	X
LOW	LOW	HIGH
LOW	HIGH	LOW
HIGH	LOW	LOW
HIGH	HIGH	LOW

$X = \overline{A + B}$

FIGURE 8.34 CMOS NOR gate.

8.15 CMOS SERIES CHARACTERISTICS

The first CMOS logic series was produced by RCA and is called the 4000 series; ICs in this series are numbered 4000, 4001, etc. Some manufacturers use this same numbering system for their CMOS devices, while others have developed their own closely related systems. For example, Motorola's CMOS devices belong to the MC14000 and MC14500 series. In addition, National Semiconductor has developed the 74C00 series which are pin-for-pin equivalents of the 7400 series of TTL devices. For example, the 74C04 is a hex-INVERTER chip that has the same pin assignments as the 7404 and is logically equivalent to the 7404.

More recently, an improved version of the 4000 series has been developed. Called the 4000B series, it provides improvements in switching speed and output drive characteristics over the original 4000 series (which is often referred to as the 4000A series).

The following paragraphs will investigate some of the important characteristics that are common to most CMOS devices.

Power-Supply Voltage   Most CMOS ICs in the 4000A series will operate with V_{DD} voltages ranging from 3 to 15 V, and most 4000B-series devices can operate with V_{DD} values from 3 to 18 V, so that power-supply regulation is not a critical consideration. When CMOS and TTL are being used together in the same system, the V_{DD} supply is usually made 5 V so that one 5-V supply can serve both types of ICs. In some cases the CMOS devices will be operated at a supply voltage greater than 5 V. In these situations, special steps have to be taken to permit the CMOS and TTL devices to work together. We will discuss these steps later.

Voltage Levels   The output levels of a CMOS IC will be very close to 0 V for the LOW state, and very close to V_{DD} for the HIGH state. The reason is that the resistance of a CMOS input is so large (10^{12} ohms) that it does not load down a CMOS output that is driving it. The input voltage requirements for the two logic states are specified as a percentage of the V_{DD} supply voltage as shown in Table 8.7. The highest LOW-state input voltage is specified as 30 percent of V_{DD}, and the lowest HIGH-state input voltage is specified as 70 percent of V_{DD}. For example, with $V_{DD} = 5$ V, any CMOS input less than $V_{IL}(\text{max}) = 1.5$ V will be accepted as a LOW, and any input greater than $V_{IH}(\text{min}) = 3.5$ V will be accepted as a HIGH.

TABLE 8.7   CMOS VOLTAGE LEVELS*

	Minimum	Maximum
V_{DD}	3 V	18 V
V_{OH}	V_{DD}	—
V_{IH}	70% V_{DD}	—
V_{OL}	—	0 V
V_{IL}	—	30% V_{DD}

*Assumes outputs are driving only CMOS inputs

Sec. 8.15   **CMOS Series Characteristics**   **335**

Noise Margins   The CMOS dc noise margins can be determined from Table 8.7 as follows:

$$V_{NH} = V_{OH}(\text{min}) - V_{IH}(\text{min})$$

$$= V_{DD} - 70\% \ V_{DD}$$

$$= 30\% \ V_{DD}$$

$$V_{NL} = V_{IL}(\text{max}) - V_{OL}(\text{max})$$

$$= 30\% \ V_{DD} - 0$$

$$= 30\% \ V_{DD}$$

The noise margins are the same in both states and depend on V_{DD}. At $V_{DD} = 5$ V, the noise margins are both 1.5 V. This is substantially better than TTL and ECL. This makes CMOS attractive for applications that are exposed to a high-noise environment. Of course, the noise margins can be made even better by using a larger V_{DD}. This improvement in noise immunity, however, would be obtained at the expense of a higher power drain because of the larger supply voltage.

Power Dissipation   When a CMOS logic circuit is in a static state (not changing), its power dissipation is extremely low. We can see the reason by examining each of the circuits shown in Figures 8.32–8.34. Note that regardless of the state of the output, there is always a very high resistance between the V_{DD} terminal and ground because there is always an OFF MOSFET in the current path. This results in a typical CMOS dc power dissipation of only 2.5 nW per gate when $V_{DD} = 5$ V; even at $V_{DD} = 10$ V this power increases to only 10 nW. With these values for P_D, it is easy to see why CMOS is widely used in applications where power drain is a prime concern.

P_D Increases with Frequency   The power dissipation of a CMOS IC will be very low as long as it is in a dc condition. Unfortunately, P_D will increase in proportion to the frequency at which the circuits are switching states. For example, a CMOS NAND gate that has $P_D = 10$ nW under dc conditions will have $P_D = 0.1$ mW at a frequency of 100 kpps, and 1 mW at 1 MHz. The reason for this dependence on frequency is illustrated in Figure 8.35.

Each time a CMOS output switches from LOW to HIGH, a transient charging current has to be supplied to the load capacitance. This capacitance consists of the combined input capacitances of any loads being driven and the device's own output capacitance. These narrow spikes of current are supplied by V_{DD} and can have a typical amplitude of 5 mA and a duration of 20 to 30 ns. Clearly, as the switching frequency increases, there will be more of these current spikes occurring per second, and the average current drawn from V_{DD} will increase.

Thus, at higher frequencies, CMOS begins to lose some of its advantage over other logic families. As a general rule, a CMOS gate will have the same average P_D as an LS-TTL gate at frequencies around 5 MHz. For MSI chips, the situation is somewhat more complex than stated here, and a logic designer must do a detailed analysis to determine whether or not CMOS has a power-dissipation advantage at a particular frequency of operation.

FIGURE 8.35 Current spikes are drawn from the V_{DD} supply each time the output switches from LOW to HIGH. This is due to charging current of load capacitance.

Fan-Out Like N-MOS and P-MOS, CMOS inputs have an extremely large resistance (10^{12} ohms) that draws essentially no current from the signal source. Each CMOS input, however, typically presents a 5-pF load to ground. This input capacitance limits the number of CMOS inputs that one CMOS output can drive (see Figure 8.36). The CMOS output has to charge and discharge the parallel combination of each input capacitance, so that the output switching time will be increased in proportion to the number of loads being driven. Typically, each CMOS load increases the driving circuit's propagation delay by 3 ns. For example, NAND gate 1 in Figure 8.36 might have a t_{PLH} of 25 ns if it were driving no loads; this would increase to 25 ns + 20(3 ns) = 85 ns if it were driving *twenty* loads.

Thus, CMOS fan-out depends on the permissible maximum propagation delay. Typically, CMOS outputs are limited to a fan-out of 50 for low-frequency operation ($\leq$ 1 MHz). Of course, for higher-frequency operation the fan-out would have to be less.

Gate 1 output drives a total C_{LOAD} of N × 5 pF

To other loads

FIGURE 8.36 Each CMOS input adds to the total load capacitance seen by the driving gate's output.

Switching Speed   Although CMOS, like N-MOS and P-MOS, has to drive relatively large load capacitances, its switching speed is somewhat faster because of its low output resistance in each state. Recall that an N-MOS output has to charge the load capacitance through a relatively large (100 kΩ) resistance. In the CMOS circuit, the output resistance in the HIGH state is the R_{ON} of the P-MOSFET, which is typically 1 kΩ or less. This allows more rapid charging of load capacitance.

A CMOS NAND gate will typically have an average t_{pd} of 50 ns at V_{DD} = 5 V, and 25 ns at V_{DD} = 10 V. The reason for the improvement in t_{pd} as V_{DD} is increased is that the R_{ON} of the MOSFETs decreases significantly at higher supply voltages. Thus, it appears that V_{DD} should be made as large as possible for operation at higher frequencies. Of course, the larger V_{DD} will result in increased power dissipation.

Unused Inputs   *CMOS inputs should never be left disconnected. All CMOS inputs have to be tied either to a fixed voltage level (0 V or V_{DD}) or to another input.* This rule applies even to the inputs of extra unused logic gates on a chip. An unconnected CMOS input is susceptible to noise and static charges that could easily bias both the *P*- and *N*-channel MOSFETs in the conductive state, resulting in increased power dissipation and possible overheating.

Static-Charge Susceptibility   The high input resistance of CMOS inputs makes them especially prone to static-charge buildup that can produce voltages large enough to break down the dielectric insulation between the MOSFET gate and channel. This static charge can result from improper handling such as pushing a chip into a styrofoam carrier. CMOS and MOS chips require careful handling precautions such as (1) storing them in conductive foam or metal containers and (2) using a grounded-tip soldering iron when soldering them into a circuit.

Most of the CMOS-series devices are now protected against static-charge damage by the inclusion of protective diodes on each input. Even so, it is best to observe the rules stated above to prevent static-charge damage.

Input Transition Times   Like TTL, the CMOS and MOS devices require reasonably fast input-signal transitions for reliable operation, although the requirement is not as stringent as for TTL. In general, CMOS devices will work best with input-signal transitions that are less than 15 μs at V_{DD} = 5 V and less than 4 μs at V_{DD} = 10 V.

REVIEW QUESTIONS

1. How do the CMOS power-supply voltage requirements compare to TTL?
2. Why is a CMOS output voltage so close to V_{DD} in the HIGH state?
3. How do the CMOS noise margins compare to TTL?
4. Why are CMOS power requirements so low under dc conditions?
5. Why does CMOS power drain go up with frequency?
6. What limits CMOS fan-out?
7. What should be done with CMOS unused inputs?
8. What precautions should be taken when handling CMOS devices?

8.16 CMOS TRISTATE LOGIC

Conventional CMOS outputs should never be connected together. Figure 8.37 illustrates what can happen when two CMOS INVERTER outputs are shorted together. The output on the left is trying to go to the HIGH state and its *P*-channel MOSFET is conducting with $R_{ON} = 1$ kΩ. The other output is trying to go to the LOW state and its *N*-channel MOSFET is conducting with $R_{ON} = 1$ kΩ. The common output terminal *X* will be at a voltage of approximately $V_{DD}/2$ because of the voltage-divider action between the two 1-kΩ resistances.

FIGURE 8.37 When CMOS outputs are shorted together, the common output terminal will be at approximately $V_{DD}/2$ if the outputs are trying to be at different levels.

This voltage is in the indeterminate range (recall that $V_{IL}(\text{max}) = 30\%\ V_{DD}$ and $V_{IH}(\text{min}) = 70\%\ V_{DD}$) and is therefore unacceptable for driving other devices. Furthermore, the current through the two conducting MOSFETs will be much greater than normal, especially at higher values of V_{DD}, and can damage the ICs.

It is relatively easy to recognize when two CMOS outputs have been unintentionally wired or shorted together because the signals at the outputs will have three different levels: (1) HIGH when both outputs are trying to be HIGH; (2) LOW when both are trying to be LOW; and (3) approximately $V_{DD}/2$ when the outputs are trying to be at different levels. This is illustrated in Figure 8.38.

FIGURE 8.38 When two ordinary CMOS outputs are shorted together, the shorted outputs will be at $\approx V_{DD}/2$ when the outputs are trying to be different.

Sec. 8.16   CMOS Tristate Logic

Tristate A few CMOS ICs have tristate outputs that can be connected together, provided that only one output is enabled at one time. The 4502 and 4503 ICs are tristate inverting buffers and noninverting buffers, respectively. They operate just like the TTL tristate devices described earlier.

1. Describe what can happen when ordinary CMOS outputs are connected together.

8.17 CMOS TRANSMISSION GATE

A special CMOS circuit that has no TTL or ECL counterpart is the *transmission gate* or *bilateral switch*, which acts essentially as a single-pole, single-throw switch controlled by an input logic level. This transmission gate will pass signals in both directions and is useful for digital and analog applications.

Figure 8.39 shows the basic arrangement for the bilateral switch. It consists of a P-MOSFET and N-MOSFET in parallel so that both polarities of input voltage can be switched. The CONTROL input and its inverse are used to turn the switch ON (closed) and OFF (open). When the CONTROL is HIGH, both MOSFETS are turned ON and the switch is closed. When CONTROL is LOW, both MOSFETs are turned OFF and the switch is open. Ideally, this circuit operates like an electromechanical relay. In practice, however, it is not a perfect short circuit when the switch is closed; the switch resistance R_{ON} is typically 200 Ω. In the open state, the switch resistance is very large, typically 10^{12} Ω, which for most purposes is an open circuit. The symbol in Figure 8.39(b) is used to represent the bilateral switch.

This circuit is called a *bilateral* switch because the input and output terminals can be interchanged. The signals applied to the switch input can be either digital or analog signals, provided that they stay within the limits of 0 to V_{DD} volts.

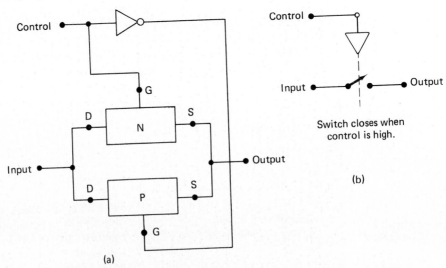

FIGURE 8.39   CMOS bilateral switch (transmission gate).

FIGURE 8.40   Logic diagram for the 4016 quad bilateral switch.

Control Input	Switch State
0 V	Open ($R_{OFF} = 10^{12}$ Ω)
+V_{DD}	Closed ($R_{ON} = 200$ Ω)

Figure 8.40 shows the logic diagram for the 4016 quad bilateral switch. It contains four bilateral switches that operate as described above. The four switches are independently controlled by their individual control inputs. A typical application is shown in Figure 8.41.

FIGURE 8.41

Explain the operation of the circuit in Figure 8.41.

Solution: Here we have two of the 4016's bilateral switches connected so that a common input signal v_{IN} can be switched to either output terminal, depending on the state of the OUTPUT SELECT input. When OUTPUT SELECT = 0, v_{IN} is routed to output X; when OUTPUT SELECT = 1, v_{IN} is routed to output Y. This v_{IN} signal can be any type of signal ranging between 0 V and V_{DD}.

8.18   IC INTERFACING

Interfacing means connecting the output(s) of one circuit or system to the input(s) of another circuit or system that has different electrical characteristics. Often a direct connection cannot be made because of the difference in the electrical characteristics of the *driver* circuit that is providing the output signal and the *load* circuit that is receiving the signal.

An *interface* circuit is a circuit connected between the driver and the load; its function is to take the driver output signal and condition it so that it is compatible with requirements of the load.

In the following sections we will address the problems involved in interfacing devices from one logic family to those of a different logic family. This type of interfacing occurs quite often in the more complex digital systems, where designers utilize different logic families for different parts of the system in order to take advantage of the strong points of each family. For example, high-speed TTL (S-TTL) might be used in those parts of the system that are operating at the highest frequency, LS-TTL in the slower parts of the system, and N-MOS for the LSI and VLSI parts of the system.

8.19   TTL DRIVING CMOS

Since the resistance of a CMOS input is so very high, it will not load down the output of a TTL driver. It would seem, then, that there would be no problem with a TTL output driving a CMOS input, provided that the CMOS was operating at $V_{DD} = 5$ V. There is a problem, however, because a TTL totem-pole output has a HIGH-state voltage, V_{OH}, that is very close to the V_{IH}(min) of a CMOS input.

Recall that V_{IH}(min) = 70% × V_{DD} = 3.5 V for CMOS. Also, recall that a TTL HIGH output is typically at 3.6 V but can go as low as V_{OH}(min) = 2.4 V if it is driving other TTL loads. Thus, something has to be done to raise the TTL output to an acceptable level for CMOS.

The most common solution to this interface problem is shown in Figure 8.42, where the TTL output is connected to +5 V with a pull-up resistor. The presence of the pull-up resistor will cause the TTL output to rise to approximately 5 V in the HIGH state, thereby providing an adequate CMOS input. We determine the value of the pull-up resistor in the same way as we did for the open-collector pull-up resistor. If the TTL output is driving only CMOS, a value from 1 kΩ to 10 kΩ is normally used.

FIGURE 8.42 External pull-up resistor is used when TLL drives CMOS.

TTL Driving High-Voltage CMOS If the CMOS IC is operating with V_{DD} greater than 5 V, the situation becomes somewhat more difficult. For example, with $V_{DD} = 10$ V, the CMOS input requires $V_{IH}(min) = 7$ V. The outputs of many TTL devices cannot be operated at more than 5 V, so a pull-up resistor connected to +10 V is prohibited. The LS-TTL series from certain manufacturers (e.g., Fairchild) can operate with an output pull-up to 10 V. In general, the device's data sheet should be checked before using a pull-up to more than 5 V.

When the TTL output cannot be pulled up to V_{DD}, there are some alternatives. One common solution is shown in Figure 8.43, where a 7407 open-collector buffer is used as the interface between a TTL totem-pole output and CMOS operating at $V_{DD} > 5$ V. The 7407 is the noninverting counterpart of the 7406 and has an output voltage rating of 30 V.

Another solution is to utilize a *level-translator* circuit such as the 40104. This is a CMOS chip that is designed to take a low-voltage input (e.g., from TTL) and translate it to a high-voltage output for CMOS.

REVIEW QUESTIONS

1. What has to be done to interface a conventional TTL output to a 5-V CMOS load?
2. What are the various ways to interface TTL to high-voltage CMOS?

FIGURE 8.43 A 7407 open-collector buffer can be used to interface TTL to high-voltage CMOS.

8.20 TTL DRIVING N-MOS

This situation is basically the same as TTL driving CMOS, since N-MOS operates with a positive supply voltage. N-MOS devices are usually LSI chips such as memories, microprocessors, and microcomputer interfaces that are designed to operate at 5 V. When a TTL output is to drive one of these devices, a pull-up resistor *might* be necessary, depending on the input requirement of the N-MOS device. Many N-MOS LSI chips are fully *TTL compatible*, which means that they will accept the full range of TTL voltage levels without any need for a pull-up resistor. Of course, the N-MOS data sheet should be consulted.

8.21 CMOS DRIVING TTL

Before we consider the problem of interfacing CMOS outputs to TTL inputs, it will be helpful to review the CMOS output characteristics for the two logic states.

Figure 8.44(a) shows the equivalent output circuit in the HIGH state. The R_{ON} of the P-MOSFET connects the output terminal to V_{DD} (remember, the N-MOSFET is OFF). Thus, the CMOS output circuit acts like a V_{DD} source with a source resistance of R_{ON}. The value of R_{ON} typically ranges from 200 to 1000 ohms.

Figure 8.44(b) shows the equivalent output circuit in the LOW state. The R_{ON} of the N-MOSFET connects the output terminal to ground (remember, the P-MOSFET is OFF). Thus, the CMOS output acts as a low resistance to ground; that is, it acts as a current sink.

CMOS Driving TTL in the HIGH State A TTL input does not require very much current in the HIGH state. Recall that $I_{IH}(\text{max}) = 40 \ \mu\text{A}$ for a standard TTL and only 20 μA for LS-TTL. Thus, a CMOS output can easily supply the HIGH-state input current to many TTL loads even if $R_{ON} = 1 \ \text{k}\Omega$.

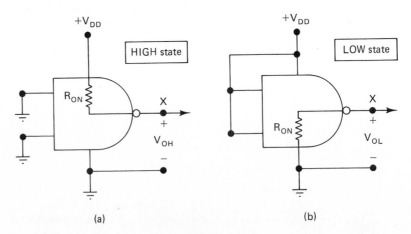

FIGURE 8.44 Equivalent CMOS output circuits for both logic states.

EXAMPLE 8.14

Calculate V_{OH} for a CMOS output that is driving 10 UL of TTL. Assume $R_{ON} = 1\ k\Omega$.

Solution: Refer to Figure 8.44(a). With X connected to 10 UL of TTL, there will be a maximum of $10 \times 40\ \mu A = 400\ \mu A$ that has to be supplied through R_{ON}. This produces a voltage drop of $400\ \mu A \times 1\ k\Omega = 0.4\ V$ across R_{ON}, so that $V_{OH} = 5\ V - 0.4\ V = 4.6\ V$, which is well above the $V_{IH}(min) = 2V$ for TTL.

CMOS Driving TTL in the LOW State   The LOW state is a problem because of the relatively high I_{IL} required by TTL. The CMOS output resistance has to sink this current, as shown in Figure 8.45. The voltage produced across R_{ON} can raise V_{OL} to well above the $V_{OL}(max)$ at which TTL normally operates.

Recall that $V_{OL}(max) = 0.4\ V$ for TTL and 0.5 V for LS-TTL. It is easy to see that an R_{ON} of $1\ k\Omega$ will produce much more than 0.4 V if it has to sink the $I_{IL}(max) = 1.6\ mA$ of a standard TTL input. On the other hand, it will produce only 0.4 V if it has to sink $I_{IL}(max) = 0.4\ mA$ of an LS-TTL input.

Most CMOS 4000-series devices can drive only one LS-TTL load (0.25 UL) and *cannot* drive even one TTL load because of the I_{IL} requirement. Thus, there is a need for some type of interface when a standard CMOS output signal is to drive a TTL load or more than one LS-TTL load. Some of the common alternatives are shown in Figure 8.46.

In Figure 8.46(a) the CMOS output drives only a single LS-TTL input. The double 74LS04 INVERTERs provide the proper logic level and drive the TTL loads. In Figure 8.46(b) the two INVERTERs are replaced by a single tristate buffer. The buffer is permanently enabled so that it simply passes the CMOS output signal to the TTL loads. The arrangement in Figure 8.46(c) uses the CMOS 4050B, a noninverting buffer that can drive 2 UL of TTL.

The 4000B series includes several ICs that can drive more than 0.25 UL of TTL. The individual device data sheets should be consulted.

FIGURE 8.45   CMOS output driving a TTL input must sink I_{IL} without causing V_{OL} to exceed 0.4 V.

(a)

(b)

(c)

FIGURE 8.46   Various methods for interfacing standard CMOS to TTL loads.

High-voltage CMOS Driving TTL   IC manufacturers have produced many LS-TTL ICs that can withstand input voltages as high as 15 V. These devices can easily be driven by CMOS outputs operating at up to $V_{DD} = 15$ V.

For those situations where high-voltage CMOS has to drive a standard TTL input, the 4050B buffer of Figure 8.45(c) will act as a level translator. It will accept input levels as high as 15 V and convert them to 5-V outputs for TTL.

_____ **REVIEW QUESTIONS**

1. True or false: CMOS has no problem driving TTL in the HIGH state.
2. True or false: A standard CMOS output can drive one TTL load.
3. Describe three methods for interfacing 5-V CMOS outputs to TTL inputs.
4. How can high-voltage CMOS be interfaced to standard TTL?

PROBLEMS

(Data sheets for the ICs referred to in these problems may be found in Appendix II)

8.1   Two different logic circuits have the following characteristics:

	Circuit A	Circuit B
V_{supply}	6 V	5 V
V_{IH}	1.6 V	1.8 V
V_{IL}	0.9 V	0.7 V
V_{OH}	2.2 V	2.5 V
V_{OL}	0.4 V	0.3 V
t_{PLH}	10 ns	18 ns
t_{PHL}	8 ns	14 ns
P_D	16 mW	10 mW

    (a) Which circuit has the best LOW-state dc noise immunity? The best HIGH-state dc noise immunity?

    (b) Which circuit can operate at higher frequencies?

    (c) Which circuit draws the most supply current?

8.2   Determine the values of $I_{IL}(max)$, $I_{IH}(max)$, $I_{OL}(max)$, and $I_{OH}(max)$ for each of the following TTL NAND gates:

    (a) 7420   (b) 74S20   (c) 74H20   (d) 74LS20

8.3   Compare the average P_D and average t_{pd} for each of the following TTL OR gates:

    (a) 7432   (b) 74S32   (c) 74LS32

8.4   Refer to the data sheet for the 74LS112 J-K FF.

    (a) Determine the input loading factor at the J and K inputs.

    (b) Determine the input loading factor at the clock and clear inputs.

    (c) How many other 74LS112s can the output of one 74LS112 drive at the clock input?

8.5   Figure 8.47(a) shows a 74107 J-K FF whose output is required to drive a total of 14 UL. Since this exceeds the fan-out of the 74107, a buffer of some type is needed. Figure 8.47(b) shows one possibility using one of the NAND gates from the 7437 quad NAND buffer, which has a much higher fan-out than the 74107. Note that $\bar{Q}$ is used since the NAND is acting as an INVERTER. Refer to the data sheet for the 7437 and determine

    (a) Its fan-out.

    (b) Its maximum sink current in the LOW state.

(a)

(b)

FIGURE 8.47

8.6 Buffer gates are generally more expensive than ordinary gates, and sometimes there are several unused gates available that can be used to solve a loading problem such as that in Figure 8.47(a). Show how 7400 NAND gates can be used to solve this problem.

8.7 Refer to the logic diagram of Figure 8.48, where the 7486 exclusive-OR output is driving several 7420 inputs. Determine whether the fan-out of the 7486 is being exceeded, and explain.

FIGURE 8.48

8.8 For the circuit of Figure 8.48 determine the longest time it will take for a change in the *A* input to be felt at output *W*. Use all worst-case conditions and maximum values of gate propagation delays. (*Hint:* Remember that NAND gates are inverting gates.)

8.9 Figure 8.49 shows a 74121 one-shot being triggered by the closing of the switch. What maximum value of *R* should be used to ensure that the *B* input is biased LOW while the switch is open?

8.10 Figure 8.50(a) shows a circuit that is used to convert a 60-Hz sinewave to a 60-pps signal that can reliably trigger FFs and counters, etc. This type of circuit might be used in a digital clock.
(a) Explain the circuit operation.
(b) A technician is testing this circuit and observes that the 74LS14 output stays LOW. He checks the waveform at the INVERTER input and it appears as shown in Figure

FIGURE 8.49

FIGURE 8.50

8.50(b). Thinking that the INVERTER is faulty, he replaces the chip and observes the same results. What do you think is causing the problem, and how can it be fixed? (*Hint:* Examine the v_x waveform carefully.)

8.11 For each waveform in Figure 8.51, determine *why* it will *not* reliably trigger a 74LS107 FF at its *CLK* input.

FIGURE 8.51

8.12 A technician breadboards a logic circuit for testing. As she tests the circuit's operation, she finds that many of the FFs and counters are triggering erratically. Like any good technician, she checks the V_{CC} line with a dc meter and reads 4.97 V, which is acceptable for TTL. She then checks all circuit wiring and replaces each IC one by one, but the problem persists. Finally she decides to observe V_{CC} on the scope and sees the waveform shown in Figure 8.52.

    What is the probable cause of the noise on V_{CC}? What did the technician forget to include when she breadboarded the circuit?

V_{CC} 4.97 V   1.3 V

FIGURE 8.52

8.13 The 7409 TTL IC is a quad two-input AND with open collector outputs. Show how 7409s can be used to implement the operation $x = A \cdot B \cdot C \cdot D \cdot E \cdot F \cdot G \cdot H \cdot I \cdot J \cdot K \cdot M$.

8.14 Determine the logic expression for output X in Figure 8.53.

FIGURE 8.53

NAND gates are 7401s (open-collector).

8.15 Determine a value for R_c in Figure 8.53 if output X is driving the clear inputs of four 74LS112 FFs.

8.16 Figure 8.54 shows a 7406 open-collector inverting buffer used to control the ON/OFF status of an LED to indicate the state of FF output Q. The LED's nominal specification is $V_F = 2.4$ V @ $I_F = 20$ mA, and $I_F(\max) = 30$ mA.

(a) What voltage will appear at the 7406 output when $Q = 0$?

(b) Choose an appropriate value for the series resistor for proper operation.

8.17 Figure 8.55 shows how two tristate buffers can be used to construct a *bidirectional transceiver* that allows digital data to be transmitted in either direction (A to B, or B to A). Describe the circuit operation for the two states of the DIRECTION input.

8.18 The circuit of Figure 8.56 is used to provide the enable inputs for the circuit of Figure 8.25.

(a) Determine which of the data inputs (A, B, or C) will appear on the bus for each combination of inputs X and Y.

(b) Explain why the circuit will not work if the NOR is changed to an EX-NOR.

FIGURE 8.54

FIGURE 8.55

FIGURE 8.56

8.19 Which of the following are advantages that ECL has over conventional TTL (7400 series)?

(a) Lower power dissipation.

(b) Shorter t_{pd}.

(c) Greater fan-out.

(d) Greater noise immunity.

(e) Complementary outputs.

(f) No current spikes during switching.

8.20 The circuit of Figure 8.57 is an N-MOS logic gate. Determine what type of gate it is. Assume that $+16$ V = logic 1, 0 V = logic 0.

FIGURE 8.57

8.21 Which of the following are advantages that P-MOS and N-MOS logic have over TTL?
(a) Greater packing density.
(b) Greater operating speed.
(c) Greater fan-out.
(d) More suitable for LSI.
(e) Lower P_D.
(f) Complementary outputs.
(g) Greater noise immunity.
(h) Simpler fabrication process.
(i) More SSI and MSI functions.
(j) Uses lower supply voltage.

8.22 Which of the following are advantages that CMOS has over P-MOS/N-MOS?
(a) Greater packing density.
(b) Higher speed.
(c) Greater fan-out.
(d) Lower output impedance.
(e) Simpler fabrication process.
(f) More suited for LSI.
(g) Lower P_D.
(h) Uses transistors as only circuit element.
(i) Lower input capacitance.

8.23 Repeat Problem 8.22 for CMOS advantages over TTL.

8.24 Which of the following operating conditions will probably result in the lowest average P_D for a CMOS logic system? Explain.
(a) $V_{DD} = 10$ V, switching frequency $f_{max} = 1$ MHz
(b) $V_{DD} = 5$ V, $f_{max} = 10$ kHz
(c) $V_{DD} = 10$ V, $f_{max} = 10$ kHz

8.25 What are the dc noise margins of CMOS logic that is operating at $V_{DD} = 12$ V?

FIGURE 8.58

8.26 Determine the waveform at output X in Figure 8.58 for the given input waveforms. Assume $R_{ON} \approx 200\ \Omega$ for the bilateral switch.

8.27 Determine the gain of the op-amp circuit of Figure 8.59 for the two states of the GAIN SELECT input. This circuit shows the basic principle of digitally controlled signal amplification.

FIGURE 8.59

8.28 Refer to Figure 8.60(a) where a standard TTL output, Q, is driving a CMOS IN-VERTER operating at $V_{DD} = 10$ V. The waveforms at Q and X appear as shown in Figure 8.60(b). Which of the following is a possible reason why X stays HIGH?

(a) The 10-V supply is faulty.

(b) The pull-up resistor is too large.

(c) The 74107 output breaks down at well below 10 V and maintains a 5.5-V level in the HIGH state. This is in the indeterminate range for the CMOS input.

(d) The CMOS input is loading down the TTL output.

FIGURE 8.60

8.29 Figure 8.61 shows a logic circuit that was poorly designed. It contains at least seven instances where the specifications and characteristics of the ICs have not been used properly. Find as many of these as you can.

FIGURE 8.61

9
MSI LOGIC CIRCUITS

Digital systems contain binary-coded data and information that is continuously being operated on in some manner. Some of the operations include: (1) *decoding and encoding*—changing the data from one type of code to another, (2) *multiplexing*—selecting one out of several groups of data, (3) *demultiplexing*—distributing data to one of several destinations, and (4) *data busing*—transmitting data among several devices on a common bus. All these operations and others have been facilitated by the availability of numerous ICs in the MSI (medium-scale-integration) category.

In this chapter we will study many of the common types of MSI devices. For each type, we will start with a brief discussion of its basic operating principle, then introduce specific ICs. We will then show how they can be used alone or in combination with other ICs in various applications.

9.1 DECODERS

A *decoder* is a logic circuit that converts an N-bit binary input code into $M*$ output lines such that each output line will be activated for only one of the possible combinations of inputs. Figure 9.1 shows the general decoder diagram with N inputs and M outputs. Since each of the N inputs can be 0 or 1, there are 2^N possible input combinations or codes. For each of these input combinations only one of the M outputs will be active HIGH; all the other outputs are LOW. Many decoders are designed to produce active LOW outputs, where only the selected output is LOW while all others are HIGH. This is always indicated by the presence of small circles on the output lines in the decoder diagram.

Some decoders do not utilize all of the 2^N possible input codes but only certain ones. For example, a BCD-to-decimal decoder has a 4-bit input code and *ten* output lines that correspond to the *ten* BCD code groups 0000 through 1001. Decoders of this type are often designed such that if any of the unused codes are applied to the input, *none* of the outputs will be activated.

In Chapter 7 we saw how decoders are used in conjunction with counters to detect the various states of the counter. In that application it was the FFs in the counter that provided the binary code inputs for the decoder. The same basic decoder circuitry

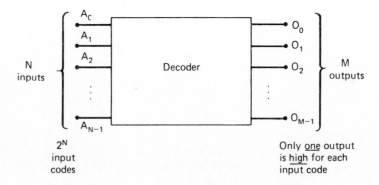

FIGURE 9.1 General decoder diagram.

*N can be any integer and M is an integer that is less than or equal to 2^N.

is used no matter where the inputs come from. Figure 9.2 shows the circuitry for a decoder with three inputs and $2^3 = 8$ outputs. It uses all AND gates, so the outputs are active HIGH. For active LOW outputs, NAND gates would be used. Note that for a given input code, the only output which is active (HIGH) is the one corresponding to the decimal equivalent of the binary input code (e.g., output O_6 goes HIGH when $C\overline{B}\overline{A} = 110_2 = 6_{10}$).

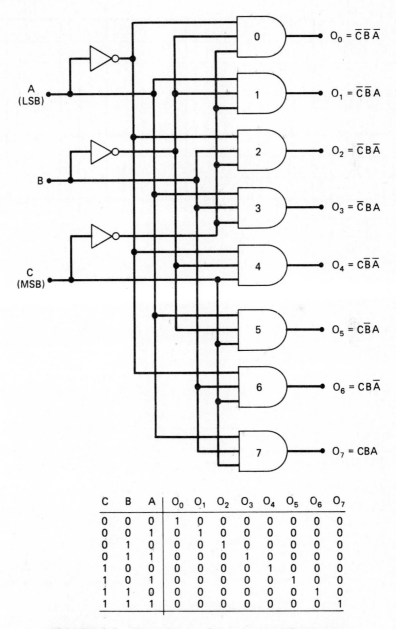

C	B	A	O_0	O_1	O_2	O_3	O_4	O_5	O_6	O_7
0	0	0	1	0	0	0	0	0	0	0
0	0	1	0	1	0	0	0	0	0	0
0	1	0	0	0	1	0	0	0	0	0
0	1	1	0	0	0	1	0	0	0	0
1	0	0	0	0	0	0	1	0	0	0
1	0	1	0	0	0	0	0	1	0	0
1	1	0	0	0	0	0	0	0	1	0
1	1	1	0	0	0	0	0	0	0	1

FIGURE 9.2   Three-line-to-8-line (or 1-of-8) decoder.

FIGURE 9.3 (a) Logic diagram for 74LS138 decoder; (b) truth table;
(c) logic symbol. (Courtesy of Fairchild, a Schlumberger company)

This decoder can be referred to in several ways. It can be called a 3-*line-to-*8-*line decoder,* because it has three input lines and eight output lines. It could also be called a *binary-to-octal decoder or convertor* because it takes a 3-bit binary input code and activates the one of the eight (octal) outputs corresponding to that code. It is also referred to as a 1-*of-*8 *decoder,* because only 1 of the 8 outputs is activated at one time.

ENABLE Inputs   Some decoders have one or more ENABLE inputs that are used to control the operation of the decoder. For example, refer to the decoder in Figure 9.2 and visualize having a common ENABLE line connected to a fourth input of each gate. Whith this ENABLE line held HIGH the decoder will function normally and the A, B, C input code will determine which output is HIGH. With ENABLE held LOW, however, *all* the outputs will be forced to the LOW state regardless of the levels at the A, B, C inputs. Thus, the decoder is ENABLED only if ENABLE is HIGH.

Figure 9.3(a) shows the logic diagram for the 74LS138 decoder as it appears in the Fairchild *TTL Data Book.* By examining this diagram carefully, we can determine exactly how this decoder functions. First, notice that it has NAND-gate outputs, so that its outputs are active LOW. Another indication is the labeling of the outputs as $\overline{O}_7, \overline{O}_6, \overline{O}_5$, etc.; the inverting overbar indicates active LOW outputs.

The input code is applied at $A_2, A_1,$ and A_0, where A_2 is the MSB. With three inputs and eight outputs, this is a 3-to-8 decoder or, equivalently, a 1-of-8 decoder.

Inputs $\overline{E}_1, \overline{E}_2,$ and E_3 are separate enable inputs that are combined in the AND gate. In order to enable the output NAND gates to respond to the input code at $A_2A_1A_0$, this AND-gate output has to be HIGH. This will occur only when $\overline{E}_1 = \overline{E}_2 = 0$ and $E_3 = 1$. In other words, $\overline{E}_1$ and $\overline{E}_2$ are active LOW, E_3 is active HIGH, and all three have to be in their active states to activate the decoder outputs. If one or more of the enable inputs is in its inactive state, the AND output will be LOW, which will force all NAND outputs to their inactive HIGH state regardless of the input code. This operation is summarized in the truth table in Figure 9.3(b). Recall that "x" represents the "don't care" condition.

The logic symbol for the 74LS138 is shown in Figure 9.3(c). Note how the active LOW outputs are represented, and how the enable inputs are represented.

EXAMPLE 9.1

Figure 9.4 shows how four 74LS138s and an INVERTER can be arranged to function as a 1-of-32 decoder. The decoders are labeled Z1–Z4 for easy reference, and the eight outputs from each one are combined into 32 outputs. Z1's outputs are $\overline{O}_0$–$\overline{O}_7$, Z2's $\overline{O}_0$–$\overline{O}_7$ outputs are renamed $\overline{O}_8$–$\overline{O}_{15}$, respectively, Z3's outputs are renamed $\overline{O}_{16}$–$\overline{O}_{23}$, and Z4's are renamed $\overline{O}_{24}$–$\overline{O}_{31}$. A 5-bit input code $A_4A_3A_2A_1A_0$ will activate only one of these 32 outputs for each of the 32 possible input codes.
(a)   Which output will be activated for $A_4A_3A_2A_1A_0 = 01101$?
(b)   What range of input codes will activate the Z4 chip?

Solution:
(a)   The 5-bit input code has two distinct portions. The A_4 and A_3 bits determine which one of the decoder chips Z1–Z4 will be enabled, while $A_2A_1A_0$ determines

---

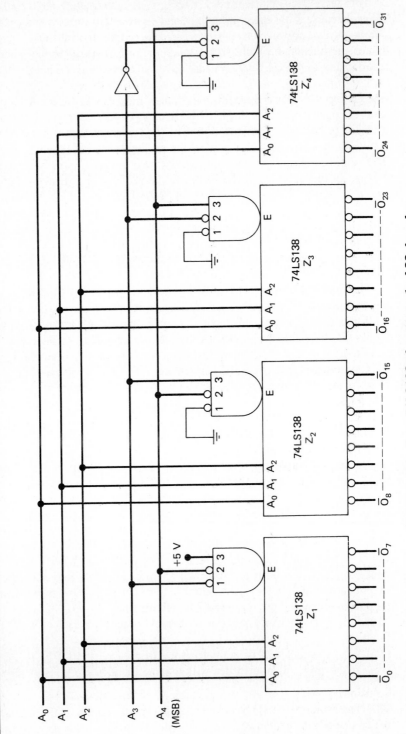

FIGURE 9.4 Four 74LS138s forming a 1-of-32 decoder.

which output of the enabled chip will be activated. With $A_4A_3 = 01$, only Z2 has all its enable inputs activated. Thus Z2 responds to the $A_2A_1A_0 = 101$ code and activates its $\overline{O}_5$ output, which has been renamed $\overline{O}_{13}$. Thus the input code 01101, which is the binary equivalent of decimal 13, will cause output $\overline{O}_{13}$ to go LOW, while all others stay HIGH.

(b) To enable Z4, both A_4 and A_3 have to be HIGH. Thus all input codes ranging from 11000(24_{10}) to 11111(31_{10}) will activate Z4. This corresponds to outputs $\overline{O}_{24}$ to $\overline{O}_{31}$.

Open-Collector Outputs Some decoders such as the TTL 7445 have open-collector outputs. For these types of decoders each output is normally OFF, providing a high-resistance path to ground; the output becomes a low-resistance path to ground when its corresponding input code is applied to the decoder inputs. These open-collector outputs are usually designed to operate at higher current and voltage limits than a normal TTL device. For example, the 7445 outputs can sink up to 80 mA in the LOW state and can tolerate voltages up to 30 V in the HIGH state. This makes them suitable for directly driving loads such as indicator lights or relays.

Applications of Decoders Decoders are used whenever an output or group of outputs is to be activated only upon the occurrence of a unique input code. When the input code is coming from a counter that is being clocked at a certain rate, the decoder outputs will be activated sequentially, and they can be used as timing or sequencing signals to control the occurrence of events such as turning motors or heaters on and off in a digitally controlled appliance.

Decoders are extremely valuable in the memory system of a computer. There they are used to select one bank of memory chips out of many in response to an address-code input from the control unit of the computer. We will investigate this in more detail in our study of memories.

Another valuable application of decoders is in the displaying of data on decimal readouts. We will elaborate on this in the following sections.

REVIEW QUESTIONS

1. Is it possible to activate more than one output of a decoder at a time?
2. True or false: When a decoder's ENABLE input is activated, all the decoder outputs will be activated.
3. The 74154 is a 1-of-16 decoder chip with two active LOW enable inputs. What is the total number of inputs and outputs for this chip? (*Ans.* Six inputs, 16 outputs)

9.2 BCD-TO-DECIMAL DECODERS

Figure 9.5(a) shows the logic diagram for a 7442 *BCD-to-decimal decoder*. Each output goes LOW when its corresponding BCD input is applied. For example, $\overline{O}_5$ will go LOW only when inputs $DCBA = 0101$, and $\overline{O}_8$ will go LOW only when $DCBA = 1000$. For input codes that are not BCD, none of the outputs will be activated. This decoder is also called a *4-to-10 decoder* or a *1-of-10 decoder*. The logic symbol for the 7442 is shown in Figure 9.5(b).

FIGURE 9.5 (a) Logic diagram for the 7442 BCD-to-decimal decoder; (b) logic symbol. (Courtesy of Fairchild, a Schlumberger company)

BCD-to-Decimal Decoder/Driver A BCD-to-decimal decoder/driver has an open-collector transistor on each output so that it can drive lamps, relays, etc. One of its major applications is in driving cold-cathode display tubes (nixie tubes), which display the decimal numerals corresponding to the BCD input code. Figure 9.6 shows a typical arrangement where a 74141 BCD-to-decimal decoder/driver is driving a nixie-tube readout. The *DCBA* inputs usually come from a counter or a storage register. For a given input BCD code, one of the outputs $\bar{O}_0$–$\bar{O}_9$ will be LOW (saturated driver transistor). This LOW essentially *grounds* the appropriate filament of the nixie tube so that current flows from the 170-V supply, through the tube, and to this filament, causing it to glow in the shape of the decimal digit corresponding to the input code. In this way the tube displays the decimal digit that is presented at the decoder inputs in BCD form. For any input codes greater than 1001, all the decoder outputs will be HIGH (open driver transistor) and none of the tube filaments will glow.

FIGURE 9.6 Typical application of a BCD-to-decimal decoder/driver with Nixie tube display.

9.3 BCD-TO-7-SEGMENT DECODER/DRIVERS

Many numerical displays use a 7-segment configuration [Figure 9.7(a)] to produce the decimal characters 0–9 and sometimes the hex characters A–F. Each segment is made up of a material that emits light when current is passed through it. Most commonly used materials include light-emitting diodes (LEDs) and incandescent filaments. Figure 9.7(b) shows the patterns of segments which are used to display the various digits.

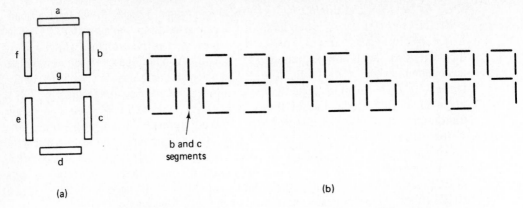

b and c
segments

(a)   (b)

FIGURE 9.7   (a) Seven-segment arrangement; (b) active segments for each digit.

For example, to display "6," segments c, d, e, f, and g are bright while segments a and b are dark.

A *BCD-to-7-segment decoder/driver* is used to take a 4-bit BCD input and provide the outputs that will pass current through the appropriate segments to display the decimal digit. The logic for this decoder is more complicated than those we looked at previously because each output is activated for more than one combination of inputs. For example, the e segment must be activated for any of the digits 0, 2, 6, and 8, which means whenever any of the codes 0000, 0010, 0110, or 1000 occurs.

Figure 9.8(a) shows a BCD-to-7-segment decoder/driver (TTL 7446 or 7447) being used to drive a 7-segment LED readout. Each segment consists of one or two LEDs. The anodes of the LEDs are all tied to V_{CC} (+5 V). The cathodes of the LEDs are connected through current-limiting resistors to the appropriate outputs of the decoder/driver. The decoder/driver has active LOW outputs which are open-collector driver transistors that can sink a fairly large current. This is because LED readouts may require 10 mA to 40 mA per segment, depending on their type and size.

To illustrate the operation of this circuit, let us suppose that the BCD input is $D = 0$, $C = 1$, $B = 0$, $A = 1$, which is BCD for 5. With these inputs the decoder/driver outputs $\bar{a}$, $\bar{f}$, $\bar{g}$, $\bar{c}$, and $\bar{d}$ will be driven LOW (connected to ground), allowing current to flow through the a, f, g, c, and d LED segments and thereby displaying the numeral 5. The $\bar{b}$ and $\bar{e}$ outputs will be HIGH (open), so LED segments b and e cannot conduct.

The 7446 and 7447 decoder/drivers are designed to activate specific segments even for input codes greater than 1001 (9). Figure 9.8(b) shows which segments are activated for each of the input codes from 0000 to 1111 (15). Note that an input code of 1111 will blank out all the segments.

The LED display used in Figure 9.8 is a *common-anode* type because the anodes of each segment are tied together to V_{CC}. Another type of 7-segment LED display uses a *common-cathode* arrangement where the cathodes of each segment are tied together and connected to ground. This type of display has to be driven by a BCD-to-7-segment decoder/driver with active-HIGH outputs that apply a HIGH voltage to the anodes of

FIGURE 9.8 (a) BCD-to-7-segment decoder/driver driving a
common-anode 7-segment LED display; (b) segment patterns for
all possible input codes.

those segments that are to be activated. The 7448 is a decoder/driver that can be used
for this purpose.

EXAMPLE 9.2

Each segment of a typical 7-segment LED display is rated to operate at 20 mA
@ 2.8 V for normal brightness. Calculate the value of the current-limiting
resistor needed to produce approximately 20 mA per segment.

Solution: Referring to Figure 9.8(a), we can see that the series resistor will have
to have a voltage drop equal to the difference between $V_{CC} = 5$ V and the
segment voltage of 2.8 V. This 2.2 V across the resistor must produce a current
of 20 mA. Thus we have

$$R_S = \frac{2.2 \text{ V}}{20 \text{ mA}} = 110 \text{ } \Omega$$

A standard resistor value close to this can be used. A 100-Ω resistor would be
a good choice.

Sec. 9.3   **BCD-to-7-Segment Decoder/Drivers**   **367**

1. How does a BCD-to-decimal decoder/driver differ from a BCD-to-7-segment decoder/driver?

9.4 LIQUID CRYSTAL DISPLAYS (LCDs)

Basically, LCDs operate from a low-voltage (typically 3 to 15 V RMS), low-frequency (25 to 60 Hz) ac signal and draw very little current. They are often arranged as 7-segment displays for numerical readouts as shown in Figure 9.9. The ac voltage needed to turn ON a segment is applied between the segment and the *backplane,* which is common to all segments. The segment and backplane form a capacitor that draws very little current as long as the ac frequency is kept low. It is generally not lower than 25 Hz, because this would produce flicker.

LCD display

Backplane

FIGURE 9.9   Liquid crystal display.

    LCDs draw much less current than LED displays and are widely used in battery-powered devices such as calculators and watches. An LCD does not emit light energy like an LED, and so it requires an external source of light.

Driving an LCD   An LCD segment will turn ON when an ac voltage is applied between the segment and the backplane, and will turn OFF when there is no voltage between the two. Rather than generating an ac signal, it is common practice to produce the required ac voltage by applying out-of-phase squarewaves to the segment and backplane. This is illustrated in Figure 9.10 for one segment. A 40-Hz squarewave is applied to the backplane and also to the input of a CMOS 4070 EXCLUSIVE-OR. The other input to the EX-OR is a CONTROL input that will control whether the segment is ON or OFF.

FIGURE 9.10   Method for driving an LCD segment. When CONTROL is LOW, segment is OFF. When CONTROL is HIGH, segment is ON.

When the CONTROL input is LOW, the EX-OR output will be exactly the same as the 40-Hz squarewave, so that the signals applied to the segment and backplane are equal. Since there is no difference in voltage, the segment will be OFF. When the CONTROL input is HIGH, the EX-OR output will be the INVERSE of the 40-Hz squarewave, so that the signal applied to the segment is out of phase with the signal applied to the backplane. As a result, the segment voltage will alternately be at $+5$ V and at -5 V relative to the backplane. This ac voltage will turn ON the segment.

This same idea can be extended to a complete 7-segment LCD display as shown in Figure 9.11. Here the CMOS 4511 BCD-to-7-segment decoder/driver supplies the CONTROL signals to each of seven EX-ORs for the seven segments. The 4511 has active HIGH outputs, since a HIGH is required to turn ON a segment.

In general, CMOS devices are used to drive LCDs for two reasons: (1) they require much less power than TTL and are more suited to the battery-operated applications where LCDs are used; (2) the TTL LOW-state voltage is not exactly 0 V and can be as much as 0.4 V. This will produce a dc component of voltage between the segment and backplane that considerably shortens the life of an LCD.

FIGURE 9.11 Method for driving 7-segment LCD.

1. Which of the following types of display devices is not arranged as seven segments? (a) LEDs, (b) nixie tubes, (c) LCDs.
2. Which display device requires a BCD-to-decimal decoder/driver that can withstand very high voltages?
3. Which display device requires an ac voltage?
4. What is the major advantage of an LCD over an LED display?

9.5 ENCODERS

A decoder accepts an N-bit input code and produces a HIGH (or LOW) at *one and only one* output line. In other words, we can say that a decoder identifies, recognizes, or detects a particular code. The opposite of this decoding process is called *encoding* and is performed by a logic circuit called an *encoder*. An encoder has a number of input lines, only one of which is activated at a given time, and produces an N-bit *output* code, depending on which input is activated. Figure 9.12 is the general diagram for an encoder with M inputs and N outputs. Here the inputs are active-HIGH, which means they are normally LOW.

A_0 O_0
A_1 O_1
A_2 O_2
Encoder
A_{M-1} O_{N-1}

M inputs
only one HIGH
at a time

N-bit
output code

FIGURE 9.12 General encoder diagram.

We saw that a *binary-to-octal decoder* accepts a 3-bit binary input code and activates one of eight output lines. An *octal-to-binary encoder* operates in the opposite manner. It accepts eight input lines and produces a 3-bit binary output code. Its circuitry is shown in Figure 9.13. It is assumed that only *one* of the input lines is made HIGH at one time, so there are only *eight* possible input conditions. The circuit is designed so that when A_0 is HIGH,* the binary code 000 is generated at the output; when A_1 is HIGH, the binary code 001 is generated; when A_2 is HIGH, the code 010 is generated; and so on (see the accompanying truth table). The circuit design is very simple, since it only involves looking at each output bit and determining for which input cases that bit is HIGH and then ORing the results. For example, the truth table shows that O_0 (LSB of output code) must be 1 whenever inputs A_1, A_3, A_5, or A_7 are HIGH. Thus, we have

---

*Actually, A_0 is not connected to the circuit because the output will automatically be 000 if inputs A_1–A_7 are all LOW.

Only one
HIGH input
at a time.

A_0	A_1	A_2	A_3	A_4	A_5	A_6	A_7	O_2	O_1	O_0
1	0	0	0	0	0	0	0	0	0	0
0	1	0	0	0	0	0	0	0	0	1
0	0	1	0	0	0	0	0	0	1	0
0	0	0	1	0	0	0	0	0	1	1
0	0	0	0	1	0	0	0	1	0	0
0	0	0	0	0	1	0	0	1	0	1
0	0	0	0	0	0	1	0	1	1	0
0	0	0	0	0	0	0	1	1	1	1

{ Other input cases }
{ not allowed. }

FIGURE 9.13   Octal-to-binary encoder (8-line-to-3-line encoder).

$$O_0 = A_1 + A_3 + A_5 + A_7$$

The other outputs are designed accordingly. This design is made simple by the fact that only *eight* out of the total of 2^8 possible input conditions are used. If more than one input is made HIGH at a given time, the output results will be erroneous. If none of the inputs are HIGH, the outputs will read 000. The encoder of Figure 9.13 is also called an 8-*line-to-3-line encoder*.

Priority Encoders   As stated above, the encoder in Figure 9.13 will produce erroneous results if two or more inputs are simultaneously activated. For example, if we follow through the logic gates in this circuit for the case where A_5 and A_3 are both HIGH, we see that the output code will be 111, which is the code for input A_7.

A *priority encoder* is a modified version of the basic encoder circuit that contains the logic circuitry needed to ensure that when two or more inputs are activated, the output code will correspond to the highest-numbered input. For example, the 74148, which is an 8-line-to-3-line priority encoder, will produce the output code 101 when both A_5 and A_3 are activated. Likewise, it will produce the output code 110 for A_6 if inputs A_0, A_2, and A_6 are activated.

Decimal-to-BCD Encoder Figure 9.14 shows the logic symbol for the 74147 decimal-to-BCD priority encoder. It has nine active LOW inputs representing the decimal digits 1–9 and produces the *inverted* BCD code corresponding to the activated input. For example, if input $\overline{A}_4$ is LOW while all other inputs are HIGH, the output code will be $O_3\overline{O}_2O_1O_0 = 1011$; this is the inverse of 0100, which is the BCD code for decimal 4.

Ten inputs

Inverted BCD

FIGURE 9.14 74147 decimal-to-BCD priority encoder.

The 74147 outputs will be normally HIGH when none of the inputs are activated. This corresponds to the decimal 0 input condition. There is actually no $\overline{A}_0$ input, since the encoder assumes the decimal 0 input state when all other inputs are HIGH. The 74147 inverted BCD outputs can be converted to normal BCD by putting each one through an INVERTER.

Switch Encoder Figure 9.15 shows how a 74147 can be used as a *switch encoder*. The 10 switches might be the keyboard switches on a calculator representing digits 0–9. The switches are normally open types, so that the encoder inputs are all normally HIGH, and the BCD output is 0000 (note the INVERTERs). When a digit key is depressed, the circuit will produce the BCD code for that digit. Since the 74147 is a priority encoder, simultaneous key depressions will produce the BCD code for the higher-numbered key.

The switch encoder of Figure 9.15 can be used whenever BCD data have to be manually entered into a digital system. A prime example would be in an electronic calculator, where the operator depresses several keyboard switches in succession to enter a decimal number. In a simple, basic calculator the BCD code for each decimal digit is entered into a 4-bit storage register. In other words, when the first key is depressed, the BCD code for that digit is sent to a 4-bit FF register; when the second

FIGURE 9.15   Decimal-to-BCD switch encoder.

switch is depressed, the BCD code for that digit is sent to *another* 4-bit FF register, and so on. Thus, a calculator that can handle eight digits will have eight 4-bit registers to store the BCD codes for these digits. Each 4-bit register drives a decoder/driver and numerical display so that the eight-digit number can be displayed.

The operation described above can be accomplished with the circuit in Figure 9.16. This circuit will take three decimal digits entered from the keyboard in sequence, encode them in BCD, and store the BCD in three FF output registers. The 12 D-type FFs Q_0–Q_{11} are used to receive and store the BCD codes for the digits. Q_8–Q_{11} stores the BCD code for the most significant digit (MSD), which is the first one entered on the keyboard. Q_4–Q_7 stores the second entered digit and Q_0–Q_3 stores the third entered digit. The X, Y, and Z FFs form a ring counter (Chapter 7) which controls the transfer of data from the encoder outputs to the appropriate output register. The OR gate produces a HIGH output any time one of the keys is depressed. This output may be affected by switch contact bounce, which would produce several pulses before settling down to the HIGH state. The OS is used to neutralize the switch bounce by triggering

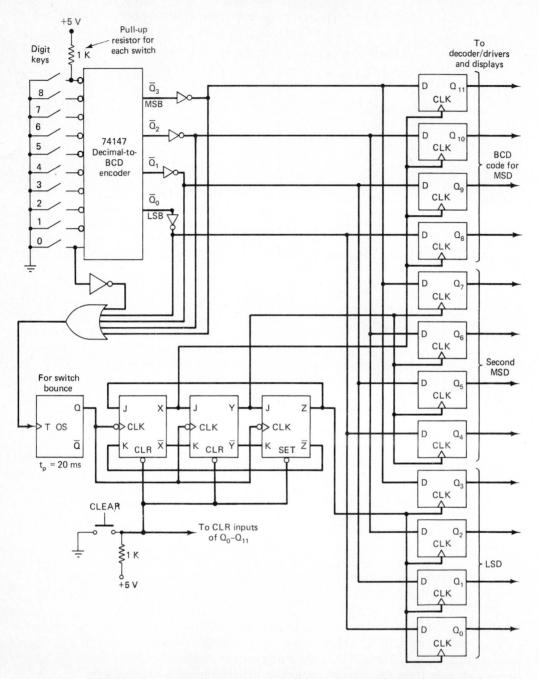

FIGURE 9.16 Circuit for keyboard entry of three-digit number into storage register.

on the first positive transition from the OR gate and remaining HIGH for 20 ms, well past the time duration of the switch bounce. The OS output clocks the ring counter.

The circuit operation is described as follows for the case where the decimal number 309 is being entered:

1. The CLEAR key is depressed. This clears all the Q_0–Q_{11} storage FFs to 0. It also clears FFs X and Y and presets FF Z to 1, so the ring counter begins in the 001 state.

2. The CLEAR key is released and the "3" key is depressed. The encoder outputs 1100 are inverted to produce 0011, the BCD code for 3. These binary values are sent to the D inputs of the three 4-bit output registers.

3. The OR output goes HIGH (since two inputs have gone HIGH) and triggers the OS output $Q = 1$ for 20 ms. After 20 ms Q returns LOW and clocks the ring counter to the 100 state (X goes HIGH). The positive transition at X is fed to the CLK inputs of FFs Q_8–Q_{11}, so the encoder outputs are transferred to these FFs. That is, $Q_{11} = 0$, $Q_{10} = 0$, $Q_9 = 1$, and $Q_8 = 1$. Note that FFs Q_0–Q_7 are not affected because their *CLK* inputs have not received a positive transition.

4. The "3" key is released and the OR gate output returns LOW. The "0" key is then depressed. This produces the BCD code of 0000, which is fed to the inputs of the storage FFs.

5. The OR output goes HIGH in response to the "0" key (note the inverter) and triggers the OS for 20 ms. After 20 ms the ring counter shifts to the 010 state (Y goes HIGH). The positive transition at Y is fed to the *CLK* inputs of Q_4–Q_7 and transfers 0000 to these FFs. Note that FFs Q_0–Q_3 and Q_8–Q_{11} are not affected by the Y transition.

6. The "0" key is released and the OR output returns LOW. The "9" key is depressed, producing BCD outputs 1001, which are fed to the storage FFs.

7. The OR output goes HIGH again, triggering the OS, which in turn clocks the ring counter to the 001 state (Z goes HIGH). The positive transition at Z is fed to the *CLK* inputs of Q_0–Q_3 and transfers the 1001 into these FFs. The other storage FFs are unaffected.

8. At this point the storage register contains 0011 0000 1001, beginning with Q_{11}. This is the BCD code for 309. These register outputs feed decoder/drivers which drive appropriate displays for indicating the decimal digits 309.

9. The storage FF outputs are also fed to other circuits in the system. In a calculator, for example, these outputs would be sent to the arithmetic section to be processed.

Several problems at the end of the chapter will deal with some other aspects of this circuit.

REVIEW QUESTIONS

1. How does an encoder differ from a decoder?
2. How does a priority encoder differ from an ordinary encoder?
3. What will be the outputs of the 74147 encoder if all its inputs are LOW? (*Ans.* 0110)

9.6 MULTIPLEXERS (DATA SELECTORS)

A *multiplexer* or *data selector* is a logic circuit that accepts several data inputs and allows only *one* of them at a time to get through to the output. The routing of the desired data input to the output is controlled by SELECT inputs (sometimes referred

FIGURE 9.17 Symbol for a digital multiplexer (MUX).

to as ADDRESS inputs). Figure 9.17 shows the symbol for a general multiplexer (MUX). In this diagram the inputs and outputs are drawn as large arrows to indicate that they may be one or more lines.

The multiplexer acts like a digitally controlled multiposition switch* where the digital code applied to the SELECT inputs controls which data inputs will be switched to the output. For example, output Z will equal data input I_0 for some particular SELECT input code; Z will equal I_1 for another particular SELECT input code; and so on. Stated another way, a multiplexer selects 1 out of N input data sources and transmits the selected data to a single output channel. This is called *multiplexing*.

Basic Two-Input Multiplexer   Figure 9.18 shows the logic circuitry for a two-input multiplexer with data inputs I_0 and I_1 and SELECT input S. The logic level applied to the S input determines which AND gate is enabled so that its data input passes through the OR gate to output Z. Looking at it another way, the Boolean expression for the output is

FIGURE 9.18 Two-input multiplexer.

*This is indicated by the dotted lines in Figure 9.17. These lines are not part of the MUX symbol but are shown just for illustration of the multiplexing function.

$$Z = I_0\bar{S} + I_1 S$$

With $S = 0$, this expression becomes

$$Z = I_0 \cdot 1 + I_1 \cdot 0$$

$$= I_0$$

which indicates that Z will be identical to input signal I_0, which can be a fixed logic level or a time-varying logic signal. With $S = 1$, the expression becomes

$$Z = I_0 \cdot 0 + I_1 \cdot 1 = I_1$$

showing that output Z will be identical to input signal I_1.

EXAMPLE 9.3

Show how multiplexers of the type shown in Figure 9.18 can be used to take two 3-bit binary numbers (X_2, X_1, X_0 and Y_2, Y_1, Y_0) and transmit one or the other number to outputs Z_2, Z_1, and Z_0, depending on an input select level.

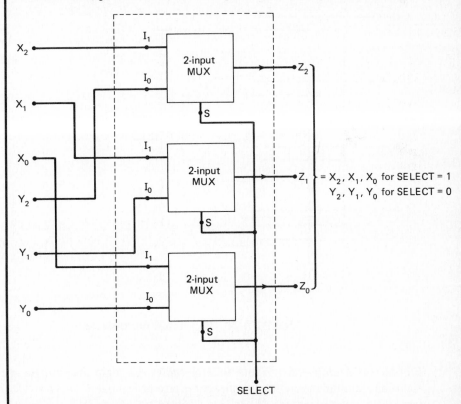

FIGURE 9.19   Example 9.3

Solution: Figure 9.19 shows 3 two-input multiplexers being used to perform the desired operation. Note that the S inputs of each multiplexer are connected together as one common select input. When $S = 1$, the X inputs to each individual multiplexer are routed through to the Z outputs. When $S = 0$, the Y inputs are routed through to the outputs.

Four-Input Multiplexer The same basic idea can be used to form the four-input multiplexer shown in Figure 9.20. Here there are four inputs, which are selectively transmitted to the output based on the four possible combinations of the $S_1 S_0$ select inputs. Each data input is gated with a different combination of select input levels. I_0 is gated with $\bar{S}_1 \bar{S}_0$ so that I_0 will pass through its AND gate to output Z only when $S_1 = 0$ and $S_0 = 0$. The table in the figure gives the outputs for the other three input select codes.

Two-, four-, eight-, and 16-input multiplexers are readily available in the TTL and CMOS logic families. These basic ICs can be combined for multiplexing a larger number of inputs.

S_1	S_0	Output
0	0	$Z = I_0$
0	1	$Z = I_1$
1	0	$Z = I_2$
1	1	$Z = I_3$

FIGURE 9.20 Four-input multiplexer.

Eight-Input Multiplexer Figure 9.21(a) shows the logic diagram for the 74151 eight-input multiplexer. This multiplexer has an enable input, $\bar{E}$, and provides both the normal and inverted outputs. When $\bar{E} = 0$, the select inputs $S_2 S_1 S_0$ will select one data input (I_0–I_7) for passage to output Z. When $\bar{E} = 1$, the multiplexer is disabled so that $Z = 0$ regardless of the select input code. This operation is summarized in Figure 9.21(b), and the 74151 logic symbol is shown in Figure 9.21(c).

(a)

Truth Table

Inputs				Outputs	
$\bar{E}$	S_2	S_1	S_0	$\bar{Z}$	Z
H	X	X	X	H	L
L	L	L	L	$\bar{I}_0$	I_0
L	L	L	H	$\bar{I}_1$	I_1
L	L	H	L	$\bar{I}_2$	I_2
L	L	H	H	$\bar{I}_3$	I_3
L	H	L	L	$\bar{I}_4$	I_4
L	H	L	H	$\bar{I}_5$	I_5
L	H	H	L	$\bar{I}_6$	I_6
L	H	H	H	$\bar{I}_7$	I_7

(b)

(c)

FIGURE 9.21 (a) Logic diagram for 74151 multiplexer; (b) truth table; (c) logic symbol. (Courtesy of Fairchild, a Schlumberger company)

Sec. 9.6   Multiplexers (Data Selectors)

EXAMPLE 9.4

The circuit in Figure 9.22 uses two 74151s, an INVERTER, and an OR gate. Describe this circuit's operation.

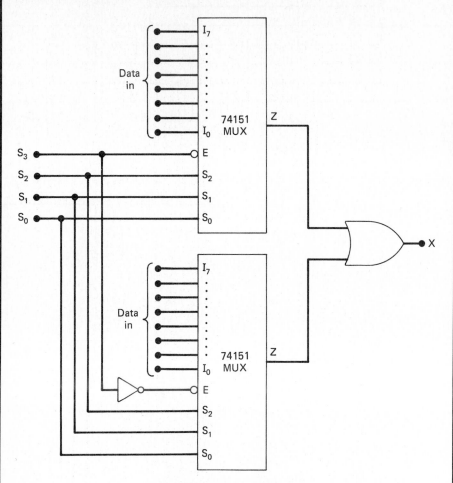

FIGURE 9.22   Example 9.4

Solution: This circuit has a total of 16 data inputs, eight applied to each multiplexer. The two multiplexer outputs are combined in the OR gate to produce a single output X. The circuit functions as a 16-input multiplexer. The four select inputs $S_3 S_2 S_1 S_0$ will select one of the 16 inputs to pass through to X.

The S_3 input determines which multiplexer is enabled. When $S_3 = 0$, the top multiplexer is enabled, and the $S_2 S_1 S_0$ inputs determine which of its data inputs will appear at its output and pass through the OR gate to X. When $S_3 = 1$, the bottom multiplexer is enabled and the $S_2 S_1 S_0$ inputs select one of its data inputs for passage to output X.

Quad Two-Input Multiplexer (74157) This is a very useful multiplexer IC that contains four two-input multiplexers like the one in Figure 9.18. The logic diagram for the 74157 is shown in Figure 9.23(a). Note the manner in which the data inputs and outputs are labeled.

(a)

(b)

FIGURE 9.23 (a) Logic diagram for 74157 multiplexer; (b) logic symbol. (Courtesy of Fairchild, a Schlumberger company)

EXAMPLE 9.5

Determine the input conditions required for each Z output to take on the logic level of its corresponding I_0 input. Repeat for I_1.

Solution: First of all, the enable input has to be active; that is, $\bar{E} = 0$. In order for Z_a to equal I_{0a}, the select input has to be LOW. These same conditions will produce $Z_b = I_{0b}$, $Z_c = I_{0c}$, and $Z_d = I_{0d}$.

With $\bar{E} = 0$ and $S = 1$, the Z outputs will follow the set of I_1 inputs; that is, $Z_a = I_{1a}$, $Z_b = I_{1b}$, $Z_c = I_{1c}$, and $Z_d = I_{1d}$.

All the outputs will be disabled (LOW) when $\bar{E} = 1$.

It is helpful to think of this multiplexer as being a simple two-input multiplexer, but where each input is four lines and the output is four lines. The four output lines switch back and forth between the two sets of four input lines under the control of the select input. This operation is represented by the 74157's logic symbol in Figure 9.23(b).

REVIEW QUESTIONS

1. What determines which input will appear at a multiplexer's output?
2. How many select inputs are required for a 32-input multiplexer?

9.7 MULTIPLEXER APPLICATIONS

Multiplexer circuits find numerous and varied applications in digital systems of all types. These applications include data selection, data routing, operation sequencing, parallel-to-serial conversion, waveform generation, and logic-function generation. We shall look at some of these applications here and several more in the problems at the end of the chapter.

Data Routing Multiplexers can route data from one of several sources to one destination. One typical application uses 74157 multiplexers to select and display the contents of either of two BCD counters using a *single* set of decoder/drivers and LED displays. The circuit arrangement is shown in Figure 9.24.

Each counter consists of two cascaded BCD stages, and each one is driven by its own clock signal. When the COUNTER SELECT line is HIGH, the outputs of counter 1 will be allowed to pass through the multiplexers to the decoder/driver to be displayed on the LED readouts. When COUNTER SELECT = 0, the outputs of counter 2 will pass through the multiplexers to the displays. In this way the decimal contents of one counter or the other will be displayed under the control of the COUNTER SELECT input. A common situation where this might be used is in a digital watch. The digital watch circuitry contains many counters and registers that keep track of seconds, minutes, hours, days, months, alarm settings, and so on. A multiplexing scheme such as this one allows different data to be displayed on the limited number of decimal readouts.

The purpose of the multiplexing technique, as it is used here, is to *time-share* the decoder/drivers and display circuits between the two counters rather than have a

FIGURE 9.24   System for displaying two multidigit BCD counters one at a time.

separate set of decoder/drivers and displays for each counter. This results in a significant saving in the number of wiring connections, especially when more BCD stages are added to each counter. Even more importantly, it represents a significant decrease in power consumption, because decoder/drivers and LED readouts typically draw relatively large amounts of current from the V_{CC} supply. Of course, this technique has the limitation that only one counter contents can be displayed at a time. However, in many applications this is not a drawback. A mechanical switching arrangement

FIGURE 9.25 (a) Parallel-to-serial converter; (b) waveforms for
$X_7X_6X_5X_4X_3X_2X_1X_0 = 10110101$.

could have been used to perform the function of switching first one counter and then the other to the decoder/drivers and displays, but the number of required switch contacts, the complexity of wiring, and the physical size could all be disadvantages over the completely logic method of Figure 9.24.

Parallel-to-Serial Conversion   Many digital systems process binary data in parallel form (all bits simultaneously) because it is faster. When these data are to be transmitted over relatively long distances, however, the parallel arrangement is undesirable because it requires a large number of transmission lines. For this reason, binary data or information that are in parallel form are often converted to serial form before being transmitted to a remote destination. One method for performing this *parallel-to-serial conversion* uses a multiplexer, as illustrated in Figure 9.25.

The data are present in parallel form at the outputs of the X register and are fed to the eight-input multiplexer. A 3-bit (MOD-8) counter is used to provide the select code bits $S_2S_1S_0$ so that they cycle through from 000 to 111 as clock pulses are applied. In this way, the output of the multiplexer will be X_0 during the first clock period, X_1 during the second clock period, and so on. The output Z is a waveform which is a serial representation of the parallel input data. The waveforms in the figure are for the case where $X_7X_6X_5X_4X_3X_2X_1X_0 = 10110101$. This conversion process takes a total of eight clock cycles. Note that X_0 (the LSB) is transmitted first and the X_7 (MSB) is transmitted last.

Operation Sequencing   The circuit of Figure 9.26 uses an eight-input multiplexer as part of a control sequencer that steps through seven steps, each of which actuates some portion of the physical process being controlled. This process could, for example, be a large high-temperature oven which is energized by seven different heaters, which must be activated one at a time. The circuit also uses a 3-line-to-8-line decoder and a MOD-8 binary counter. The operation is described as follows:

1. Initially the counter is in the 000 state. The counter outputs are fed to the select inputs of the multiplexer and to the inputs of the decoder. Thus, the decoder output $\overline{O}_0 = 0$ and the others are all 1, so all the ACTUATOR inputs of the process are LOW. The SENSOR outputs of the process all start out LOW. The multiplexer output $\overline{Z} = \overline{I}_0 = 1$, since the S inputs are 000.

2. The START pulse initiates the sequencing operation by setting FF A HIGH, bringing the counter to the 001 state. This causes decoder output $\overline{O}_1$ to go LOW, thereby activating actuator 1, which is the first step in the process (it could be the turning on of a heater).

3. Some time later SENSOR output 1 goes HIGH, indicating the completion of the first step (it could be the reaching of a certain temperature level). This HIGH is now present at the I_1 input of the multiplexer. It is inverted and reaches the $\overline{Z}$ output since the select code from the counter is 001.

4. The LOW at $\overline{Z}$ is fed to the *CLK* of FF A. This negative transition advances the counter to the 010 state.

5. Decoder output $\overline{O}_2$ now goes LOW, activating actuator 2, which is the second step in the process. $\overline{Z}$ now equals $\overline{I}_2$ (select code is 010). Since SENSOR output 2 is still LOW, $\overline{Z}$ will go HIGH.

FIGURE 9.26 Seven-step control sequencer.

6. When the second process step is complete, SENSOR output 2 goes HIGH, producing a LOW at $\bar{Z}$ and advancing the counter to 011.

7. This same action is repeated for each of the other steps. When the seventh step is completed, SENSOR output 7 goes HIGH, causing the counter to go from 111 to 000, where it will remain until another START pulse reinitiates the sequence.

Logic Function Generation   Multiplexers can be used to implement logic functions directly from a truth table without the need for simplification. When used for this purpose the select inputs are used as the logic variables and each data input is connected permanently HIGH or LOW as necessary to satisfy the truth table.

Figure 9.27 illustrates how an eight-input multiplexer can be used to implement the logic circuit that satisfies the given truth table. The input variables A, B, C are connected to S_0, S_1, S_2, respectively, so that the levels on these inputs determine which data input appears at output Z. According to the truth table, Z is supposed to be LOW when $CBA = 000$. Thus, multiplexer input I_0 should be connected LOW. Likewise, Z is supposed to be LOW for $CBA = 011$, 100, 101, and 110, so that inputs I_3, I_4, I_5, and I_6 should also be connected LOW. The other sets of CBA conditions must produce $Z = 1$, so multiplexer inputs I_1, I_2, and I_7 are connected permanently HIGH.

C	B	A	Z
0	0	0	0
0	0	1	1
0	1	0	1
0	1	1	0
1	0	0	0
1	0	1	0
1	1	0	0
1	1	1	1

(a)   (b)

FIGURE 9.27   Multiplexer used to implement a logic function described by truth table.

It is easy to see that any three-variable truth table can be implemented with this eight-input multiplexer. This method of implementation is often more efficient than the using separate logic gates. For example, if we write the sum-of-products expression for the truth table in Figure 9.27, we have

$$Z = \bar{A}\bar{B}C + \bar{A}B\bar{C} + ABC$$

This *cannot* be simplified either algebraically or by K-mapping, so its gate implementation would require three INVERTERs and four NAND gates, for a total of two ICs.

There is an even more efficient method for using multiplexers to implement logic functions, but it is beyond the scope of this text. This method can usually be found in logic design textbooks.

9.8 DEMULTIPLEXERS (DATA DISTRIBUTORS)

A multiplexer takes several inputs and transmits *one* of them to the output. A *demultiplexer* performs the reverse operation; it takes a single input and distributes it over several outputs. Figure 9.28 shows the general diagram for a demultiplexer (DEMUX). The large arrows for inputs and outputs can represent one or more lines. The select input code determines to which output the DATA input will be transmitted. In other words, the demultiplexer takes one input data source and selectively distributes it to 1 of N output channels just like a multiposition switch.

FIGURE 9.28   General demultiplexer.

1-Line-to-8-Line Demultiplexer   Figure 9.29 shows the logic diagram for a demultiplexer that distributes one input line to eight output lines. The single data input line I is connected to all eight AND gates, but only one of these gates will be enabled by the SELECT input lines. For example, with $S_2S_1S_0 = 000$, only AND gate 0 will be enabled, and data input I will appear at output O_0. Other SELECT codes cause input I to reach the other outputs. The truth table summarizes the operation.

The demultiplexer circuit of Figure 9.29 is very similar to the 3-line-to-8-line decoder circuit in Figure 9.2 except that a fourth input (I) has been added to each gate. It was pointed out earlier that many IC decoders have an ENABLE input, which is an extra input added to the decoder gates. This type of decoder chip can therefore be used as a demultiplexer, with the binary code inputs (e.g., A, B, C in Figure 9.2) serving as the SELECT inputs and the ENABLE input serving as the data input I. For this reason, IC manufacturers often call this type of device a *decoder/demultiplexer*, and it can be used for either function.

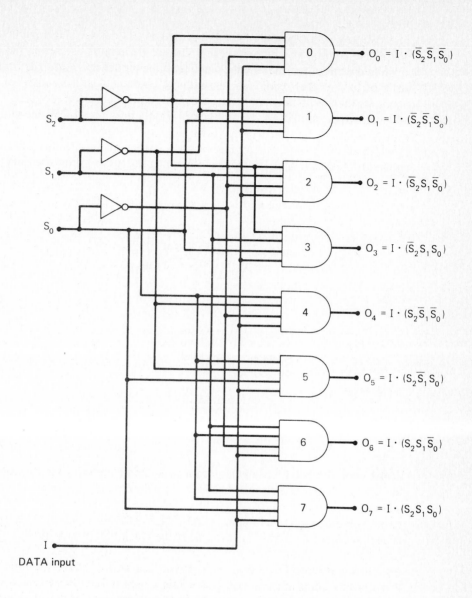

SELECT code			Outputs							
S_2	S_1	S_0	O_7	O_6	O_5	O_4	O_3	O_2	O_1	O_0
0	0	0	0	0	0	0	0	0	0	I
0	0	1	0	0	0	0	0	0	I	0
0	1	0	0	0	0	0	0	I	0	0
0	1	1	0	0	0	0	I	0	0	0
1	0	0	0	0	0	I	0	0	0	0
1	0	1	0	0	I	0	0	0	0	0
1	1	0	0	I	0	0	0	0	0	0
1	1	1	I	0	0	0	0	0	0	0

FIGURE 9.29 One-line-to-8-line demultiplexer.

Sec. 9.8   Demultiplexers (Data Distributors)

We saw earlier how the 74LS138 is used as a 1-of-8 decoder. Figure 9.30 shows how it can be used as a demultiplexer. The enable input $\overline{E}_1$ is used as the data input I, while the other two enable inputs are held in their active states. The $A_2A_1A_0$ inputs are used as the select code. To illustrate the operation, let's assume that the select inputs are 000. With this input code, the only output that can be activated is $\overline{O}_0$, while all other outputs are HIGH. $\overline{O}_0$ will go LOW only if $\overline{E}_1$ goes LOW and will be HIGH if $\overline{E}_1$ goes HIGH. In other words, $\overline{O}_0$ will follow the signal on $\overline{E}_1$ (i.e., the data input, I) while all other outputs stay HIGH. In a similar manner, a different select code applied to $A_2A_1A_0$ will cause the corresponding output to follow the data input, I.

FIGURE 9.30   The 74LS138 decoder used as a demultiplexer.

Clock Demultiplexer   Many applications of the demultiplexing principle are possible. Figure 9.31 shows the 74LS138 demultiplexer being used as a *clock demultiplexer*. Under control of the SELECT lines, the clock signal is routed to the desired destination. For example, with $S_2S_1S_0 = 000$, the clock signal applied to I will appear at output $\overline{O}_0$. With $S_2S_1S_0 = 101$, the clock will appear at $\overline{O}_5$.

Synchronous Data Transmission System   Figure 9.32 shows the logic diagram for a synchronous data transmission system that is used to serially transmit four 4-bit data words from a transmitter to a remote receiver. Let's look at the transmitter circuitry first. The data words are stored in registers A, B, C, and D that are connected as recirculating shift registers with a common SHIFT (clock) input. Each register will shift right on the PGT of the SHIFT pulses from AND gate 2. The LSB of each register is connected as a data input to the 4-input multiplexer.

The two MOD-4 counters control the transmission of the data register contents to the multiplexer output Z. The *word counter* selects the register data that will appear at Z. As this counter cycles from 00 to 11, the data from each register will sequentially appear at Z. The *bit counter* makes sure that 4 data bits from each register are transmitted through the multiplexer before advancing to the next register. The bit counter advances one count for each SHIFT pulse, so that after four SHIFT pulses,

FIGURE 9.31 Clock demultiplexer.

it recycles to 00. The NGT at the Q_1 output of the bit counter will cause the word counter to be incremented to the next count to select the next data register for transmission. In this way, the contents of each of the data registers will be transmitted to Z, one bit at a time, starting with register A (for $S_1S_0 = 00$) and proceeding through each register as the word counter advances one count for every four SHIFT pulses. The Z signal will thus contain 16 bits of serial data, 4 bits from each register. These data are said to be *time-multiplexed* because four different sets of data are appearing on the same output line at different times.

The transmission process is controlled by the two D-FFs, AND gates 1 and 2, and the one-shot. The operation of this control logic will be described in a later paragraph.

The Receiver   The receiver circuitry contains a 1-to-4 demultiplexer that receives the Z signal from the transmitter's multiplexer and *demultiplexes* it; that is, it separates the four different sets of data and distributes them to four different outputs so that the data that came from register A will appear serially at O_0, the data from register B at O_1, and so on. The end result is almost the same as having each transmitter data register connected to its corresponding output in the receiver, except that the data are sent from one register at a time over the serial data transmission path.

The MOD-4 counters in the receiver have the same function as their counterparts in the transmitter. The word counter selects which demultiplexer output will be receiving data, and the bit counter allows 4 bits of data to reach each output before advancing the word counter to its next state. The functions of the FF, OS, and AND gate are described below.

FIGURE 9.32 Synchronous data transmission system.

Complete Operation It should be clear that in order for this data transmission to work properly, there has to be some means for synchronizing the selection of the multiplexer inputs in the transmitter with the selection of the demultiplexer outputs in the receiver. We will go through a complete operation cycle to see how this synchronization is accomplished. For this illustration we will assume the following register data:

$$[A] = 0110, \quad [B] = 1001, \quad [C] = 1011, \quad [D] = 0100$$

As we go through the following steps, we will refer to the waveforms in Figure 9.33.

FIGURE 9.33   Waveforms during one complete transmission cycle.

1. FFs W and X in the transmitter and FF Y in the receiver are normally LOW. The LOWs from X and Y will keep both sets of counters in the zero state. The LOW at W prevents CLOCK pulses from getting through AND gate 1.

2. With both word counters at 00, the LOW at A_0 passes through the multiplexer to Z into the demultiplexer to output O_0. All other demultiplexer outputs are LOW, since they are not selected.

3. This is the situation prior to time t_0. At t_0, the TRANSMIT pulse sets $W = 1$ to enable AND gate 1 to pass CLOCK pulses. These CLOCK pulses also become the

TRANSMITTED CLOCK that is sent to the receiver along with the TRANS-MITTED DATA.

4. The NGT of the first CLOCK pulse out of AND gate 1 will set X and Y both HIGH at time t_2. This removes the resets from the counters and also enables AND gates 2 and 3 to pass CLOCK pulses starting at t_3. The pulses out of AND gate 2 are the SHIFT pulses. The pulses out of AND gate 3 are exactly the same as the SHIFT pulses out of AND gate 2.

5. The PGTs of the SHIFT pulses at t_3, t_4, and t_5 will shift A_1, A_2, and A_3 into the multiplexer, out of Z, into the demultiplexer, and out of output O_0. These three PGTs are also counted by both bit counters.

6. The PGT at t_6 will shift all the registers back to their original data and will recycle the bit counters to 00. The NGT at Q_1 of the bit counters will increment the transmitter and receiver word counters to 01 to select I_1 and O_1. Thus, the HIGH at B_0 will pass through the multiplexer into the demultiplexer and out of O_1.

7. The SHIFT pulses at t_7, t_8, and t_9 will shift B_1, B_2, and B_3 into the multiplexer and out of O_1. At t_{10} the bit counters will recycle and increment the word counters to 10 to select I_2 and O_2. This places the HIGH from C_0 at Z and at O_2.

8. The SHIFT pulses at t_{11}, t_{12}, and t_{13} will shift C_1, C_2, and C_3 into the multiplexer and out of O_2. At t_{14} the bit counters recycle and increment the word counters to 11 to select I_3 and O_3. This places the LOW from D_0 at Z and at O_3.

9. The SHIFT pulses at t_{15}, t_{16}, and t_{17} will shift D_1, D_2, and D_3 into the multiplexer and out of O_3. At t_{18} the bit counters recycle and increment the word counters to 00. The NGT at Q_1 of the word counters will trigger their respective one-shots to produce narrow clearing pulses for FFs W, X, and Y. With these FFs all LOW, all CLOCK pulses and SHIFT pulses are inhibited, and all the counters remain in the zero state.

10. The circuit conditions are back to their original state. No more data will be transmitted until the next TRANSMIT pulse occurs.

The waveforms at Z and O_0–O_3 show how the register data are multiplexed onto the Z signal and then demultiplexed so that each output receives the correct data.

REVIEW QUESTIONS

1. How does a demultiplexer differ from a multiplexer?
2. How many select inputs will a 1-to-16 demultiplexer have?

9.9 TRISTATE REGISTER

In many modern computers, especially microcomputers, the transfer of data takes place over a common set of connecting lines called a *data bus*. In these bus-organized computers, many different devices can have their outputs and inputs tied to the common data bus lines. Because of this, the devices that are tied to the data bus will often have tristate outputs, or they will be tied to the data bus through tristate buffers as we saw in Figure 8.25.

Some of the devices that are commonly connected to a data bus are (1) microprocessors, which will be discussed in Chapter 12; (2) semiconductor memory chips,

Truth Table

	Inputs				Output
MR	CP	$\overline{IE}_1$	$\overline{IE}_2$	D_n	Q_{N+1}
H	X	X	X	X	L
L	L	X	X	X	Q_N
L	$\int$	H	X	X	Q_N
L	$\int$	X	H	X	Q_N
L	$\int$	L	L	L	L
L	$\int$	L	L	H	H

When either $\overline{OE}_1$ or $\overline{OE}_2$ is HIGH the output is in the OFF
state (high impedance); however this does not affect the
contents or sequential operating of the register

H = HIGH voltage level
L = LOW voltage level
X = immaterial

Logic Diagram

FIGURE 9.34 Truth table and logic diagram for 74173 tristate register.
(Courtesy of Fairchild, a Schlumberger company)

Sec. 9.9   Tristate Register

which will be covered in Chapter 11; (3) digital-to-analog (D/A) converters and analog-to-digital (A/D) converters, to be covered in Chapter 10.

Almost always, the devices connected to a data bus will contain registers that hold their data. The outputs of these registers will have tristate buffers to allow data busing. There are many IC registers available that include the tristate buffers on the same chip. One of these is the TTL 74173 (and its CMOS counterparts, the 4076B and 74C173), whose logic diagram and truth tables are shown in Figure 9.34.

The 74173 is a 4-bit register with parallel in/parallel out capability. Note that the FF outputs are connected to tristate buffers that provide outputs O_0–O_3. Also note that the data inputs D_0–D_3 are connected to the D inputs of the register FFs through logic circuitry. This logic allows two modes of operation: (1) *load*, where the data at the D_0–D_3 inputs are transferred into the FFs on the PGT of the clock pulse at CP; (2) *hold*, where the data in the register does not change when the PGT of CP occurs.

EXAMPLE 9.6

(a) What input conditions will produce the load operation?
(b) What input conditions will produce the hold operation?

Solution:
(a) First of all, a PGT at CP is required for the load operation. By following through the logic circuitry, or by looking at the last two entries in the truth table, we see that both INPUT ENABLE inputs, $\overline{IE}_1$ and $\overline{IE}_2$, have to be LOW.
(b) If either of the conditions above is *not* met, the register data will not change, but will hold at their current value.

EXAMPLE 9.7

What input conditions are needed for the register data to appear at O_0–O_3?

Solution: The tristate buffers have an active-HIGH enable input that is driven by the output AND gate. The AND-gate output will be HIGH only when both OUTPUT ENABLE inputs, $\overline{OE}_1$ and $\overline{OE}_2$, are LOW. When either of these is HIGH, the output terminals will be in their Hi-Z state.

Note that the load operation can take place even if the outputs are Hi-Z.

Figure 9.35 shows the logic symbol for the 74173. We will use this symbol in the following section on data busing.

9.10 DATA BUSING

The data bus is very important in computer systems, and its significance will not be appreciated until our later studies of memories and microprocessors. For now, we will

FIGURE 9.35 **Logic symbol for 74173.**

illustrate the data-bus concept for register-to-register data transfer. Figure 9.36 shows a bus-organized system for three 74173 tristate registers. Note that each register has its pair of $\overline{OE}$ inputs tied together as one $\overline{OE}$ input, and likewise for the $\overline{IE}$ inputs. Also note that the registers will be referred to as registers *A, B,* and *C* from top to bottom. This is indicated by the subscripts on each input and output.

In this arrangement, the data bus consists of four lines labeled DB_0–DB_3. Corresponding outputs of each register are connected to the same data-bus line (e.g., O_{3A}, O_{3B}, and O_{3C} are connected to DB_3). Since the three registers have their outputs connected together, it is imperative that only one register has its outputs enabled while the other two register outputs are in the Hi-*Z* state. Otherwise, there will be "bus contention" (two or more sets of outputs fighting each other) and possible chip damage.

Corresponding register inputs are also tied to the same bus line (e.g., D_{3A}, D_{3B}, and D_{3C} are tied to DB_3). Thus, the levels on the bus will always be ready to be transferred to one or more of the registers depending on the $\overline{IE}$ inputs.

Data Transfer Operation The contents of any one of the three registers can be parallel-transferred over the data bus to one of the other registers through the proper application of logic levels to the register enable inputs. In a typical system, the control unit of a computer (i.e., the CPU) will generate the signals that select which register will put its data on the data bus, and which one will take the data from the data bus. The following example will illustrate this.

EXAMPLE 9.8

Describe the input signal requirements for transferring $[A] \rightarrow [C]$.

Solution: First of all, only register *A* should have its outputs enabled. That is, we need

FIGURE 9.36 Tristate registers connected to data bus.

$$\overline{OE}_A = 0, \qquad \overline{OE}_B = \overline{OE}_C = 1$$

This will place the contents of register A onto the data-bus lines.

Next, only register C should have its inputs enabled. For this, we want

$$\overline{IE}_C = 0, \qquad \overline{IE}_A = \overline{IE}_B = 1$$

This will allow only register C to accept data from the data bus when the PGT of the clock signal occurs.

Finally, a clock pulse is required to actually transfer the data from the bus into the register-C FFs.

It is easy to see how we can add more registers to the data bus so that a greater number of register-to-register data transfers are possible. Of course, each register adds two more enable inputs that have to be controlled for each data transfer. Each register also adds data inputs that have to be driven by the outputs of the register that is putting data on the bus. In systems where a large number of registers (or other devices) are connected to the bus, the register outputs may have to be buffered by circuits that are called *bus drivers*. Even when the devices are all MOS or CMOS, the total capacitive load connected to each bus line will deteriorate the transition times of the level changes on each bus line. A bus-driver IC will have very low output resistance in either logic state and will be able to charge and discharge the bus capacitances more rapidly than a normal IC.

Simplified Bus Representation   Usually many devices are connected to the same data bus. On a circuit schematic this can produce a confusing array of lines and connections. For this reason, a more simplified representation of data-bus connections is often used on block diagrams and in all but the most detailed circuit schematics. This simplified representation is shown in Figure 9.37 for an eight-line data bus.

The connections to and from the data bus are represented by wide arrows. The numbers in brackets indicate the number of bits that each register contains, as well as the number of lines connecting the register inputs and outputs to the bus.

Bidirectional Busing   Each register in Figure 9.36 has both its inputs and its outputs connected to the data bus, so that corresponding inputs and outputs are shorted together. For example, each register has output O_2 connected to input D_2 because of their common connection to DB_2. This, of course, would not be true if bus drivers were connected between the register outputs and the data bus.

Because inputs and outputs are often connected together in bus sysems, IC manufacturers have developed ICs that connect inputs and outputs together *internal* to the chip in order to reduce the number of IC pins. Figure 9.38 illustrates this for a 4-bit register. The separate data input lines (D_0–D_3) and output lines (O_0–O_3) have been replaced by input/output lines (I/O_0–I/O_3).

Each I/O line will function as either an input or an output depending on the states of the enable inputs. Thus, they are called *bidirectional data lines*. The 74LS299 is an 8-bit register with common I/O lines. Many memory ICs and microprocessors have bidirectional transfer of data.

Sec. 9.10   Data Busing   **399**

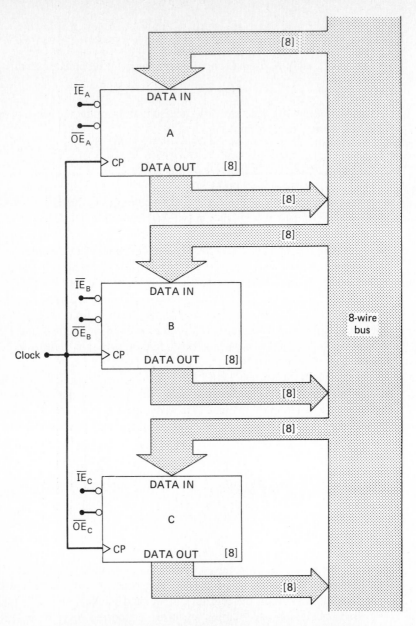

FIGURE 9.37 Simplified drawing of bus arrangement.

1. What is meant by the term "data bus"?
2. What is "bus contention," and what must be done to prevent it from occurring?
3. What is the function of a bus driver?
4. What is the principal reason for having registers with common I/O lines?

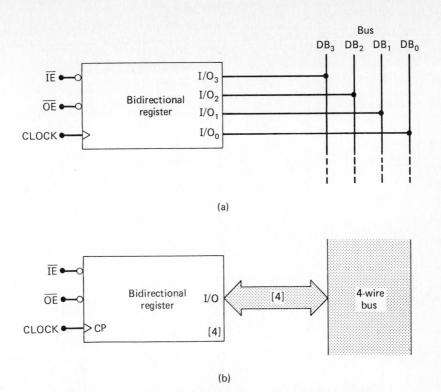

(a)

(b)

FIGURE 9.38 Bidirectional register; (b) simplified diagram.

PROBLEMS

9.1 What is the number of inputs and outputs of a decoder that accepts 64 different input combinations?

9.2 What set of input conditions will produce a LOW at $\overline{O}_6$ of a 74LS138?

9.3 Show how to use 74LS138s to form a 1-of-16 decoder.

9.4 Figure 9.39 shows how a decoder can be used in the generation of control signals. Assume that a RESET pulse has occurred at time t_0, and determine the CONTROL waveform in relation to t_0.

9.5 Modify the circuit of Figure 9.39 to generate a CONTROL waveform that goes LOW from t_{20} to t_{24}. (*Hint:* The modification does not require additional logic.)

9.6 The 7442 decoder of Figure 9.5 does not have an ENABLE input. However, we can operate it as a 1-of-8 decoder by not using outputs $\overline{O}_8$ and $\overline{O}_9$ and by using the D input as an ENABLE. This is illustrated in Figure 9.40. Describe how this arrangement works as an enabled 1-of-8 decoder, and state how the level on D either enables or disables the outputs.

9.7 Determine the various input codes which would activate output $\overline{g}$ in Figure 9.8.

9.8 Repeat Problem 9.7 for output $\overline{d}$.

9.9 Show how to connect BCD-to-7-segment decoder/drivers and LED 7-segment displays to the clock circuit of Figure 7.35 to display minutes and hours. Assume each segment is to operate at approximately 10 mA @ 2.5 V.

(a)

RESET

CLOCK

(b)

FIGURE 9.39

ENABLE →

Input code

D C B A

7442

$\bar{O}_9$ $\bar{O}_8$, $\bar{O}_7$ $\bar{O}_6$ $\bar{O}_5$ $\bar{O}_4$ $\bar{O}_3$ $\bar{O}_2$ $\bar{O}_1$ $\bar{O}_0$

Not
used

FIGURE 9.40

9.10 (a) Refer to Figure 9.10 and draw the segment and backplane waveforms relative to ground for CONTROL = 0. Then draw the waveform of segment voltage relative to backplane voltage.

(b) Repeat for CONTROL = 1.

9.11 Determine the output levels for the 74147 encoder when $\bar{A}_8 = \bar{A}_4 = 0$ and all other inputs are HIGH.

9.12 Figure 9.41 shows the block diagram of a logic circuit used to control the number of copies made by a copy machine. The machine operator selects the number of desired copies by closing one of the selector switches S_0–S_9. This number is encoded in BCD by the encoder and sent to a comparator circuit. The operator then hits a momentary-

FIGURE 9.41

contact START switch, which clears the counter and initiates a HIGH OPERATE output that is sent to the machine to signal it to make copies. As the machine makes each copy, a copy pulse is generated and fed to the BCD counter. The counter outputs are continually compared with the switch encoder outputs in the comparator. When the two BCD numbers match, indicating that the desired number of copies have been made, the comparator output $\overline{X}$ goes LOW; this causes the operate level to return LOW and stop the machine so that no more copies are made. Activating the START switch will cause this process to be repeated. Design the complete logic circuitry for the comparator and control sections of this system.

9.13 The keyboard circuit of Figure 9.16 is designed to accept a three-digit decimal number. What would happen if *four* digit keys were activated (e.g., 3095)? Design the necessary logic to be added to this circuit so that after three digits have been entered, any further digits will be ignored until the CLEAR key is depressed. In other words, if 3095 is entered on the keyboard, the output registers will display 309 and ignore the 5 and any subsequent digits until the circuit is CLEARED.

9.14 A technician breadboards the keyboard entry circuit of Figure 9.16 and tests its operation by trying to enter a series of three-digit numbers. He finds that sometimes the digit "0" is entered instead of the digit he pressed. He also observes that it happens with all the keys more or less randomly, although it is worse for some keys than others. He replaces all the ICs, and the malfunction persists. Which of the following circuit faults would explain his observations? Explain each choice.

(a) The technician forgot to ground the unused inputs of the OR gate.

(b) He has mistakenly used $\overline{Q}$ instead of Q from the one-shot.

(c) The switch bounce from the digit keys lasts longer than 20 ms.

(d) The Y and Z outputs are shorted together.

9.15 Repeat Problem 9.14 with the following symptom: the registers and displays stay at 0 no matter how many times a key is pressed.

9.16 What circuit malfunction could cause the wrong digit to be entered only when an even-numbered key is pressed? (For example, pressing the "2" key causes a "3" to be entered, as does pressing the "3" key; pressing the "4" causes a "5" to be entered, and so on.)

9.17 The circuit in Figure 9.42 uses 3 two-input multiplexers (Figure 9.18). Determine the function performed by this circuit.

FIGURE 9.42

9.18 Use the idea from the last problem to arrange several 74151 1-of-8 multiplexers to form a 1-of-64 multiplexer.

9.19 Show how two 74157s and a 74151 can be arranged to form a 1-of-16 multiplexer with no other required logic. Label the inputs I_0-I_{15} to show how they correspond to the select code.

9.20 (a) Expand the circuit of Figure 9.24 to display the contents of 2 three-stage BCD counters.

(b) Count the number of connections in this circuit and compare it to the number required if a separate decoder/driver and display were used for each stage of each counter.

9.21 Figure 9.43 shows how a multiplexer can be used to generate logic waveforms with any desirable pattern. The pattern is programmed using eight SPDT switches, and the waveform is repetitively produced by pulsing the MOD-8 counter. Draw the waveform at Z for the given switch positions.

FIGURE 9.43

9.22 Change the MOD-8 counter in Figure 9.43 to a MOD-16 counter and connect the MSB to the multiplexer $\overline{E}$ input. Draw the Z waveform.

9.23 Design the additional logic needed for the parallel-to-serial converter (Figure 9.25) so that it operates under the control of an input CONVERT level as follows:
(a) The CONVERT input starts LOW and the counter holds in the 000 state.
(b) When the CONVERT input goes HIGH, the counter goes through one complete cycle and stops at 000.
(c) This operation is repeated when CONVERT returns LOW and then goes HIGH.

9.24 Show how a 74151 can be used to generate the logic function: $Z = AB + BC + AC$.

9.25 Show how a 16-input multiplexer can be used to generate the function $Z = \overline{ABC}D + BCD + A\overline{BD} + AB\overline{C}D$.

9.26 Show how the 7442 decoder can be used as a 1-to-8 demultiplexer. (*Hint:* See Problem 9.6.)

9.27 Draw the waveforms at Z, O_0, O_1, O_2, and O_3 in Figure 9.32 for the following register data: $[A] = 0011$, $[B] = 0110$, $[C] = 1001$, $[D] = 0111$.

9.28 Suppose the synchronous data transmission system of Figure 9.32 is malfunctioning as follows: the Z waveform is correct, but the O_0 waveform is identical to the Z waveform at all times while the other outputs are constantly LOW. Which of the following circuit faults could cause this malfunction? Assume the circuitry is TTL, and explain each choice.
(a) There is a broken connection to the $\overline{MR}$ of the receiver word counter.
(b) There is a broken connection to the Y FF's *CLR* input.
(c) The transmitted CLOCK line is open.
(d) The D input of FF Y is shorted to ground.
(e) The D input of FF W is shorted to ground.

9.29 Refer to the bus arrangement in Figure 9.36. Describe the input-signal requirements for simultaneously transferring the contents of register C to both of the other registers.

9.30 Assume that the registers in Figure 9.36 are initially at $[A] = 1011$, $[B] = 1000$, and

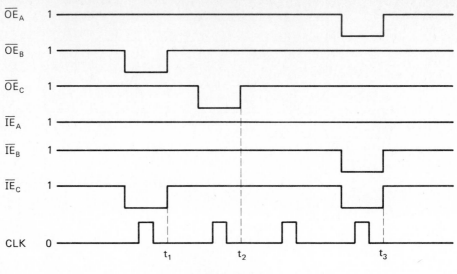

FIGURE 9.44

$[C] = 0111$. The signals in Figure 9.44 are applied to the register inputs. Determine the contents of each register at times t_1, t_2, and t_3.

9.31 Figure 9.45 shows two more devices that are to be added to the data bus of Figure 9.36. One is a set of buffered switches that can be used to manually enter data into any of the bus registers. The other device is an output register that is used to latch any data that are on the bus during a data transfer operation and display them on a set of LEDs.

(a) Assume all registers are at 0000. Outline the sequence of operations needed to load the registers with the following data from the switches: $[A] = 1011$, $[B] = 0001$, $[C] = 1110$.

(b) What will be the state of the LEDs at the end of this sequence?

9.32 Now that the circuitry of Figure 9.45 has been added to Figure 9.36, a total of five devices are connected to the data bus. The circuit in Figure 9.46(a) will now be used to generate the enable signals needed to perform the different data transfers over the data bus. It uses a 74LS139 chip that contains two identical independent 1-of-4 decoders with active LOW enable. The top decoder is used to select the device that will put data on the data bus (output select), and the bottom decoder is used to select the device that is to take the data from the data bus (input select). Assume that this circuit's outputs are connected to the corresponding enable inputs of the devices tied to the data bus. Also assume that all registers are initially at 0000 at time t_0, and the switches are in the positions shown in Figure 9.45.

(a) Determine the contents of each register at times t_1, t_2, and t_3 in response to the waveforms in Figure 9.46(b).

(b) Can "bus contention" ever occur with this circuit?

9.33 In theory, the answer to Problem 9.32(b) is "no" because only one decoder output can go LOW at one time, and so only one device's outputs can be enabled. In practice, however, there will be a very short overlap interval where two devices' outputs are enabled as the output select code changes from one code to another. This is usually caused by the fact that the newly enabled device becomes enabled faster than the previously enabled device becomes disabled. What can be done to prevent "bus contention" during the transition from one output select code to another?

FIGURE 9.45

(a)

(b)

FIGURE 9.46

10
INTERFACING
with the ANALOG WORLD

O U T L I N E

Digital systems perform all their internal operations in binary or some type of binary code. Any information that is to be input to a digital system must be put into binary form before it can be processed by the digital circuits. On the other hand, the outputs of a digital system are in some type of binary code and very often must be converted to a different form depending on how the outputs are to be used.

Many devices are used on the input and/or output sides of digital systems to serve as the communications link to the outside world. (Refer to Figure 10.1) Process-related I/O devices provide the means by which a digital system (i.e., computer) monitors and controls a physical process. On the *input* side, measurements of process parameters that are *analog* in nature are usually transduced (changed to a proportional electronic voltage or current) and sent to an analog-to-digital converter, ADC, which converts the analog quantity to a corresponding digital representation. Some process parameters are already digital in nature but must be changed to the appropriate form for the digital system. For example, digital input circuitry is used to count pulses from magnetic flow and turbine meters or to sense the ON/OFF status of switches or valves within the process.

FIGURE 10.1   ADC and DAC are used to interface a purely digital system with the analog outside world.

Process-related *output* devices translate the computer's outputs to appropriate actuating signals needed to control the process. These actuating signals might be simply opening and closing of switch contacts or pulses to a stepping motor. Many times, however, the required actuating signal has to be analog in nature, such as a voltage for controlling the speed of a dc motor. In these cases, a digital-to-analog converter, DAC, is needed to convert the digital system output to the required analog form.

Thus, we see that ADCs and DACs function as the *interface* between a completely digital system or device, such as a computer, and the analog outside world. This function is becoming increasingly more important as inexpensive microcomputers move into areas of process control where computer control was previously not feasible.

10.1 DIGITAL-TO-ANALOG CONVERSION

The two most important process-related I/O operations are digital-to-analog (D/A) and analog-to-digital (A/D) conversion. Since many A/D conversion methods utilize the D/A conversion process, we will examine D/A conversion first.

Basically, D/A *conversion* is the process of taking a value represented in *digital* code (such as straight binary or BCD) and converting it to a voltage or current which is proportional to the digital value. This voltage or current is an *analog* quantity, since it can take on many different values over a given range. Figure 10.2(a) shows the block diagram of a typical 4-bit D/A converter. We will not concern ourselves with the internal circuitry until later. For now, we will examine the various input/output relationships.

The digital inputs D, C, B, and A are usually derived from the output register of a digital system. The $2^4 = 16$ different binary numbers represented by this 4 bits

(a)

D	C	B	A	V_{OUT}	
0	0	0	0	0	volts
0	0	0	1	1	
0	0	1	0	2	
0	0	1	1	3	
0	1	0	0	4	
0	1	0	1	5	
0	1	1	0	6	
0	1	1	1	7	
1	0	0	0	8	
1	0	0	1	9	
1	0	1	0	10	
1	0	1	1	11	
1	1	0	0	12	
1	1	0	1	13	
1	1	1	0	14	
1	1	1	1	15	volts

(b)

FIGURE 10.2 Four-bit D/A converter with voltage output.

are listed in Figure 10.2(b). For each input number, the D/A converter output voltage is a different value. In fact, for this case, the analog output voltage V_{OUT} is equal in volts to the binary number. It could also have been twice the binary number or some other proportionality factor. The same idea would hold true if the D/A output were a current I_{OUT}.

EXAMPLE 10.1

A 5-bit D/A converter has a current output. For a digital input of 10100, an output current of 10 mA is produced. What will I_{OUT} be for a digital input of 11101?

Solution: The digital input 10100 is the binary equivalent of 20_{10}. Since $I_{OUT} = 10$ mA for this case, the proportionality factor is 0.5; that is, $I_{OUT} = 0.5 \times$ binary value. Thus, the binary input 11101 is equivalent to 29_{10}, so $I_{OUT} = 0.5 \times 29 = 14.5$ mA.

Input Weights   For the DAC of Figure 10.2 it should be noted that each digital input contributes a different amount to the analog output. This is easily seen if we examine the cases where only one input is HIGH:

D	C	B	A		V_{OUT} (V)
0	0	0	1	→	1
0	0	1	0	→	2
0	1	0	0	→	4
1	0	0	0	→	8

The contributions of each digital input are *weighted* according to their position in the binary number. Thus, A, which is the LSB, has a *weight* of 1 V, B has a weight of 2 V, C has a weight of 4 V, and D, the MSB, has the largest weight, 8 V. The weights are successively doubled for each bit, beginning with the LSB. Thus, we can consider V_{OUT} to be the weighted sum of the digital inputs. For instance, to find V_{OUT} for the digital input 0111 we can add the weights of the C, B, and A bits to obtain 4 V + 2 V + 1 V = 7 V.

EXAMPLE 10.2

A 5-bit D/A converter produces $V_{OUT} = 0.2$ V for a digital input of 00001. Find the value of V_{OUT} for a 11111 input.

Solution: Obviously, 0.2 V is the weight of the LSB. Thus, the weights of the other bits must be 0.4 V, 0.8 V, 1.6 V, and 3.2 V, respectively. For a digital input of 11111, then, the value of V_{OUT} will be 3.2 V + 1.6 V + 0.8 V + 0.4 V + 0.2 V = 6.2 V.

Resolution (Step Size)   *Resolution* of a D/A converter is defined as the smallest change that can occur in the analog output as a result of a change in the digital input. Referring to the table in Figure 10.2, we can see that the resolution is 1 V, since V_{OUT}

can change by no less than 1 V when the input code is changed. The resolution is always equal to the weight of the LSB and is also referred to as the *step size*, since it is the amount V_{OUT} will change as the input code goes from one step to the next. This is illustrated more graphically in Figure 10.3, where the digital inputs are being derived from the outputs of a 4-bit binary counter. The counter is being continuously cycled through its 16 states by the clock input. The waveform at the D/A output is a repetitive staircase which goes up 1 V per step as the counter advances from 0000 to 1111. When the counter returns to 0000, the D/A output returns to 0 V. The resolution or step size is the size of the jumps in the staircase waveform. In this example each step is 1 V.

Although resolution can be expressed as the amount of voltage or current per step, it is more useful to express it as a percentage of the *full-scale output*. To illustrate, the D/A converter of Figure 10.3 has a maximum full-scale output of 15 V (when the digital input is 1111). The step size is 1 V, which gives a percentage resolution of

$$\% \text{ resolution} = \frac{\text{step size}}{\text{full scale (F.S.)}} \times 100\% \tag{10.1}$$

$$= \frac{1 \text{ V}}{15 \text{ V}} \times 100\% = 6.67\%$$

EXAMPLE 10.3

A 10-bit D/A converter has a step size of 10 mV. Determine the full-scale output voltage and the percentage resolution.

Solution: With 10 bits, there will be $2^{10} - 1 = 1023$ steps of 10 mV each. The full-scale output will therefore be 10 mV $\times$ 1023 = 10.23 V and

$$\% \text{ resolution} = \frac{10 \text{ mV}}{10.23 \text{ V}} \times 100\% \approx 0.1\%$$

Example 10.3 helps to illustrate the fact that the percentage resolution becomes smaller as the number of input bits is increased. In fact, the percentage resolution can also be calculated from

$$\% \text{ resolution} = \frac{1}{\text{total \# steps}} \times 100\% \tag{10.2}$$

For an N-bit binary input code the total number of steps is $2^N - 1$. Thus, for the previous example

$$\% \text{ resolution} = \frac{1}{2^{10} - 1} \times 100\%$$

$$= \frac{1}{1023} \times 100\%$$

$$\approx 0.1\%$$

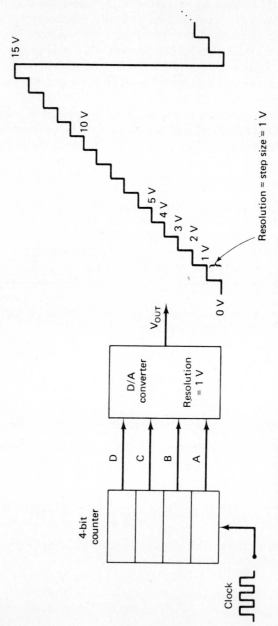

FIGURE 10.3 Output waveforms of D/A converter as inputs are provided by a binary counter.

This means that it is *only the number of bits* which determines the *percentage* resolution. Increasing the number of bits increases the number of steps to reach full scale, so each step is a smaller part of the full-scale voltage. Most DAC manufacturers specify resolution as the number of bits.

BCD Input Code   The D/A converters we have considered thus far have used a binary input code. Many D/A converters use a BCD input code where 4-bit code groups are used for each decimal digit. Figure 10.4 shows the diagram of an 8-bit (two-digit) converter of this type. Each 4-bit code group can range from 0000 to 1001, so the BCD inputs can represent any decimal number from 00 to 99. *Within* each code group the weights of the different bits are proportioned the same as in the binary code, but the *group* weights are different by a factor of 10. For example, A_0, the LSB of the least significant digit (LSD), could have a weight of 0.1 V. Thus, B_0, C_0, and D_0 would be 0.2, 0.4, and 0.8 V, respectively. The weight of A_1, the LSB of the MSD, would be 1 V (10 times A_0). Similarly, B_1, C_1, and D_1 would be 2, 4, and 8 V, respectively.

FIGURE 10.4   D/A converter using BCD input code.

EXAMPLE 10.4

If the weight of A_0 is 0.1 V in Figure 10.4, find:
(a) Step size.
(b) Full-scale output and percentage resolution.
(c) V_{OUT} for $D_0C_0B_0A_0 = 1000$ and $D_1C_1B_1A_1 = 0101$.

Solution:
(a) Step size is the weight of the LSB of the LSD, 0.1 V.
(b) There are 99 steps since there are two BCD digits. Thus, full-scale output is $99 \times 0.1 = 9.9$ V. The resolution is [using equation (10.1)]

$$\frac{\text{step size}}{\text{F.S.}} \times 100\% = \frac{0.1}{9.9} \times 100\% \approx 1\%$$

We could also have used equation (10.2) to calculate percentage resolution, since the total number of steps is 99.

(c) To find V_{OUT}, add the weights of all the bits that are 1s. Thus,

$$V_{OUT} = \overbrace{0.8 \text{ V}}^{D_0} + \overbrace{4 \text{ V}}^{C_1} + \overbrace{1 \text{ V}}^{A_1} = \textbf{5.8 V}$$

Alternatively, $D_1 C_1 B_1 A_1 = 5_{10}$ and $D_0 C_0 B_0 A_0 = 8_{10}$, so the input code is 58, multiplied by 0.1 V per step, to again give **5.8 V**.

REVIEW QUESTIONS

1. Define D/A conversion.
2. Define resolution (step size).
3. True or false: Percentage resolution of a DAC increases in proportion to the number of input bits.

10.2 D/A-CONVERTER CIRCUITRY*

There are several methods and circuits for producing the D/A operation which has been described. We shall examine only one of the basic schemes, to give an insight into the ideas used. It is not important to be familiar with all the various circuit schemes because D/A converters are available as ICs or as encapsulated packages that do not require any circuit knowledge. Instead, it is important to know the significant performance characteristics of D/A converters, in general, so that they can be used intelligently. These will be covered in Section 10.3.

Figure 10.5(a) shows the basic circuit for one type of 4-bit D/A converter. The inputs A, B, C, and D are binary inputs which are assumed to have values of either 0 V or 5 V. The *operational amplifier* is employed as a summing amplifier, which produces the weighted sum of these input voltages. It may be recalled that the summing amplifier multiplies each input voltage by the ratio of the feedback resistor R_F to the corresponding input resistor R_{IN}. In this circuit $R_F = 1$ kΩ and the input resistors range from 1 to 8 kΩ. The D input has $R_{IN} = 1$ kΩ, so the summing amplifier passes the voltage at D with no attenuation. The C input has $R_{IN} = 2$ kΩ, so it will be attenuated by 0.5. Similarly, the B input will be attenuated by 0.25 and the A input by 0.125. The amplifier output can thus be expressed as

$$V_{OUT} = -(V_D + 0.5V_C + 0.25V_B + 0.125V_A) \tag{10.3}$$

The negative sign is present because the summing amplifier is a polarity-inverting amplifier, but it will not concern us here.

Clearly, the summing amplifier output is an analog voltage which represents a weighted sum of the digital inputs, as shown by the table in Figure 10.5(b). This table lists all the possible input conditions and the resultant amplifier output voltage. The output is evaluated for any input condition by setting the appropriate inputs to either 0 V or 5 V. For example, if the digital input is 1010, then $V_D = V_B = 5$ V and $V_c = V_A = 0$ V. Thus, using (10.3),

---

*This section can be omitted by those not concerned with the internal circuitry of D/A converters.

Input code				V_{OUT} (volts)
D	C	B	A	
0	0	0	0	0
0	0	0	1	−0.625 ← LSB
0	0	1	0	−1.250
0	0	1	1	−1.875
0	1	0	0	−2.500
0	1	0	1	−3.125
0	1	1	0	−3.750
0	1	1	1	−4.375
1	0	0	0	−5.000
1	0	0	1	−5.625
1	0	1	0	−6.250
1	0	1	1	−6.875
1	1	0	0	−7.500
1	1	0	1	−8.125
1	1	1	0	−8.750 Full-
1	1	1	1	−9.375 ← scale

(a)

(b)

FIGURE 10.5   Simple D/A converter using op-amp summing amplifier.

$$V_{OUT} = -(5 \text{ V} + 0 \text{ V} + 0.25 \times 5 \text{ V} + 0 \text{ V})$$

$$= -6.25 \text{ V}$$

The resolution of this D/A converter is equal to the weighting of the LSB, which is $0.125 \times 5 \text{ V} = 0.625 \text{ V}$. As shown in the table, the analog output increases by 0.625 V as the binary input number advances one step.

EXAMPLE 10.5

(a) Determine the weights of each input bit of Figure 10.5(a).
(b) Change R_F to 250 Ω and determine the full-scale output.

Solution:
(a) The MSB passes with gain = 1, so its weight in the output is 5 V. Thus,

$$\text{MSB} \rightarrow 5 \text{ V}$$

$$\text{2nd MSB} \rightarrow 2.5 \text{ V}$$

$$\text{3rd MSB} \rightarrow 1.25 \text{ V}$$

$$\text{4th MSB} = \text{LSB} \rightarrow 0.625 \text{ V}$$

(b) If R_F is reduced by a factor of four to 250 Ω, each input weight will be four times *smaller* than the values above. Thus, the full-scale output will be reduced by this same factor and becomes $-9.375/4 = -2.344 \text{ V}$.

Conversion Accuracy The table in Figure 10.5(b) gives the *ideal* values of V_{OUT} for the various input cases. How close the circuit comes to producing these values depends primarily on two factors: (1) the precision of the input and feedback resistors, and (2) the precision of the input voltage levels. The resistors can be made very accurate (within 0.01 percent of the desired values) by trimming, but the input voltage levels must be handled differently. It should be clear that the digital inputs cannot be taken directly from the outputs of FFs or logic gates because the output logic levels of these devices are not precise values like 0 V and 5 V but vary over a given range. For this reason, it is necessary to insert a *precision-level amplifier* in between each logic input and its input resistor to the summing amplifier. This is shown in Figure 10.6.

The level amplifiers produce precise output levels of 5 V and 0 V, depending on whether the digital inputs are HIGH or LOW. A very stable, precise 5-V *reference* supply is required to produce the accurate 5-V level.

DAC with Current Output Many other circuits can be used to produce D/A conversion, and it is not our intention to present them all here. We will, however, look at another DAC method that uses an analog *current* output instead of voltage output. The basic scheme is shown in Figure 10.7(a) for a 4-bit DAC.

FIGURE 10.6 Complete 4-bit D/A converter including precision-level amplifiers.

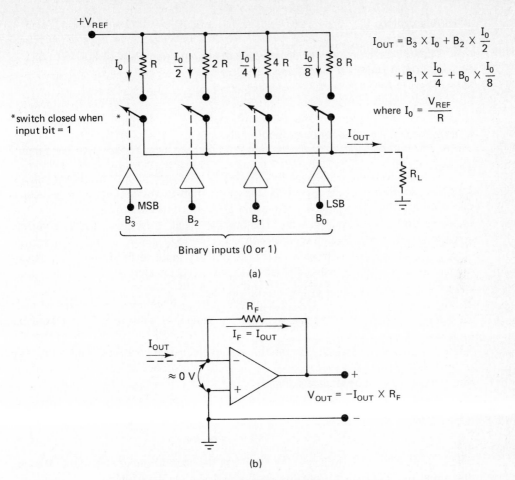

$$I_{OUT} = B_3 \times I_0 + B_2 \times \frac{I_0}{2}$$

$$+ B_1 \times \frac{I_0}{4} + B_0 \times \frac{I_0}{8}$$

where $I_0 = \dfrac{V_{REF}}{R}$

*switch closed when input bit = 1

Binary inputs (0 or 1)

(a)

$I_F = I_{OUT}$

$\approx 0\ V$

$V_{OUT} = -I_{OUT} \times R_F$

(b)

FIGURE 10.7 (a) Basic current-output DAC; (b) connected to an op-amp current-to-voltage converter.

The circuit uses four parallel current paths, each controlled by a semiconductor switch (such as the CMOS bilateral switch covered in Chapter 8). The state of each switch is controlled by the voltage levels applied to the binary inputs. The current through each parallel path is determined by V_{REF} and the resistance in the path. The resistors are binarily weighted, and the total current, I_{OUT}, will depend on which switches are open or closed according to the binary inputs. This output current can be made to flow through a load, R_L, but R_L must be made much smaller than R so as to have no effect on the value of current. Ideally, R_L should be a short to ground.

EXAMPLE 10.6

Assume $V_{REF} = 10$ V and $R = 10$ kΩ. Determine the resolution and full-scale output for this DAC.

Solution: $I_0 = V_{REF}/R = 1$ mA. This is the weight of the MSB. The other three currents will be 0.5, 0.25, and 0.125 mA. The LSB is 0.125 mA, which is also the resolution.

The full-scale output will occur when the binary inputs are all HIGH so that each current switch is closed and

$$I_{OUT} = 1 + .5 + .25 + .125 = \textbf{1.875 mA}$$

Note that the output current is proportional to V_{REF}. If V_{REF} is increased or decreased, I_{OUT} will change proportionally.

In order for I_{OUT} to be accurate, R_L should be a short to ground. One common way to accomplish this is to use an op-amp as a current-to-voltage converter, as shown in Figure 10.7(b). Here the I_{OUT} from the DAC is connected to the op-amp's "−" input, which is virtually at ground. The op-amp negative feedback forces a current equal to I_{OUT} to flow through R_F to produce $V_{OUT} = -I_{OUT} \times R_F$. Thus, V_{OUT} will be an analog voltage that is proportional to the binary input to the DAC. This analog output can drive a wide range of loads without being affected.

Multiplying DACs Many DACs require a reference voltage source that is used internally to help generate the analog output. In some DACs this reference voltage has to be a fixed, precise value in order to produce an accurate output. In other types the reference voltage can actually be variable and even change polarities. This latter type of DAC is called a *multiplying DAC* because the analog output is the product of the binary input and the reference voltage.

For example, we saw that the output current from the DAC of Figure 10.7(a) is proportional to V_{REF}. That is,

$$I_{OUT} = k(V_{REF} \times B) \tag{10.4}$$

where B represents the binary input, and k is the proportionality constant. We can determine the value of k from the results of Example 10.6, where we found that $I_{OUT} = 1.875$ mA for $V_{REF} = 10$ V and $B = 1111_2 = 15_{10}$. Thus, we have

$$1.875 \text{ mA} = k(10 \text{ V} \times 15)$$

or

$$k = 1.875 \text{ mA}/150 \text{ V} = 0.0125 \text{ mA/V}$$

Once k is known, I_{OUT} can be found for any values of V_{REF} and binary input.

EXAMPLE 10.7

Find I_{OUT} for the DAC of Figure 10.7(a) when $V_{REF} = 3.8$ V and the binary input is 0110 = 6_{10}.

Solution: Using equation (10.4) with $k = 0.0125$ mA/V

$$I_{OUT} = (0.0125 \text{ mA/V})(3.8 \text{ V} \times 6)$$

$$= \textbf{0.285 mA}$$

A multiplying DAC can be used to multiply an analog voltage (V_{REF}) by a binary value to produce an analog output. Some multiplying DACs allow V_{REF} to be of either polarity so that the output product can have either polarity. These are called *two-quadrant* multiplying DACs. Some DACs also allow the binary input to be of either polarity by using the MSB as a sign bit. These are called *four-quadrant* multiplying DACs.

_____ **REVIEW QUESTIONS**

1. What factors determine the accuracy of the summing-amplifier DAC?
2. True or false: A current-output DAC will produce an output current proportional to the binary input.
3. What is a multiplying DAC?

10.3   DAC SPECIFICATIONS

A wide variety of DACs are currently available as ICs or as self-contained encapsulated packages. One should be familiar with the more important manufacturers' specifications in order to evaluate a DAC for a particular application.

Resolution   As mentioned earlier, the percentage resolution of a DAC is dependent solely on the number of bits. For this reason, manufacturers usually specify a DAC resolution as the number of bits. A 10-bit DAC has a finer (smaller) resolution than an 8-bit DAC.

Accuracy   DAC manufacturers have several ways of specifying accuracy. The two most common are called *full-scale error* and *linearity error*, which are normally expressed as a percentage of the converter's full-scale output (%F.S.)

Full-scale error is the maximum deviation of the DAC's output from its expected (ideal) value. For example, assume that the DAC of Figure 10.5 has an accuracy of $\pm 0.01\%$ F.S. Since this converter has a full-scale output of 9.375 V, this percentage converts to

$$\pm 0.01\% \times 9.375 \text{ V} = \pm 0.9375 \text{ mV}$$

This means that the output of this DAC can, at any time, be off by as much as 0.9375 mV from its expected value.

Linearity error is the maximum deviation in step size from the ideal step size. For example, the DAC of Figure 10.5 has an expected step size of 0.625 V. If this converter has a linearity error of $\pm 0.01\%$ F.S., this would mean that the actual *step size* could be off by as much as 0.9375 mV.

Some of the more expensive DACs have full-scale and linearity errors as low as 0.001% F.S. General-purpose DACs usually have accuracies in the 0.01–0.1% range.

It is important to understand that accuracy and resolution of a D/A converter must be compatible. It is illogical to have a resolution of, say, 1 percent and an accuracy of 0.1 percent, or vice versa. To illustrate, a D/A converter with a resolution of 1 percent and a F.S. output of 10 V can produce an output analog voltage within 0.1 V of any desired value, assuming perfect accuracy. It makes no sense to have a costly accuracy of 0.01 percent of F.S. (or 1 mV) if the resolution already limits the closeness to the desired value to 0.1 V. The same can be said for having a resolution that is very small (many bits) while the accuracy is poor; it is a waste of input bits.

Settling Time   The operating speed of a DAC is specified as *settling time*, which is the time required for the output to go from zero to full-scale as the binary input is changed from all 0s to all 1s. Typical values of settling time will range from 50 ns to 10 μs. In general, DACs with a current output will have shorter settling times than those with a voltage output. For example, the DAC1280 can operate as either current output or voltage output. Its output settling time is 300 ns when current output is used, and 2.5 μs when voltage output is used. The main reason for this difference is the response time of the op-amp that is used as the current-to-voltage converter.

Offset Voltage   Ideally, the output of a DAC will be zero volts when the binary input is all 0s. In practice, there will be a small output voltage caused by the offset error in the output op-amp. This offset is typically 0.05% F.S. Almost all voltage DACs will have an external offset adjustment capability that allows you to zero the offset error.

REVIEW QUESTIONS

1.   True or false: Voltage DACs usually are slower responding than current DACs.
2.   True or false: It makes no sense to have an 8-bit DAC with a full-scale error of 0.02 percent.

10.4   DAC APPLICATIONS

DACs are used whenever the output of a digital circuit has to provide an analog voltage or current to drive an analog device. Some of the most common applications are described in the following paragraphs.

Control   The digital output from a computer can be converted to an analog control signal to adjust the speed of a motor, the temperature of a furnace, or to control almost any physical variable.

Automatic Testing   Computers can be programmed to generate the analog signals (through a DAC) needed to test analog circuitry. The test circuit's analog output response will normally be converted to a digital value by an ADC and fed into the computer to be stored, displayed, and sometimes analyzed.

Digital Amplitude Control   A multiplying DAC can be used to digitally adjust the amplitude of an analog signal. Recall that a multiplying DAC produces an output that is the product of a reference voltage and the binary input. If the reference voltage is a time-varying signal, the DAC output will follow this signal but with an amplitude determined by the binary input code. A typical application of this is digital "volume control," where the output of a digital circuit or computer can adjust the amplitude of an audio signal. We will examine this in Problem 10.14.

A/D Converters Several types of A/D converters utilize DACs as part of their circuitry. We will see this in the next section.

10.5 ANALOG-TO-DIGITAL CONVERSION

An *A/D converter* takes an analog input voltage and after a certain amount of time produces a digital output code which represents the analog input. The A/D conversion process is generally more complex and time consuming than the D/A process, and many different methods have been developed and used. We shall examine several of these methods in detail, even though it may never be necessary to design or construct A/D converters (they are available as completely packaged units). However, the techniques that are used provide an insight into what factors determine an A/D converter's performance.

Several important types of ADC utilize a D/A converter as part of their circuitry. Figure 10.8 is a general block diagram for this class of ADC. The timing for the operation is provided by the input clock signal. The control unit contains the logic circuitry for generating the proper sequence of operations in response to the START COMMAND, which initiates the conversion process. The comparator has two *analog* inputs and a *digital* output that switches states, depending on which analog input is greater.

The basic operation of A/D converters of this type consists of the following steps:

1. The START COMMAND goes HIGH, starting the operation.
2. At a rate determined by the clock, the control unit continually modifies the binary number that is stored in the register.

FIGURE 10.8 General diagram of one class of A/D converters.

Sec. 10.5   Analog-to-Digital Conversion **423**

3. The binary number in the register is converted to an analog voltage $V_{A'}$, by the D/A converter.

4. The comparator compares $V_{A'}$ with the analog input V_A. As long as $V_{A'} < V_A$, the comparator output stays HIGH. When $V_{A'}$ exceeds V_A by at least an amount $= V_T$ (threshold voltage), the comparator output goes LOW and stops the process of modifying the register number. At this point, $V_{A'}$ is a close approximation to V_A, and the digital number in the register, which is the digital equivalent of $V_{A'}$, is also the digital equivalent of V_A, within the resolution and accuracy of the system.

The several variations of this A/D conversion scheme differ mainly in the manner in which the control section continually modifies the numbers in the register. Otherwise, the basic idea is the same, with the register holding the required digital output when the conversion process is complete.

10.6 DIGITAL-RAMP A/D CONVERTER

One of the simplest versions of the general A/D converter of Figure 10.8 uses a binary counter as the register and allows the clock to increment the counter one step at a time until $V_{A'} \geq V_A$. It is called a *digital-ramp A/D converter* because the waveform at $V_{A'}$ is a step-by-step ramp (actually a staircase) like the one shown in Figure 10.3.

Figure 10.9 shows the complete diagram for a digital-ramp A/D converter. Its operation proceeds as follows:

FIGURE 10.9   Digital-ramp A/D converter.

1. A positive START pulse is applied, which resets the counter to zero. It also inhibits the AND gate so that no clock pulses get through to the counter while the START pulse is HIGH.

2. With the counter at zero, $V_{A'} = 0$, so the comparator output is HIGH (assume that V_A is some positive voltage).

3. When the START pulse returns to LOW, the AND gate is enabled and pulses are allowed into the counter.

4. As the counter advances, the D/A output $V_{A'}$ increases in steps of voltage equal to its resolution.

5. This continues until $V_{A'}$ reaches a step that exceeds V_A by an amount equal to V_T or greater.* At this point, the comparator output goes LOW, inhibiting the pulses to the counter so that the counter has stopped at a count that is the desired digital representation of V_A. The conversion process is now complete. The HIGH-to-LOW transition at the comparator output signals the end of the conversion. This signal is often called EOC (end of conversion).

EXAMPLE 10.8

Assume the following values for the A/D converter of Figure 10.9: clock frequency = 1 MHz; V_T = 0.1 mV; D/A converter has F.S. output = 10.23 V and a 10-bit input. Determine: (a) the digital equivalent obtained for V_A = 3.728 V; (b) the conversion time; and (c) the resolution of this converter.

Solution:
(a) The DAC has a 10-bit input and a 10.23-V F.S. output. Thus, the number of total possible steps is $2^{10} - 1 = 1023$, so the step size is

$$\frac{10.23 \text{ V}}{1023} = 10 \text{ mV}$$

This means that $V_{A'}$ increases in steps of 10 mV as the counter counts up from zero. Since V_A = 3.728 V and V_T = 0.1 mV, then $V_{A'}$ has to reach 3.7281 V or more before the comparator switches low. This will require

$$\frac{3.2781 \text{ V}}{10 \text{ mV}} = 372.81 = 373 \text{ steps}$$

At the end of the conversion, then, the counter will hold the binary equivalent of 373, which is 0101110101. This is the desired digital equivalent of V_A = 3.728 V, as produced by this A/D converter.

(b) Three hundred seventy-three steps were required to complete the conversion. Thus, 373 clock pulses occurred at the rate of one per microsecond. This gives a total conversion time of 373 μs.

(c) The resolution of this converter is equal to the step size of the D/A converter, which is 10 mV. In percent it is 10 mV/10.23 V × 100 percent ≈ 0.1 percent.

---

*IC comparators typically have threshold sensitivities of 10–100 μV.

A/D Resolution and Accuracy As pointed out in Example 10.8, the resolution of the A/D converter is equal to the resolution of the D/A converter which it contains. The D/A output voltage $V_{A'}$ is a staircase waveform that goes up in discrete steps until it exceeds V_A. Thus, $V_{A'}$ is an approximation to the value of V_A, and the best we can expect is that $V_{A'}$ is within 10 mV of V_A if the resolution (step size) is 10 mV. We can think of the resolution as being an inherent error which is often referred to as *quantization error*. This quantization error, which can be reduced by increasing the number of bits in the counter and D/A converter, is sometimes specified as an error of $+1$ LSB, indicating that the result could be off by that much due to the finite (nonzero) step size. In a later problem we will see how this quantization error can be modified so that it is $\pm\frac{1}{2}$ LSB, which is a more common situation.

Looking at it from a different aspect, the input V_A can take on an *infinite* number of values from 0 V to F.S. The approximation $V_{A'}$, however, can take on only a finite number of discrete values. This means that a small range of V_A values will have the same digital representation. To illustrate, in Example 10.8 any value of V_A from 3.720 V to 3.7299 V will require 373 steps, thereby resulting in the same digital representation. In other words, V_A must change by 10 mV (the resolution) to produce a change in digital output.

As in the D/A converter, *accuracy* is not related to the resolution but is dependent on the accuracy of the circuit components, such as the comparator, the D/A precision resistors and level amplifiers, the reference supplies, and so on. An error specification of 0.01 percent F.S. indicates that the A/D converter result may be off by 0.01 percent of F.S., owing to nonideal components. This error is *in addition* to the quantization error due to the resolution. These two sources of error are usually of the same order of magnitude for a given ADC.

EXAMPLE 10.9

A certain 8-bit ACD has a full-scale input of 2.55 V (i.e., $V_A = 2.55$ V produces a 11111111 digital output). It has a specified error of 1% F.S. Determine the maximum amount by which the digital output can differ from the analog input.

Solution: The step size is 2.55 V/$(2^8 - 1)$, which is exactly 10 mV. This means that even if the DAC has no inaccuracies, the digital output could be off by as much as 10 mV because $V_{A'}$ can change only in 10-mV steps; this is the quantization error. The specified error of 1% F.S. is 1% $\times$ 2.55 V = 2.55 mV. This means that the $V_{A'}$ value can be off by as much as 2.55 mV because of component inaccuracies. Thus, the total possible error could be as much as 10 mV + 2.55 mV = **12.55 mV**.

Conversion Time, t_C In the digital-ramp converter, the counter starts at zero and counts up until $V_{A'} > V_A$. Clearly, then, the time it takes to complete the conversion will depend on the value of V_A. A larger value of V_A will require more steps before the staircase voltage exceeds V_A. The maximum conversion time occurs when V_A is slightly lower than full-scale, so the staircase has to reach the full-scale step in order for $V_{A'} > V_A$. Thus,

$$t_C(\text{max}) = 2^N - 1 \text{ clock cycles}$$

For example, the ADC in Example 10.8 would have a maximum conversion time of

$$t_C(\text{max}) = (2^{10} - 1) \times 1 \ \mu s = 1023 \ \mu s$$

Sometimes, average conversion time is specified; it is half of the maximum conversion time. For the digital-ramp converter, this would be

$$t_C(\text{avg}) = \frac{t_C(\text{max})}{2} \approx 2^{N-1} \text{ clock cycles}$$

The major disadvantage of the digital-ramp method is that conversion time essentially doubles for each bit that is added to the counter, so that resolution can be improved only at the cost of a longer t_C. This makes this type of ADC unsuitable for applications where repetitive A/D conversions of a fast-changing analog signal have to be made. For low-speed applications, however, the relative simplicity of the digital-ramp converter is an advantage over the more complex, higher-speed ADCs.

REVIEW QUESTIONS

1. Describe the basic operation of the digital-ramp ADC.
2. Explain what is meant by "quantization error."
3. Why does conversion time increase with the value of the analog input voltage?
4. True or false: Everything else being equal, an 8-bit digital-ramp ADC will have a better resolution, but a longer conversion time, than a 10-bit ADC.
5. Give one advantage and one disadvantage of a digital-ramp ADC.

10.7 DATA ACQUISITION

There are many applications in which analog data have to be *digitized* (converted to digital) and transferred into a computer's memory. The process by which the computer acquires these digitized analog data is referred to as *data acquisition*. The computer can do several different things with the data, depending on the application. In an analog storage application, such as a digital oscilloscope, the computer will store the data and then transfer them to a DAC at a later time to reproduce the analog data. In a process control application, the computer can examine the data or perform computations on them to determine what control outputs to generate.

Figure 10.10(a) shows how a microcomputer is connected to a digital-ramp ADC for the purpose of data acquisition. The computer generates the START pulses that initiate each new A/D conversion. The EOC (end-of-conversion) signal from the ADC is fed to the computer. The computer examines EOC to find out when the current A/D conversion is complete; then it transfers the digital data from the ADC output into its memory.

The waveforms in Figure 10.10(b) illustrate how the computer acquires a digital version of the analog signal, V_A. The $V_{A'}$ staircase waveform that is generated internal to the ADC is shown superimposed on the V_A waveform for purposes of illustration. The process begins at t_0 when the computer generates a START pulse to start an A/D conversion cycle. The conversion is completed at t_1 when the staircase first exceeds

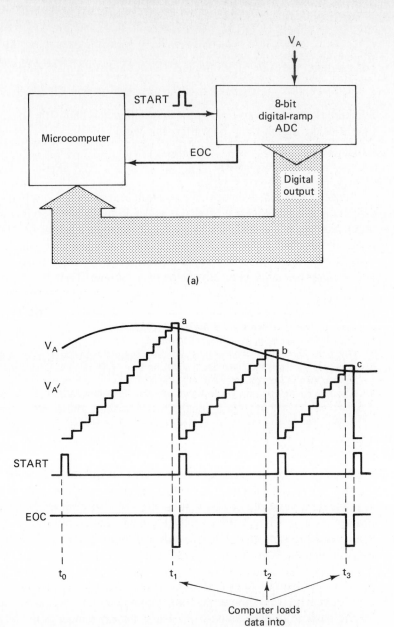

(a)

(b)

FIGURE 10.10 **(a) Typical computer data acquisition system; (b) waveforms showing how computer initiates each new conversion cycle and then loads digital data into memory at end of conversion.**

V_A, and EOC goes LOW. This NGT at EOC signals the computer that the ADC has a digital output that now represents the value of V_A at point a, and the computer will load these data into its memory.

The computer generates a new START pulse shortly after t_1 to initiate a second conversion cycle. Note that this resets the staircase to zero and EOC back HIGH because the START pulse resets the counter in the ADC. The second conversion ends at t_2 when the staircase again exceeds V_A. The computer then loads the digital data corresponding to point b into its memory. These steps are repeated at t_3, t_4, and so on.

The process whereby the computer generates a START pulse, examines EOC, and loads ADC data into memory is done under the control of a program that the computer is executing. This data acquisition program will determine how many data points from the analog signal will be stored in the computer memory.

REVIEW QUESTIONS

1. What does the term "digitize" mean?
2. Briefly describe the process of computer data acquisition.

10.8 THE SUCCESSIVE-APPROXIMATION ADC

This is one of the most widely used types of ADC. It has more complex circuitry than the digital-ramp ADC but a much shorter conversion time. In addition, successive-approximation converters (SAC) have a fixed value of conversion time that is not dependent on the value of the analog input.

The basic arrangement, as shown in Figure 10.11, is similar to the digital-ramp ADC. The SAC, however, does not use a counter to provide the input to the D/A converter block but uses a register instead. The control logic modifies the contents of the register bit-by-bit until the register data are the digital equivalent of the analog input V_A (within the resolution of the converter). The process takes place as follows:

1. The control logic sets the MSB of the register HIGH and all other bits LOW. This produces a value of $V_{A'}$ at the DAC output equal to the weight of the MSB. If $V_{A'}$ is now greater than V_A, the comparator output, COMP, goes LOW and causes the control logic to clear the MSB back to LOW. Otherwise, the MSB is kept HIGH.

2. The control logic sets the next bit of the register to 1. This produces a new value of $V_{A'}$. If this value is greater than V_A, COMP goes LOW to tell the control logic to clear the bit back to 0. Otherwise, the bit is kept at 1.

3. This process is continued for each of the bits in the register. This trial-and-error process requires one clock cycle per bit. After all bits have been tried, the register holds the digital equivalent of V_A.

We can better illustrate the successive-approximation process with a specific example. For simplicity, we will use a 4-bit converter with a step size of 1 V. Table 10.1 shows the sequence of steps for converting $V_A = 10.4$ V. There are four steps, one per clock cycle. Note how each step generates a better approximation to V_A. At the end of the fourth step the conversion is complete and the required digital value 1010 is in the register. This value is equivalent to 10 V and is *less than* V_A; this is a characteristic of the successive-approximation method.

FIGURE 10.11 **Block diagram of a successive-approximation ADC.**

TABLE 10.1

	Steps	Register	$V_{A'}$ (V)	V_A (V)	Comparator
	Initial status	0000	0	10.4	HIGH
I.	A. Set MSB to 1	1000	8	10.4	HIGH
	B. Leave it at 1 since $V_{A'} < V_A$.				
II.	A. Set second MSB to 1	1100	12	10.4	LOW
	B. Reset it to 0 since $V_{A'} > V_A$.	1000	8	10.4	HIGH
III.	A. Set third MSB to 1	1010	10	10.4	HIGH
	B. Leave it at 1 since $V_{A'} < V_A$.				
IV.	A. Set LSB to 1.	1011	11	10.4	LOW
	B. Reset it to 0 since $V_{A'} > V_A$.	1010	10	10.4	HIGH
	Digital number now in the REGISTER is the final result.	1010			

Conversion Time The same operation occurs for more practical SACs that use a greater number of bits for better resolution. In general, an *N*-bit SAC will require *N* clock cycles to perform a conversion *regardless of the value of* V_A. This is because the control circuitry has to try a 1 in each bit position to see whether it is needed or not.

EXAMPLE 10.10

Compare the maximum conversion times of a 10-bit digital-ramp A/D converter and a 10-bit successive approximation A/D converter if both utilize a 500-kHz clock frequency.

Solution: For the digital-ramp converter, the maximum conversion time is

$$(2^N - 1) \times (1 \text{ clock cycle}) = 1023 \times 2 \ \mu s = 2046 \ \mu s$$

For a 10-bit successive approximation converter, the conversion time is always 10 clock periods or

$$10 \times 2 \ \mu s = 20 \ \mu s$$

Thus, it is about 100 times faster than the digital-ramp converter.

Since SACs have relatively fast conversion times, their use in data acquisition applications (i.e., Figure 10.10) will permit more data values to be acquired in a given time interval. This can be very important when the analog data are changing at a relatively fast rate.

Because many SACs are available as ICs, it is rarely necessary to design the control logic circuitry, so we will not cover it here. For those who are interested in the details of the control logic, many manufacturers' data books should provide sufficient detail.

REVIEW QUESTIONS

1. Give one advantage and one disadvantage of the successive-approximation converter when compared to the digital-ramp converter.
2. Why is the SAC conversion time independent of the analog input value?

10.9 FLASH ADCs

This type is the highest-speed ADC available, but it requires much more circuitry than the other types. For example, a 6-bit flash ADC requires 63 analog comparators, an 8-bit unit requires 255 comparators, and a 10-bit converter requires 1023 comparators. The large number of comparators needed effectively limits *discrete* flash converters to the 6-bit level. IC flash converters are currently available in 8-bit units, and most manufacturers predict that 9- and 10-bit units will hit the market in the near future.

The principle of operation will be described for a 3-bit flash converter in order to keep the circuitry at a workable level. Once the 3-bit converter is understood, it should be easy to extend the basic idea to higher-bit flash converters.

The flash converter in Figure 10.12(a) has a 3-bit resolution and a step size of 1 V. The voltage divider sets up reference levels for each comparator so that there are seven levels corresponding to 1 V (weight of LSB), 2 V, 3 V, . . . , and 7 V (full-scale). The analog input, V_A, is connected to the other input of each comparator.

With $V_A < 1$ V, all the comparator outputs C_1–C_7 will be HIGH. With

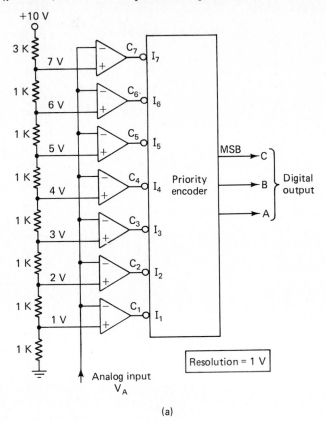

(a)

V_A	C_1	C_2	C_3	C_4	C_5	C_6	C_7	C	B	A
<1 V	1	1	1	1	1	1	1	0	0	0
>1 V, <2 V	0	1	1	1	1	1	1	0	0	1
>2 V, <3 V	0	0	1	1	1	1	1	0	1	0
>3 V, <4 V	0	0	0	1	1	1	1	0	1	1
>4 V, <5 V	0	0	0	0	1	1	1	1	0	0
>5 V, <6 V	0	0	0	0	0	1	1	1	0	1
>6 V, <7 V	0	0	0	0	0	0	1	1	1	0
>7 V	0	0	0	0	0	0	0	1	1	1

(b)

FIGURE 10.12   (a) Three-bit flash ADC; (b) truth table.

$V_A > 1$ V, one or more of the comparator outputs will be LOW. The comparator outputs are fed into an active LOW priority encoder that generates a binary output corresponding to the highest-numbered comparator output that is LOW. For example, when V_A is between 3 and 4 V, outputs C_1, C_2, and C_3 will be LOW and all others will be HIGH. The priority encoder will respond only to the LOW at C_3 and will produce a binary output $CBA = 011$, which represents the digital equivalent of V_A, within the resolution of 1 V. When V_A is greater than 7 V, C_1–C_7 will all be LOW, and the encoder will produce $CBA = 111$ as the digital equivalent of V_A. The table in Figure 10.12(b) shows the responses for all possible values of analog input.

The flash converter uses no clock signal because there is no timing or sequencing required. The conversion process takes place as soon as V_A is applied, and the conversion time depends only on the propagation delays of the comparators and the encoder. A typical flash converter can have a conversion time of 50 ns.

It should now be easy to see why the flash-converter circuitry increases dramatically as the number of desired bits increases. The 3-bit converter in Figure 10.12 requires seven comparators because there are $2^3 = 8$ possible voltage levels 0 V, 1 V, 2 V, . . . , 7 V. There is no comparator required for the 0-V level. For an 8-bit flash converter there would be $2^8 = 256$ voltage levels including 0 V, so 255 comparators would be required. In general, then, an N-bit flash converter requires $2^N - 1$ comparators.

Clearly, the major advantage of the flash converter is its operating speed, but it achieves this high-speed operation at the expense of increased circuit complexity. The high cost of these ADCs restricts their use to those applications where high speed is a prime requirement.

REVIEW QUESTIONS

1. True or false: A flash ADC does not contain a DAC.
2. How many comparators would a 12-bit flash converter contain?
3. State the major advantage and disadvantage of a flash converter.

10.10   DIGITAL VOLTMETER

A digital voltmeter (DVM) converts an analog voltage to its BCD-code representation, which is then decoded and displayed on some type of readout. Figure 10.13 shows a 3-digit DVM circuit that uses a digital-ramp ADC (shown inside the dashed lines). Three cascaded BCD counters provide the inputs to a 3-digit BCD/A converter that has a step size of 10 mV and a full-scale output of 9.99 V. Each BCD counter stage also drives a 4-bit register that feeds a decoder/driver and display. The contents of the BCD counters is transferred to the registers at the end of each conversion cycle, so that the displays do not show the counters resetting and counting, but only display the final count that represents the unknown voltage.

The clock pulses are gated into the counters with the comparator output. As long as $V_A > V_{A'}$, COMP = 1 and the counter will receive pulses. As the counter advances, the $V_{A'}$ waveform goes up 10 mV per step until $V_{A'}$ exceeds V_A by 0.1 mV. At that point, COMP goes LOW and disables the AND gate so that the counter will no longer

FIGURE 10.13 Continuous-conversion DVM using a digital-ramp ADC.

advance. The negative transition at COMP also triggers one-shot OS1 which produces 1-μs pulse at Q_1. The PGT of Q_1 provides the clock transition that transfers the BCD counter outputs to their respective registers. The NGT of Q_1 triggers a second one-shot Q_2 producing a 1-μs pulse at Q_2 to reset all the counters back to zero. This brings $V_{A'}$ back to 0V and COMP returns HIGH, allowing pulses into the counter to begin a new conversion cycle.

Thus, this DVM will perform one conversion right after another. Of course, the storage registers will keep the displays from showing the conversion process. The display readings will change only if V_A changes, so that a different counter contents is transferred to the registers at the end of the conversion cycle.

A numerical example will help illustrate this circuit's operation. Assume that V_A is 6.372 V. In order to cause COMP output to switch LOW, $V_{A'}$ must exceed 6.3721 V. Since the DAC output increases by 10 mV/step, this requires

$$\frac{6.3721 \text{ V}}{10 \text{ mV}} = 637.21 \longrightarrow 638 \text{ steps}$$

Thus, the counters will count up to 638, which will be transferred to the registers and displayed. A small LED can be used to display a decimal point so that the operator sees 6.38 V.

EXAMPLE 10.11

What will happen if V_A is greater than 9.99 V?

Solution: With $V_A > 9.99$ V, the comparator output will stay HIGH, allowing clock pulses into the counter continuously. The counter will repetitively count up to 999 and recycle to 000. One-shot Q_1 will *never* be triggered and the register contents will not change from its previous value. A well-designed DVM would have some means for detecting this *over-range* condition and activating some type of over-range indicator. One possible method would simply add a 13th bit to the counter string. This bit would go HIGH only if the counter recycled from 999 to 000, and would indicate the over-range condition.

The DVM can be modified to read input voltages over several ranges by using a suitable amplifier or attenuator between V_A and the comparator. For Example, if V_A were 63.72 V, it could be attenuated by a factor of 10, so the comparator would receive 6.372 V at its + input and the counters would display 638 at the end of the conversion. The decimal-point indicator would be placed in front of the LSD, so the display would read 63.8 V.

10.11 TRISTATE ADC

When the tristate concept was first discussed, it was shown that many tristate devices could be connected to a common bus as long as only one device was activated at one time. The rapid emergence of microprocessors and their bus-type structure has brought about the development of tristate ADCs whose outputs can be tied to a system bus. Figure 10.14 shows a functional diagram for an ADC of this type.

This ADC converts the analog input voltage V_A to an 8-bit output. It has a full-scale rating of 2.55 V that translates to a resolution (step size) of 10 mV. The conversion process is initiated by a pulse applied to its START input. The EOC output from the ADC goes LOW while the conversion is taking place and returns HIGH to signal the end of the conversion.

The digital output lines D_7–D_0 come from a tristate register that is part of the converter. A HIGH on the ENABLE input will enable these outputs so that the digital representation of V_A is present on these lines. A LOW on the ENABLE input puts these output lines in their HIGH-Z state. In most situations the ENABLE input will be

FIGURE 10.14 Tristate ADC.

pulsed HIGH only after the EOC output has indicated that the conversion is complete. If the ENABLE input is made HIGH during the t_c interval, the output lines will indicate the results of the previous A/D conversion.

10.12 *SAMPLE-AND-HOLD CIRCUITS*

When an analog voltage is connected directly to the input of an ADC, the conversion process can be adversely affected if the analog voltage is changing during the conversion time. The stability of the conversion process can be improved by using a *sample-and-hold circuit* to hold the analog voltage constant while the A/D conversion is taking place. A simplified diagram of a sample-and-hold (S/H) circuit is shown in Figure 10.15.

The S/H circuit contains an input buffer amplifier *A1* that presents a high impedance to the analog signal, and has a low output impedance that can rapidly charge the hold capacitor, C_h. The capacitor will be connected to the output of *A1* when the digitally controlled switch is closed. This is called the *sample* operation. The switch will be closed long enough for C_h to charge to the current value of the analog input. For example, if the switch is closed at time t_0, the *A1* output will quickly charge C_h up to a voltage V_0. When the switch opens, C_h will *hold* this voltage so that the output of *A2* will apply this voltage to the ADC. The output buffer amplifier *A2* presents a high input impedance that will not discharge the capacitor voltage appreciably during the conversion time of the ADC, so the ADC will essentially receive a dc input voltage V_0.

In a computer-controlled data acquisition system such as the one discussed earlier, the sample-and-hold switch would be controlled by a digital signal from the computer. The computer signal would close the switch in order to charge C_h to a new sample of the analog voltage; the amount of time the switch would have to remain

* Control = 1 → switch closed → sample mode
Control = 0 → switch open → hold mode

FIGURE 10.15 Simplified diagram of a sample-and-hold circuit.

closed is called the *acquisition time* and it depends on the value of C_h and the characteristics of the S/H circuit. The LF198 is an integrated S/H circuit that has a typical acquisition time of 4 μs for $C_h = 1000$ pF, and 20 μs for $C_h = 0.01$ μF. The computer signal would then open the switch to allow C_h to hold its value and provide a relatively constant analog voltage at the A2 output. For example, with the LF198, the capacitor voltage will typically discharge at the rate of only 30 mV per second for a 1000-pF capacitor.

REVIEW QUESTIONS

1. Describe the function of a sample-and-hold circuit.
2. True or false: The amplifiers in the S/H circuit are used to provide voltage amplification.

10.13 MULTIPLEXING

When analog inputs from several sources are to be converted, a multiplexing technique can be used so that one A/D converter may be time-shared. The basic scheme is illustrated in Figure 10.16 for a three-channel acquisition system. Switches S_1, S_2, and S_3 are used to switch each analog signal sequentially to the A/D converter. The control circuitry controls the actuation of these switches, which are usually semiconductor switches, so that only one switch is closed at a time. The control circuitry also generates the START pulse for the ADC. The operation proceeds in the following manner:

1. The control circuit closes S_1, which connects V_{A1} to the ADC input.
2. A START pulse is generated and the ADC converts V_{A1} to its digital equivalent. S_1 stays closed long enough to allow the conversion to be completed.
3. The ADC outputs representing V_{A1} can now be transferred to another location. In many cases this location would be in the memory of a computer.

Sec. 10.13   **Multiplexing**

4. The control circuit opens S_1 and closes S_2 to connect V_{A2} to the ADC input.

5. Steps 2 and 3 are repeated.

6. S_2 is open and S_3 is closed to connect V_{A3} to the ADC input.

7. Steps 2 and 3 are repeated.

The multiplexing clock controls the rate at which the analog signals are sequentially switched into the ADC. The maximum rate is determined by the delay time of the switches and the conversion time of the ADC. The switch delay time can be minimized by using semiconductor switches such as the CMOS bilateral switches described in Chapter 8. It may be necessary to connect a sample-and-hold circuit at the input of the ADC if the analog inputs will change significantly during the ADC conversion time.

Many integrated ADCs contain the multiplexing circuitry on the same chip as the ADC. The ADC0808, for example, can multiplex eight different analog inputs into one ADC. It uses a 3-bit select input code to determine which analog input is connected to the ADC.

FIGURE 10.16 Conversion of several analog inputs by multiplexing through one A/D converter.

PROBLEMS

10.1 An 8-bit D/A converter produces an output voltage of 2.0 V for an input code of 01100100. What will be the value of V_{OUT} for an input code of 10110011?

10.2 Determine the weights of each input bit for the D/A converter of Problem 10.1.

10.3 What is the resolution of the D/A converter of Problem 10.1? Express it in volts and in percent.

10.4 What is the resolution in volts of a 10-bit D/A converter whose F.S. output is 5 V?

10.5 How many bits are required for a D/A converter so that its F.S. output is 10 mA and its resolution is less than 40 μA?

10.6 What is the percentage resolution of the D/A converter of Figure 10.17? What is the step size if the top step is 2 V?

10.7 What is the cause of the negative-going spikes on the V_{OUT} waveform of Figure 10.17? (*Hint:* Note that the counter is a ripple counter and that the spikes occur on every other step.)

FIGURE 10.17

10.8 True or False: The percentage resolution of a D/A converter depends *only* on the number of input bits.

10.9 A 12-bit (three-digit) D/A converter which uses the BCD input code has a full-scale output of 9.99 V. Determine the step size, percentage resolution, and the value of V_{OUT} for an input code of 0110 1001 0101.

10.10 Which will have a higher percentage resolution, a D/A converter using an 8-bit binary input code or one using an 8-bit BCD code?

10.11 The step size of the D/A converter of Figure 10.5 can be changed by changing the value of R_F. Determine the required value of R_F for a step size of 0.5 V. Will the new value of R_F change the percentage resolution?

10.12 Assume that the output of the DAC in Figure 10.7(a) is connected to the op-amp of Figure 10.7(b).
(a) With $V_{REF} = 5$ V, $R = 20$ kΩ, and $R_F = 10$ kΩ, determine the step size and full-scale voltage at V_{OUT}.
(b) Change the value of R_F so that the full-scale voltage at V_{OUT} is -2 V.
(c) Use this new value of R_F, and determine the proportionality factor, k, in the relationship $V_{OUT} = k(V_{REF} \times B)$.

10.13 The DAC from Problem 10.12 can be used as a multiplying DAC, where V_{REF} is a signal and V_{OUT} is an amplified version of V_{REF} with the amplification factor controlled by the binary input to the DAC. This binary input can come from a computer that is being used to control the amplitude of an analog signal.

Assume that V_{REF} is an audio signal with an amplitude of 5 V p-p. Determine the amplitude of V_{OUT} for each possible binary input.

10.14 An 8-bit D/A converter has a full-scale error of 0.2 percent F.S. If the DAC has a full-scale output of 10 mA, what is the most that it can be in error for any digital input? If the D/A output reads 50 μA for a digital input of 00000001, is this within the specified range of accuracy?

10.15 The control of a positioning device may be achieved using a *servomotor*, which is a motor designed to drive a mechanical device as long as an error signal exists. Figure 10.18 shows a simple servo-controlled system which is controlled by a digital input that could be coming directly from a computer or from an output medium such as magnetic tape. The lever arm is moved vertically by the servomotor. The motor rotates clockwise or counterclockwise, depending on whether the voltage from the power amplifier (PA) is positive or negative. The motor stops when the PA output is zero.

FIGURE 10.18

The mechanical position of the lever is converted to a dc voltage by the potentiometer arrangement shown. When the lever is at its zero reference point, $V_P = 0$ V. The value of V_P increases at the rate of 1 V/inch until the lever is at its highest point (10 inches) and $V_P = 10$ V. The desired position of the lever is provided as a digital code from the computer and is then fed to a D/A converter, producing V_A. The *difference* between V_P and V_A (called *error*) is produced by the *differential* amplifier and amplified by the PA to drive the motor in the direction that causes the error signal to decrease to 0—that is, moves the lever until $V_P = V_A$.

(a) If it is desired to position the lever within a resolution of 0.1 inch, what is the number of bits needed in the digital input code?

(b) In actual operation, the lever arm might oscillate slightly around the desired position, especially if a *wirewound* potentiometer is used. Can you explain why?

10.16 An 8-bit digital-ramp A/D converter with a 40-mV resolution uses a clock frequency of 2.5 MHz and a comparator with $V_T = 1$ mV. Determine:

(a) The digital output for $V_A = 6.000$ V.

(b) The digital output for 6.035 V.

(c) The maximum and average conversion times for this ADC.

10.17 Why were the digital outputs the same for parts (a) and (b) of Problem 10.16?

10.18 What would happen in the A/D converter of Problem 10.16 if an analog voltage of $V_A = 10.853$ V were applied to the input? What waveform would appear at the D/A output? Incorporate the necessary logic in this A/D converter so that an "over-scale" indication will be generated whenever V_A is too large.

10.19 An A/D converter has the following characteristics: resolution, 12 bits; full-scale error, 0.03 percent F.S.; full-scale output, +5 V.

(a) What is the quantization error in volts?

(b) What is the total possible error in volts?

10.20 The quantization error of an A/D converter such as the one in Figure 10.9 is always positive since the $V_{A'}$ value must exceed V_A in order for the COMPARATOR output to switch states. This means the value of $V_{A'}$ could be as much as 1 LSB greater than V_A. This quantization error can be modified so that $V_{A'}$ would be within $\pm \frac{1}{2}$ LSB of V_A. This can be done by adding a fixed voltage equal to $\frac{1}{2}$ LSB ($\frac{1}{2}$ of a step) to the value of $V_{A'}$. Figure 10.19 shows this symbolically for a converter that has a resolution of 10 mV/step. A fixed voltage of +5 mV is added to the D/A output in the summing amplifier and the result, $V_{A''}$, is fed to the comparator which has $V_T = 1$ mV.

For this modified converter, determine the digital output for:

(a) $V_A = 5.022$ V

(b) $V_A = 5.028$ V

Determine the quantization error in each case by comparing $V_{A'}$ and V_A. Note that the error is positive in one case and negative in the other.

10.21 Construct a table such as Table 10.1 showing the sequence of operations for a 6-bit successive-approximation converter which has a resolution of 0.1 V per step and an analog input of $V_A = 4.2$ V. Assume that $V_T = 1$ mV for the comparator.

10.22 A certain 8-bit successive-approximation converter has 2.55 V F.S. The conversion time for $V_A = 1$ V is 80 μs. What will be the conversion time for $V_A = 1.5$ V?

10.23 Figure 10.20 shows the waveform at $V_{A'}$ for a 6-bit SAC with a step size of 40 mV during a complete conversion cycle. Examine this waveform and describe what is occurring at times $t_0–t_5$. Then determine the resultant digital output.

10.24 Draw the circuit diagram for a 4-bit flash converter with BCD output and a resolution of 0.1 V. Assume that a +5-V precision supply voltage is available.

10.25 A voltage $V_A = 3.853$ V is applied to the DVM of Figure 10.13. Asume that the counters are initially reset to zero and sketch the waveforms at $V_{A'}$, COMP, Q_1, Q_2, and the AND gate output for *two* complete cycles. Use a 100-kHz clock and assume both one-shots have $t_p = 1$ μs.

FIGURE 10.19

10.26 Refer to the DVM circuit in Figure 10.13. Assume that the circuit works correctly for V_A values up to about 7.99 V. However, when V_A is increased above 8 V, the displays do not change from their previous reading, the staircase waveform at $V_{A'}$ steps up to only about 7.99 volts before jumping back down to zero, and COMP remains HIGH. What are some possible causes for this malfunction?

10.27 Refer to the sample-and-hold circuit of Figure 10.15. What circuit fault would result in V_{OUT} looking exactly like V_A? What fault would cause V_{OUT} to be stuck at zero?

10.28 Use the CMOS 4016 bilateral switches for S_1, S_2, and S_3 in Figure 10.16 and design the necessary control logic so that each analog input is converted to its digital equivalent in sequence. The A/D converter is a 10-bit successive-approximation type using a 50-kHz clock signal and requires a 10-μs duration start pulse to begin each conversion. The digital outputs are to remain stable for 100 μs after the conversion is complete before switching to the next analog input. Choose the appropriate multiplexing clock frequency.

FIGURE 10.20

11
MEMORY DEVICES

OUTLINE

A major advantage of digital over analog systems is the ability to easily store large quantities of digital information and data for short or long periods. This memory capability is what makes digital systems so versatile and adaptable to many situations. For example, in a digital computer the internal main memory stores instructions that tell the computer what to do under *all* possible circumstances so that the computer will do its job with a minimum amount of human intervention.

This chapter is devoted to a review of the most commonly used types of memory devices and systems. We have already become very familiar with the flip-flop, which is an electronic memory device. We have also seen how groups of FFs called registers can be used to store information and how this information can be transferred to other locations. FF registers are high-speed memory elements which are used extensively in the internal operations of a digital computer, where digital information is continually being moved from one location to another. Advances in LSI and VLSI technology have made it possible to obtain large numbers of FFs on a single chip arranged in various memory-array formats. These bipolar and MOS semiconductor memories are the fastest memory devices available, and their cost has been continuously decreasing as LSI technology improves.

Digital data can also be stored as charges on capacitors, and a very important type of semiconductor memory uses this principle to obtain high-density storage at low power-requirement levels. Still another means for storage is provided by magnetic devices. In particular, *magnetic core* memories have been widely used since the late 1950s, although MOS memories have gradually replaced them in most computers.

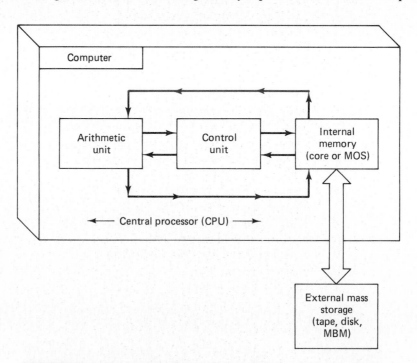

FIGURE 11.1 A computer system normally uses high-speed internal memory and slower external mass memory.

Semiconductor and magnetic core memories are used as the *internal* memory of a computer (Figure 11.1) where fast operation is important. The internal computer memory is in constant communication with the central processing unit (CPU) as a program of instructions is being executed. The program and any data used by the program are usually stored in the internal memory.

Although semiconductor and magnetic core memories are well suited for high-speed internal memory, their cost per bit of storage prohibits their use as *mass storage* devices. Mass storage refers to memory that is external to the main computer (Figure 11.1) and has the capacity to store millions of bits of data *without* the need for electrical power. Mass memory is normally much slower than internal memory and is used to store information (programs, data, etc.) not currently being used by the computer. This information is transferred to the internal memory when the computer needs it.

Magnetic tape and magnetic disk are popular mass storage devices that are much less expensive in cost per bit than internal memory devices. A newer entry into the mass memory category is the *magnetic bubble memory* (MBM), a semiconductor device that uses magnetic principles to store millions of bits on one chip. The MBM is relatively slow and cannot be used as internal memory.

We will take a more detailed look at the characteristics of these memory devices after we define some of the more important terms that are common to many memory systems.

11.1 MEMORY TERMINOLOGY

Memory cell: A device or electrical circuit used to store a single bit (0 or 1). Examples of memory cells include a flip-flop, a single magnetic core, and a single spot on magnetic tape or disk.

Memory word: A group of bits (cells) in a memory that represents instructions or data of some type. For example, a register consisting of eight FFs can be considered to be a memory that is storing an 8-bit word. Word sizes in modern computers typically range from 4 to 64 bits, depending on the size of the computer.

Byte: A special term used for an 8-bit word. A byte always consists of 8 bits, which is the most common word size in microcomputers.

Capacity: A way of specifying how many bits can be stored in a particular memory device or complete memory system. To illustrate, suppose that we have a memory which can store 4096 20-bit words. This represents a total capacity of 81,920 bits. We could also express this memory's capacity as 4096×20. When expressed this way, the first number (4096) is the number of words and the second number (20) is the number of bits per word (word size). The number of words in a memory is often a multiple of 1024. It is common to use the designation "1K" to represent 1024 when referring to memory capacity. Thus, a memory that has a storage capacity of $4K \times 20$ is actually a 4096×20 memory.

EXAMPLE 11.1

A certain semiconductor memory chip is specified as $2K \times 8$. How many words can be stored on this chip? What is the word size? How many total bits can this chip store?

Solution:

$$2K = 2 \times 1024 = \textbf{2048} \text{ words}$$

Each word is 8 bits (one byte). The total number of bits is therefore

$$2048 \times 8 = \textbf{16,384} \text{ bits}$$

Address: A number that identifies the location of a word in memory. Each word stored in a memory device or system has a unique address. Addresses are always specified as a binary number, although octal, hexadecimal, and decimal numbers are often used for convenience. Figure 11.2 illustrates a small memory consisting of eight words. Each of these eight words has a specific address represented as a 3-bit number ranging from 000 to 111. Whenever we refer to a specific word location in memory, we use its address code to identify it.

Read operation: The operation whereby the binary word stored in a specific memory location (address) is sensed and then transferred to another location. For example, if we want to use word 4 of the memory of Figure 11.2 for some purpose, we have to perform a read operation on address 100. The read operation is often called a *fetch* operation, since a word is being fetched from memory. We will use both terms interchangeably.

Write operation: The operation whereby a new word is placed into a particular memory location. It is also referred to as a *store* operation. Whenever a new word is written into a memory location, it replaces the word that was previously stored there.

Access time: A measure of a memory device's operating speed. It is the amount of time required to perform a read operation. More specifically, it is the time between the memory receiving a read command signal and the data becoming available at the memory output. The symbol t_{ACC} is used for access time.

Cycle time: Another measure of a memory device's speed. It is the amount of time required for the memory to perform a read or write operation and then return to its original state ready for the next command. Cycle time is normally longer than access time.

Volatile memory: Any type of memory that requires the application of electrical power in order to store information. If the electrical power is removed, all information stored in the memory will be lost. Many semiconductor memories are volatile, while all magnetic memories are nonvolatile.

Random-access memory (RAM): Memory in which the actual physical location of a memory word has no effect on how long it takes to read from or write into that location. In other

Addresses

0 0 0	Word 0
0 0 1	Word 1
0 1 0	Word 2
0 1 1	Word 3
1 0 0	Word 4
1 0 1	Word 5
1 1 0	Word 6
1 1 1	Word 7

FIGURE 11.2 Each word location has a specific binary address.

words, the access time is the same for any address in memory. Most semiconductor memories and magnetic core memories are RAMs.

Sequential-access memory (SAM): A type of memory in which the access time is not constant, but varies depending on the address location. A particular stored word is found by sequencing through all address locations until the desired address is reached. This produces access times which are much longer than those of random access memories. Examples of sequential-access memory devices include magnetic tape, disc, and magnetic bubble memory.

Read/write memory (RWM): Any memory that can be read from or written into with equal ease.

Read-only memory (ROM): A broad class of semiconductor memories designed for applications where the ratio of read operations to write operations is very high. Technically, a ROM can be written into (programmed) only once, and this operation is normally performed at the factory. Thereafter information can only be read from the memory. Other types of ROM are actually read-mostly memories (RMM) which can be written into more than once, but the write operation is more complicated than the read operation and it is not performed very often. The various types of ROM will be discussed later.

Static memory devices: Semiconductor memory devices in which the stored data will remain permanently stored as long as power is applied, without the need for periodically rewriting the data into memory.

Dynamic memory devices: Semiconductor memory devices in which the stored data *will not* remain permanently stored, even with power applied, unless the data are periodically rewritten into memory. The latter operation is called a *refresh* operation.

REVIEW QUESTIONS

1. Define the following terms: (a) memory cell; (b) memory word; (c) address; (d) byte; (e) access time.
2. A certain memory has a capacity of 8K × 16. How many bits are in each word? How many words are being stored? How many memory cells does this memory contain? (*Ans.* 16, 8192, 131,072)
3. Explain the difference between the read (fetch) and write (store) operations.
4. True or false: A volatile memory will lose its stored data when electrical power is interrupted.
5. Explain the difference between SAM and RAM.
6. Explain the difference between RWM and ROM.
7. True or false: A dynamic memory will hold its data as long as electrical power is applied.

11.2 GENERAL MEMORY OPERATION

Although each type of memory is different in its internal operation, certain basic operating principles are the same for all memory systems. An understanding of these basic ideas will help in our study of individual memory devices.

Every memory system requires several different types of input and output lines to perform the following functions:

1. Select the address in memory that is being accessed for a read or write operation.
2. Select either a read or write operation to be performed.
3. Supply the input data to be stored in memory during a write operation.
4. Hold the output data coming from memory during a read operation.
5. Enable (or disable) the memory so that it will (or will not) respond to the address inputs and read/write command.

Figure 11.3 illustrates these basic functions in a simplified diagram of a 32×4 memory that stores 32 4-bit words. Since the word size is 4 bits, there are four data input lines I_0–I_3 and four data output lines O_0–O_3. During a write operation the data to be stored into memory have to be applied to the data input lines. During a read operation the word being read from memory appears at the data output lines.

FIGURE 11.3 General diagram for a 32 × 4 memory.

This memory has 32 different storage locations and therefore 32 different addresses, ranging from 00000_2 to 11111_2 (0 to 31 in decimal). Thus, we need *five* address inputs, A_0–A_4, to specify one of the 32 address locations. To access one of the memory locations for a read or write operation, the 5-bit address code for that particular location has to be applied to the address inputs.

The read/write ($R/\overline{W}$) input line determines which memory operation is to take place. Some memory systems use two separate inputs, one for read and one for write. When a single $R/\overline{W}$ input is used, it is common to have $R/\overline{W} = 1$ commanding a read operation and $R/\overline{W} = 0$ a write operation.

Many memory systems have some means for completely disabling all or part of the memory so that it will not respond to the other inputs. This is represented in Figure 11.3 as the MEMORY ENABLE input, although it can have different names in the various memory systems. It is an active-HIGH input that enables the memory to operate normally when it is kept HIGH. A LOW on this input disables the memory so that it will not respond to the address and $R/\overline{W}$ inputs. This type of input is useful when several memory modules are combined to form a larger memory. We will examine this idea later.

EXAMPLE 11.2

A certain memory has a capacity of 4K × 8.
(a) How many data input and data output lines does it have?
(b) How many address lines does it have?
(c) What is its capacity in bytes?

Solution:
(a)   Eight of each, since the word size is eight.
(b)   The memory stores 4K = 4 × 1024 = 4096 words. Thus, there are 4096 memory addresses. Since 4096 = 2^{12}, it requires a 12-bit address code to specify one of 4096 addresses.
(c)   A byte is 8 bits. This memory has a capacity of 4096 bytes.

The example memory in Figure 11.3 illustrates the important input and output functions common to most memory systems. Of course, each type of memory will often have other input and output lines that are peculiar to that memory. These will be described as we discuss the individual memory types.

_____ **REVIEW QUESTION**

1.   How many address inputs, data inputs, and data outputs are required for a 16K × 12 memory? (*Ans.* 14, 12, 12)

11.3   SEMICONDUCTOR MEMORY TECHNOLOGIES

On the basis of their manufacturing process or *technology*, semiconductor memories are divided into two categories: *bipolar memories* (made with bipolar transistors) and *MOS memories* (made with MOSFETs). Figure 11.4 shows the various technologies that fall into each category.

The memory devices that are manufactured using the various technologies will generally differ in their electrical and physical characteristics, and they are usually compared according to the following criteria:

1. Density (i.e., capacity in bits per chip).
2. Speed (i.e., access time).
3. Power requirements (i.e., μW or mW per bit).
4. Cost per bit.
5. Noise immunity.

FIGURE 11.4   Semiconductor memory technologies.

Sec. 11.3   **Semiconductor Memory Technologies**   **449**

Both bipolar and MOS memory technologies have their particular strengths and weaknesses based on the above criteria. In general, though, the bipolar devices have the advantage of higher speed, while MOS memory devices are superior in all the other areas.

Bipolar Memories   All the bipolar technologies use resistors, diodes, and bipolar transistors for the memory circuitry on the chip. The relative complexity of the bipolar circuitry imposes a limitation on the memory capacity that can be placed on a single silicon chip, and bipolar memories have the lowest capacity of any technology.

Standard TTL and Schottky TTL technologies account for the major portion of bipolar memory devices in use today. These technologies use the same basic structures that were discussed in Chapter 8. The basic characteristics of standard TTL technology memories are high speed, moderate capacity, high power consumption, relatively low noise immunity, and relatively high cost. Schottky TTL memories share these same characteristics but are faster than standard TTL memories. TTL memories are suited for applications that demand relatively high speed and medium capacity and where power consumption is not a critical factor.

ECL memories are the fastest memory devices available today because they use the nonsaturating bipolar transistor operation described in Chapter 8. Unfortunately, high speed is the only advantage it has over the other technologies, and it is used only where the utmost in operating speed is required without regard to cost or power consumption.

MOS Memories   The workhorse of memory technologies is N-channel MOS technology. NMOS technology currently provides the highest capacity of all memory technologies at the lowest cost per bit, and at power levels that are much lower than bipolar memories. Their only disadvantage is their slower speed compared to bipolar. The refinements in the NMOS fabrication process, however, have not yet reached their limits in terms of speed and density, and NMOS memory devices are gradually approaching the operating speed of the lower-speed bipolar memories. *VMOS* (*vertical MOS*) and *HMOS* (*high-performance MOS*) are two of the newer MOS processes that appear to be approaching the performance of bipolar technologies.

CMOS memory devices provide the lowest power consumption of all the memory technologies. CMOS memories also have the highest noise immunity of any

Technology	Speed	Power required	Capacity	Noise immunity	Cost
TTL/S-TTL	Fast	High	Low	Low	High
ECL	Fastest	Highest	Lowest	Lowest	Highest
NMOS	Medium (approaching TTL)	Low	Highest	High	Lowest
CMOS/SOS	Slowest	Lowest	High	Highest	Low

FIGURE 11.5   Comparison of memory technologies.

technology. In capacity, on the other hand, CMOS is somewhat lower than NMOS, although it is higher than bipolar, and CMOS operating speed is the slowest of any technology. An improved CMOS process uses a *silicon-on-sapphire* (SOS) substrate instead of pure silicon to achieve faster speeds and higher densities. CMOS and SOS are best suited for applications in high-noise environments or where power consumption is extremely critical (i.e., battery-operated systems).

Figure 11.5 compares the major memory technologies.

REVIEW QUESTIONS

1. Which memory technology has the greatest capacity per chip?
2. Which memory technology would be best to use in a hand calculator?
4. Which memory technology would be best to use in a very high speed computer?
4. Which memory technology generally has the lowest cost per bit?
5. Which memory technology would be best to use in a high-noise industrial environment?

11.4 READ-ONLY MEMORIES—ROMs

This type of semiconductor memory is designed to hold data that either are permanent or will not change frequently. During normal operation, no new data can be written into a ROM but data can be read from ROM. For some ROMs the data that are stored have to be built in during the manufacturing process; for other ROMs the data can be entered electrically. The process of entering data is called *programming* or *burning* the ROM. Some ROMs cannot have their data changed once they have been programmed; others can be *erased* and reprogrammed as often as desired. We will take a detailed look later at these various types of ROMs. For now, we will assume that the ROMs have been programmed and are holding data.

ROMs are used to store data and information that is not to change during the operation of a system. A major use for ROMs is in the storage of programs in microcomputers. Since all ROMs are *nonvolatile,* these programs are not lost when the microcomputer is turned off. When the microcomputer is turned on, it can immediately begin executing the program stored in ROM. ROMs are also used for program and data storage in microprocessor-controlled equipment such as sophisticated electronic cash registers.

ROM Block Diagram A typical block diagram for a ROM is shown in Figure 11.6(a). It has three sets of signals: address inputs, control input(s), and data outputs. From our previous discussions we can determine that this ROM is storing 16 words, since it has $2^4 = 16$ possible addresses, and each word contains 8 bits, since there are eight data outputs. Thus, this is a 16×8 ROM. Another way to describe this ROM's capacity is to say that it stores 16 bytes of data.

The data outputs of most ROM ICs are either open-collector or tristate outputs to permit the connection of many ROM chips to the same data bus for memory expansion. The most common numbers of data outputs for ROMs are 4 bits and 8 bits, with 8-bit words being the most common.

The control input *CS* stands for *chip select.* This is essentially an enable input that enables or disables the ROM outputs. Some manufacturers use different labels for

(a)

Word	A₃ A₂ A₁ A₀	D₇-D₀
0	0	DE
1	1	3A
2	2	85
3	3	AF
4	4	19
5	5	7B
6	6	00
7	7	ED
8	8	3C
9	9	FF
10	A	B8
11	B	C7
12	C	27
13	D	6A
14	E	D2
15	F	5B

(c)

Word	A_3	A_2	A_1	A_0	D_7	D_6	D_5	D_4	D_3	D_2	D_1	D_0
0	0	0	0	0	1	1	0	1	1	1	1	0
1	0	0	0	1	0	0	1	1	1	0	1	0
2	0	0	1	0	1	0	0	0	0	1	0	1
3	0	0	1	1	1	0	1	0	1	1	1	1
4	0	1	0	0	0	0	0	1	1	0	0	1
5	0	1	0	1	0	1	1	1	1	0	1	1
6	0	1	1	0	0	0	0	0	0	0	0	0
7	0	1	1	1	1	1	1	0	1	1	0	1
8	1	0	0	0	0	0	1	1	1	1	0	0
9	1	0	0	1	1	1	1	1	1	1	1	1
10	1	0	1	0	1	0	1	1	1	0	0	0
11	1	0	1	1	1	1	0	0	0	1	1	1
12	1	1	0	0	0	0	1	0	0	1	1	1
13	1	1	0	1	0	1	1	0	1	0	1	0
14	1	1	1	0	1	1	0	1	0	0	1	0
15	1	1	1	1	0	1	0	1	1	0	1	1

(b)

FIGURE 11.6 (a) ROM block diagram; (b) programmed data; (c) hex data.

the control input, such as *CE* (chip enable) or *OE* (output enable). Many ROMs have more than two or more control inputs that have to be activated in order to enable the data outputs. The *CS* input shown in Figure 11.6(a) is active HIGH. If it had a small circle or an inversion symbol ($\overline{CS}$), it would be active LOW. Note that there is no *R/W* (read/write) input because the ROM cannot be written into under normal operating circumstances.

The READ Operation Let's assume that the ROM has been programmed with the data shown in the table of Figure 11.6(b). Sixteen different data words are stored at the 16 different address locations. For example, the data word stored at location 0011 is 10101111. Of course, the data are stored in binary inside the ROM, but very often we use hexadecimal notation to efficiently show the programmed data. This is done in Figure 11.6(c).

In order to read a data word from ROM, we need to do two things: apply the appropriate address inputs, and then activate the control inputs. For example, if we want to read the data stored at location 0111 of the ROM in Figure 11.6, we have to apply $A_3A_2A_1A_0 = 0111$ to the address inputs, and then apply a HIGH to CS. The address inputs will be decoded inside the ROM to select the correct data word, 11101101, that will appear at the D_7–D_0 outputs. If CS is kept LOW, the ROM outputs will be disabled and will be in the Hi-Z state.

REVIEW QUESTIONS

1. True or false: All ROMs are nonvolatile.
2. Describe the procedure for reading from ROM.

11.5 ROM ARCHITECTURE

The internal architecture (structure) of a ROM IC is very complex, and we need not be familiar with all its detail. It is instructive, however, to look at a simplified diagram of the internal architecture, such as that shown in Figure 11.7 for the 16 × 8 ROM. There are four basic parts: *row decoder, column decoder, register array, output buffers*.

Register Array The register array stores the data that have been programmed into the ROM. Each register contains a number of memory cells equal to the word size. In this case, each register stores an 8-bit word. The registers are arranged in a square matrix array that is common to many semiconductor memory chips. We can specify the position of each register as being in a specific row and specific column. For example, register 0 is in row 0/column 0, and register 9 is in row 1/column 2.

The eight data outputs of each register are connected to an internal data bus that runs through the entire circuit. Each register has two enable inputs (E); both have to be HIGH in order for the register's data to be placed on the bus.

Address Decoders The applied address code $A_3A_2A_1A_0$ determines which register in the array will be enabled to place its 8-bit data word onto the bus. Address bits A_1A_0 are fed to a 1-of-4 decoder which activates one row-select line, and address bits A_3A_2 are fed to a second 1-of-4 decoder which activates one column-select line. Only one register will be in both the row and the column selected by the address inputs, and this one will be enabled.

EXAMPLE 11.3

What register will be enabled by input address 1101?

Solution: $A_3A_2 = 11$ will cause the column decoder to activate the column 3 select line, and $A_1A_0 = 01$ will cause the row decoder to activate the row 1 select line. This will place HIGHs at both enable inputs of register 13, thereby causing its data outputs to be placed on the bus. Note that the other registers in column 3 will only have one enable input activated; likewise for the other row 1 registers.

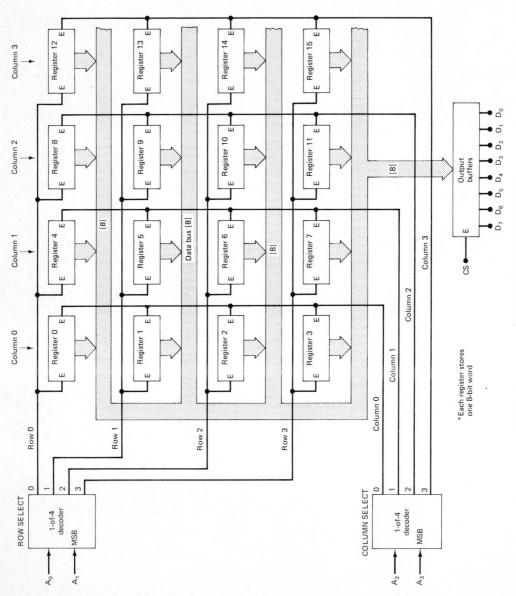

FIGURE 11.7 Architecture of 16 × 8 ROM.

EXAMPLE 11.4

What input address will enable register 10?

Solution: The enable inputs of this register are connected to the row 2 and column 2 select lines, respectively. To select row 2, the A_1A_0 inputs have to be at 10, and to select column 2, the A_3A_2 inputs have to be at 10. Thus, the required address will be $A_3A_2A_1A_0 = 1010$.

Output Buffers   The register that is enabled by the address inputs will place its data on the data bus. These data feed into the output buffers, which will pass the data to the external data outputs, provided that *CS* is HIGH. If *CS* is LOW, the output buffers are in the Hi-Z state, and D_7-D_0 will be floating.

The architecture shown in Figure 11.7 is similar to that of most IC ROMs. Depending on the number of stored data words, the registers in some ROMs will not be arranged in a square array. For example, the Intel 2708 is a MOS ROM that stores 1024 8-bit words. Its 1024 registers are arranged in a 64-by-16 array. In practice, ROMs may contain anywhere from 512 to 256K bits.

EXAMPLE 11.5

Describe the internal architecture of a ROM that stores 4K bytes and uses a square register array.

Solution: 4K is actually $4 \times 1024 = 4096$, so this ROM holds 4096 8-bit words. Each word can be thought of as being stored in an 8-bit register, and there are 4096 registers connected to a common data bus internal to the chip. Since $4096 = 64^2$, the registers are arranged in a 64-by-64 array; that is, there are 64 rows and 64 columns. This requires a 1-of-64 decoder to decode six address inputs for the row select, and a second 1-of-64 decoder to decode six other address inputs for the column select. Thus, a total of 12 address inputs are required. This makes sense, since $2^{12} = 4096$, and there are 4096 different addresses.

REVIEW QUESTIONS

1. What address inputs are required if we want to read the data from register 9 in Figure 11.7? (*Ans.* $A_3A_2A_1A_0 = 1001$)
2. Describe the function of the row-select decoder, column-select decoder, and output buffers in the ROM architecture.

11.6 ROM TIMING

There will be a propagation delay between the application of a ROM's inputs and the appearance of the data outputs during a READ operation. This time delay, called *access time,* t_{ACC}, is a measure of the ROM's operating speed. Access time is described graphically by the waveforms in Figure 11.8.

The top waveform represents the address inputs, the middle waveform is an

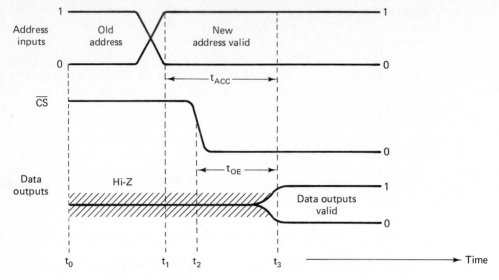

FIGURE 11.8 Typical timing for an ROM READ operation.

active LOW chip select, $\overline{CS}$, and the bottom waveform represents the data outputs. At time t_0 the address inputs are all at some specific level, some HIGH, and some LOW. $\overline{CS}$ is HIGH, so that the ROM data outputs are in their Hi-Z state (represented by hashed line).

Just prior to t_1, the address inputs are changing to a new address for a new READ operation. At t_1 the new address is valid; that is, each address input is at a valid logic level. At this point the internal ROM circuitry begins to decode the address inputs to select the register which is to send its data to the output buffers. At t_2 the $\overline{CS}$ input is activated to enable the output buffers. Finally, at t_3, the outputs change from the Hi-Z state to the valid data that represent the data stored at the specified address.

The time delay between t_1, when the new address becomes valid, and t_3, when the data outputs become valid, is the access time t_{ACC}. Typical bipolar ROMS will have access times in the 30- to 90-ns range, while MOS ROMs will range from 200 to 900 ns.

Another important timing parameter is the *output enable time*, t_{0E}, which is the delay between the $\overline{CS}$ input and the valid data output. Typical values of t_{0E} are 20 ns for bipolar ROMs and 100 ns for MOS ROMs. This timing parameter is important in situations where the address inputs are already set to their new values, but the ROM outputs have not yet been enabled. When $\overline{CS}$ goes LOW to enable the outputs, the delay will be t_{0E}.

11.7 TYPES OF ROMS

Now that we have a general understanding of the internal architecture and external operation of ROM devices, we will take a brief look at the various types of ROMs to see how they differ in the way they are programmed and in their ability to be erased and reprogrammed.

Mask-Programmed ROM   This type of ROM has its storage locations written into (programmed) by the manufacturer according to the customer's specifications. A photographic negative called a *mask* is used to control the electrical interconnections on the chip. A special mask is required for each different set of information to be stored in the ROM. Since these masks are expensive, this type of ROM is economical only if you need a large quantity of the same ROM. Some ROMs of this type are available as off-the-shelf devices preprogrammed with commonly used information or data such as certain mathematical tables and character generator codes for CRT displays. A major disadvantage of this type of ROM is that it cannot be reprogrammed in the event of a design change requiring a modification of the stored program. The ROM would have to be replaced by a new one with the desired program written into it. Several types of user-programmable ROMs have been developed to overcome this disadvantage. Mask-programmed ROMs, however, still represent the most economical approach when a large quantity of identically programmed ROMs are needed. For convenience, we will use the mnemonic MROM whenever we refer to a mask-programmed ROM.

Figure 11.9 shows a small portion of a memory-cell array for a bipolar MROM. Each memory cell consists of a single bipolar transistor connected in the common-collector configuration (input at base, output at emitter). The top row of four transis-

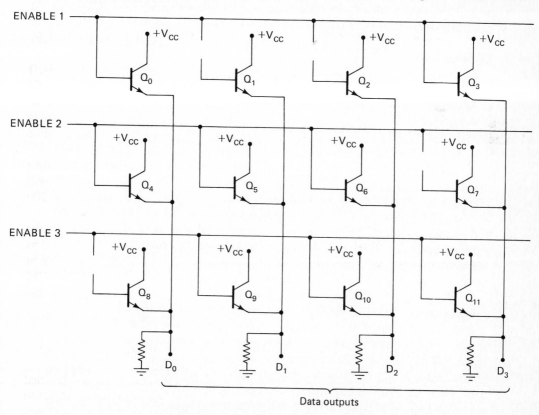

FIGURE 11.9   Typical bipolar MROM cell structure.

Sec. 11.7   **Types of ROMs**

tors constitutes a register of four cells. Note that some of the cells (Q_0 and Q_3) have their bases connected to the ENABLE 1 line, while others (Q_1 and Q_2) do not. The same is true of the registers in the middle and bottom rows. The presence or absence of these base connections determines whether a cell is storing a 1 or a 0, and the condition of each base connection is controlled by the photographic mask during production.

Note that corresponding cells in each register have their emitters connected (bused) to a common output. The same is true for all other registers that are part of the memory array. As we shall see, this presents no problem because no more than one register's cells will be enabled at one time.

If all the input ENABLE lines are held LOW, none of the transistors will be conducting, and all the data output lines will be at 0 V. The contents of the top register can be read by applying a HIGH to the ENABLE 1 line. This will turn on transistors Q_0 and Q_3, but not Q_1 and Q_2, since their base connections are open. This produces a HIGH at the emitters of Q_0 and Q_3 that appears at data outputs D_0 and D_3. Note that the other registers have no effect on the output, since their enable inputs are LOW. Thus, the data outputs would read 1001, which represents the data word stored in the top register.

Of course, we need to ensure that only one enable input can go HIGH at one time. This is automatically ensured by the address decoding circuitry (recall Figure 11.7) that enables only the one register that corresponds to the applied address code.

EXAMPLE 11.6

What data word is stored in the middle register?

Solution: The data word is 1110 because only transistor Q_7 has an open base connection.

Bipolar MROMs are available in several capacities. One of the more popular ones is the 74187, which is organized as a 256×4 memory. The 74187 has an access time of 40 ns. Another widely used bipolar MROM is the 3324A, which has a capacity of 512×8 and an access time of 70 ns.

MOS MROMs have an internal structure similar to Figure 11.9 except that each cell contains a MOSFET instead of a bipolar transistor. The 2308 is an NMOS mask-programmed ROM with a $1K \times 8$ capacity and an access time of 200 ns. The MCM 14524 is a CMOS mask-programmed ROM that stores 256 4-bit words and has an access time of over 1 μs, but has a quiescent power consumption of only 50 nW.

Programmable ROMs (PROMs) A mask-programmable ROM is very expensive and would not be used except in high-volume applications, where the cost would be spread out over many units. For lower-volume applications, manufacturers have developed *fusible-link* PROMs that are user-programmable; that is, they are not programmed during the manufacturing process but are custom programmed by the user. Once programmed, however, a PROM is like a MROM in that it cannot be erased and reprogrammed. Thus, if the program in the PROM is faulty or has to be changed, the PROM has to be thrown away.

The fusible-link PROM structure is very similar to the MROM in that certain connections either are left intact or are opened in order to program a memory cell as a 1 or 0. In the MROM of Figure 11.9 these connections were from the enable lines to the transistor bases. In a PROM each of these connections is made with a thin fuse link that comes intact from the manufacturer. The user can selectively *blow* any of these fuse links to produce the desired stored memory data. Typically this is accomplished by applying a carefully controlled voltage to the device to produce a current flow that will cause the fuse link to open in a manner similar to blowing a fuse. Once a fuse link is broken, it cannot be reconnected.

The process of programming a PROM and then verifying the programmed data can be extremely time-consuming and tedious when done manually. A number of commercial PROM programmers are available for several hundred dollars that permit keyboard entry of the program into RWM (read/write memory) and then perform the fuse blowing and verification automatically without user intervention.

A popular bipolar PROM is the 7641, which is a 512×8 memory chip. There are currently no PROMs manufactured with MOS technology.

Erasable Programmable ROM (EPROM)   An EPROM can be programmed by the user and it can also be *erased* and reprogrammed as often as desired. Once programmed, the EPROM is a *nonvolatile* memory that will hold its stored data indefinitely. The process for programming an EPROM involves the application of special voltage levels (typically in the 25- to 50-V range) to the appropriate chip inputs for a specified amount of time (typically 50 ms per address location). The programming process is usually performed by a special programming circuit that is separate from the circuit in which the EPROM will eventually be working. The complete programming process can take up to several minutes for one EPROM chip.

The storage cells in an EPROM are field-effect transistors with a silicon gate that has no electrical connections (i.e., a *floating gate*). By applying a special high-voltage programming pulse to the device, high-energy electrons are injected into the floating-gate region, thus turning the transistor "on." These electrons remain trapped in this region once the pulse is terminated, since there is no discharge path. During this programming process, the chip's address and data pins are used to determine which memory cells will be affected by the programming pulse.

Once a memory cell has been programmed, it can be erased only by exposing it to ultraviolet (UV) light applied through a window on the chip. The UV light causes a flow of photocurrent from the floating gate back to the silicon substrate, thereby restoring the gate to its initial condition. Note that there is no way to expose just a single cell to the UV light without exposing all the cells. Thus, the UV will erase the entire memory. The erasure process typically requires 15 to 30 minutes of exposure to the UV rays.

EPROMs are available in a wide selection with capacities up to $32K \times 8$ and access times down to 250 ns. We will use the popular Intel 2716 to illustrate typical EPROM operation. The 2716 is a $2K \times 8$ NMOS EPROM that operates from a single $+5$-V supply, unlike earlier EPROMs such as the 2708 which required three different supply voltages. The block symbol for the 2716 is shown in Figure 11.10. Note that it has eleven address inputs, since $2^{11} = 2048$, and eight data outputs. It has two

FIGURE 11.10   Symbol for the 2716 EPROM.

power-supply inputs, V_{cc} and V_{PP}, both of which require $+5$ V during normal operation. V_{PP}, however, has to be set at $+25$ V during the programming process.

There are two control inputs. $\overline{OE}$ is the output enable which controls the output buffers to determine whether or not EPROM data appears at the output terminals. $\overline{CE}$ is a chip enable input that has two different functions. During normal operation, it acts as a power-control input that determines whether or not the chip will operate at all. To read from the EPROM, $\overline{CE}$ has to be LOW in order for the internal circuitry to select the data and route it to the output buffers, and $\overline{OE}$ has to be LOW for the data to appear at the output pins. When $\overline{CE}$ is HIGH, the chip is in a *standby* mode where it consumes much less power (132 mW) than it does in the active mode (525 mW).

$\overline{CE}$ is also used during the programming process where it has to be pulsed from LOW to HIGH for 50 ms each time a new word is to be programmed into one of the 2048 address locations. Figure 11.11 shows the conditions needed to program a data word into one location. Assume that the EPROM has previously been erased with UV light so that it is a "clean" EPROM (all cells are 1s). Note that V_{PP} is connected to $+25$ V. The steps required to program any address location are as follows:

1. Apply the desired address to the address inputs.
2. Apply the desired 8-bit data word to the data pins D_7–D_6. These data pins will function as inputs in the program mode (note $\overline{OE}$ is HIGH).
3. Apply a 50-ms LOW-to-HIGH pulse to $\overline{CE}$. At the termination of this pulse, the selected address location should be storing the applied data word.

FIGURE 11.11 To program a 2716, the desired data are applied to the data pins, and a program pulse is applied to $\overline{CE}$.

4. In order to verify that the data word has been programmed properly, the address location should be read. This is done by applying LOWs to $\overline{CE}$ and $\overline{OE}$ and reading the levels at the data output pins. *Note:* BEFORE THIS VERIFY STEP, THE DATA INPUTS THAT WERE APPLIED DURING THE PROGRAMMING STEPS MUST BE DISCONNECTED FROM THE DATA OUTPUT PINS.

The programming process when done manually can take several hours. Numerous commercial EPROM programmers are available that can program and verify an entire 2716 in less than two minutes, once the user has entered the data into the EPROM programmer's RWM from a keyboard. Some EPROM programmers allow data to be transferred into the RWM from a computer. This process, called *downloading,* allows the user to first develop his program on a computer and then, when it is finished and working, transfer it from the computer's memory to the EPROM programmer, which will "burn it" into the EPROM.

Electrically Erasable PROM (EEPROM) As noted above, EPROMs have two major disadvantages. First, they have to be removed from their sockets in order to be erased and reprogrammed. Second, erasure removes the complete memory contents; this necessitates complete reprogramming even when only one memory word has to be changed. The *electrically erasable PROM* (EEPROM) was developed around 1980 as an improvement to the EPROM.

The EEPROM takes advantage of the same floating-gate structure as the

EPROM. It adds the feature of *electrical* erasability through the addition of a thin oxide region above the drain of the MOSFET memory cell. By applying a HIGH voltage (21 V) between the MOSFET's gate and drain, a charge can be induced onto the floating gate, where it will remain even when power is removed. Reversal of the same voltage causes a removal of the trapped charges from the floating gate and erases the cell. Since this charge-transport mechanism requires very low currents, programming and erasing of an EEPROM can usually be done *in-circuit* (i.e., without a UV light source and PROM programmer unit).

A major advantage offered by EEPROMs over EPROMs is the ability to electrically erase and reprogram *individual* words in the memory array. Another advantage is that a complete EEPROM can be erased in about 10 ms (in circuit) versus about 30 minutes for an EPROM in external UV light. An EEPROM can also be programmed more rapidly; it requires only a 10-ms programming pulse for each data word, as compared to 50 ms for an EPROM.

Because the EEPROM can be erased and reprogrammed by applying appropriate voltages, we need not remove it from the circuit it is part of, provided that additional support components are also part of the circuitry. This support circuitry includes the 21-V programming voltage (V_{PP}), which is usually generated from the +5-V supply through a DC-to-DC converter, and circuitry to control the 10-ms timing and sequencing for the erasure and programming operations.

The Intel 2816 was the original EEPROM introduced in 1981. It has a 2K × 8 capacity, an access time of 250 ns, and the characteristics described above. The 2816 has the same pin layout as the 2716 EPROM and the same logic symbol (Figure 11.10), and they can be used interchangeably once they are programmed. The requirements for erasing and programming the 2816 are described in Intel's *Memory Components Handbook*.

Improved versions of the 2816 are currently being readied for production by Intel. The 2816A, 2817, and 2817A offer many improvements, the main one being the on-chip inclusion of the support circuitry needed for erasing and programming. This will greatly reduce the amount of external circuitry needed to incorporate an EEPROM into a system.

Programmable Logic Arrays (PLAs) A PLA is really not classified as a ROM, but in programming and operation it is very similar to a fusible-link PROM. A PLA IC is a logic circuit made up of AND and OR gates interconnected to generate one or more outputs that are the sum-of-product (S-P) functions of several inputs. The user can selectively blow fusible links that interconnect the logic inputs to the AND gates, and the AND gates to the OR gates to produce any desired sum-of-products output function. Once the PLA has been programmed for particular output functions, it cannot be erased and reprogrammed. Like the fusible-link PROM, PLAs are manufactured only with bipolar technology. A typical PLA might have 16 logic inputs, 8 logic outputs, and an input-to-output propagation time of 100 ns.

PLAs provide a logic-circuit designer with a way to custom implement complex logic functions on a single chip or by combining several PLAs. This can provide more cost-efficient designs by reducing the number of required chips, chip interconnections, and board space. Reliability is also improved because there are fewer interconnections and fewer chips that can become faulty.

━━━━━━━━━━━━━━━━━━━━━━━━━━━━━━━ REVIEW QUESTIONS

1. True or false: A MROM can be programmed by the user.
2. How does a PROM differ from an MROM? Can it be erased and reprogrammed?
3. True or false: PROMs are available in both bipolar and MOS versions.
4. How is an EPROM erased?
5. True or false: There is no way to erase only a portion of an EPROM's memory.
6. Outline the steps in programming and verifying a 2716 EPROM.
7. What are the advantages of an EEPROM over an EPROM?
8. What is a PLA? How is it similar to a PROM?

11.8 ROM APPLICATIONS

ROMs can be used in any application requiring nonvolatile data storage where the data rarely or never have to change. We will briefly describe some of the most common application areas.

Microcomputer Program Storage (Firmware)   At present, this is the most wide-spread application of ROMs. Personal and business microcomputers use ROMs to store their operating system programs and their language interpreters (i.e., BASIC) so that the computer can be used immediately after power is turned on. Products that include a microcomputer to control their operation use ROMs to store the control programs. Examples are electronic games, electronic cash registers, electronic scales, and microcomputer-controlled automobile fuel injection.

The microcomputer programs that are stored in ROM are referred to as *firmware* because they are not subject to change; programs that are stored in RWM are referred to as *software* because they can be easily altered.

Implementing Combinatorial Logic   Once a ROM is programmed, the relationship between the address inputs and data outputs is fixed and will not change. This relationship can be described in a truth table like the one in Figure 11.6, which could just as easily represent the truth table for a combinatorial logic circuit. In other words, we can think of a ROM as acting like a combinatorial logic circuit that has a number of inputs equal to its address inputs and a number of outputs equal to its data outputs. The ROM can be programmed so that each data output represents a specific function of the inputs.

When used in this way, a ROM performs the same function as a PLA with the same advantages of reduced chip count, board size, and improved reliability. A ROM is generally slower than a PLA, however, and this must be taken into account when deciding which one to use.

Data Tables   ROMs are often used to store tables of data that do not change. Some examples are the trigonometric tables (i.e., sine, cosine, etc.) and code-conversion tables.

Several standard ROM "look-up" tables are available with trig functions. One, the National Semiconductor MM4220BM, stores the sine function for angles between 0 and 90 degrees. This ROM is organized as a 128 × 8, with seven address inputs and eight data outputs. The address inputs represent the angle in increments of approximately 0.7°. For example, address 0000000 is 0°, address 0000001 is 0.7°, address 0000010 is 1.41°, and so on up to address 1111111, which is 89.3°. When an address is applied to the ROM, the data outputs will represent the approximate sine of the angle. For example, with input address 1000000 (representing approximately 45°) the data outputs will be 10101101. Since the sine is less than or equal to one, these data are interpreted as a fraction; that is, .10101101, which when converted to decimal equals .707 (the sine of 45°).

A *code converter* is a circuit that takes data expressed in one type of code and produces an output expressed in another type. Code conversion is needed, for example, when a computer is outputting data in straight binary code, and we want to convert it to BCD in order to display it on 7-segment LED readouts.

One of the easiest methods of code conversion uses a ROM programmed so that the application of a particular address (the old code) produces a data output that represents the equivalent in the new code. The 74185 is a TTL ROM that stores the binary-to-BCD code conversion for 6-bit binary input. To illustrate, an address input of 100110, (decimal 38) will produce a data output of 00111000, which is the BCD code for decimal 38.

Character Generators If you have ever looked closely at the alphanumeric characters (letters, numbers, etc.) printed on a computer's video display screen, you have noticed that each is generally made up of a group of dots. Depending on the character being displayed, some dot positions are made bright while others are dark. Each character is made to fit into a pattern of dot positions, usually arranged as a 5 × 7 or 7 × 9 matrix. The pattern of dots for each character can be represented as a binary code (i.e., bright dot = 1, dark dot = 0).

A *character-generator ROM* stores the dot pattern codes for each character at an address corresponding to the ASCII code for that character. For example, the dot pattern for the letter "A" would be stored at address 1000001, where 1000001 is the ASCII code for uppercase A. Character-generator ROMs are used extensively in any application that displays or prints out alphanumeric characters.

11.9 SEMICONDUCTOR RAMS

Recall that the term RAM stands for random-access memory, meaning that any memory address location is as easily accessible as any other. Many types of memory can be classified as having random access, but when the term RAM is used with semiconductor memories it is usually taken to mean read/write memory (RWM) as opposed to ROM. Since it is common practice to use RAM to mean semiconductor RWM, we will do so throughout the following discussions.

RAMs are used in computers for the *temporary* storage of programs and data. The contents of many RAM address locations will be changing continually as the computer executes a program. This requires fast read and write cycle times for the RAM so as not to slow down the computer operation.

A major disadvantage of RAMs is that they are volatile and will lose all stored information if power is interrupted or turned off. Some RAMs, however, use such small amounts of power in the standby mode (no read or write operations taking place) that they can be powered from batteries whenever the main power is interrupted. Of course, the main advantage of RAMs is that they can be written into and read from rapidly with equal ease.

The following discussion of RAMs will draw on some of the material covered in our treatment of ROMs, since many of the basic concepts are common to both types of memory.

11.10 RAM ARCHITECTURE

As with the ROM, it is helpful to think of the RAM as consisting of a number of registers, each storing a single data word, and each having a unique address. RAMs typically come with word capacities of 1K, 4K, 8K, 16K, or 64K and word sizes of 1, 4, or 8 bits. As we will see later, the word capacity and word size can be expanded by combining memory chips.

Figure 11.12 shows the simplified architecture of a RAM which stores 64 words of 4 bits each (i.e., a 64 × 4 memory). These words have addresses ranging from 0 to 63_{10}. In order to select one of the 64 address locations for reading or writing, a binary address code is applied to a decoder circuit. Since $64 = 2^6$, the decoder requires a 6-bit input code. Each address code activates one particular decoder output, which, in turn, enables its corresponding register. For example, assume an applied address code of

$$A_5A_4A_3A_2A_1A_0 = 011010$$

Since $011010_2 = 26_{10}$, decoder output 26 will go high, selecting register 26 for either a read or a write operation.

Read Operation   The address code picks out one register in the memory chip for reading or writing. In order to *read* the contents of the selected register, the READ/WRITE $(R/\overline{W})$* input must be a 1. In addition, the CHIP SELECT (CS) input must be activated (a 1 in this case). The combination of $R/\overline{W} = 1$ and $CS = 1$ enables the output buffers so that the contents of the selected register will appear at the four data outputs. $R/\overline{W} = 1$ also *disables* the input buffers so that the data inputs do not affect the memory during a read operation.

Write Operation   To write a new 4-bit word into the selected register requires $R/\overline{W} = 0$ and $CS = 1$. This combination *enables* the input buffers so that the 4-bit word applied to the data inputs will be loaded into the selected register. The $R/\overline{W} = 0$ also *disables* the output buffers, which are tristate, so that the data outputs are in their

---

*Some IC manufacturers use the symbol $\overline{WE}$ (write enable) instead of $R/\overline{W}$. In either case the operation is the same.

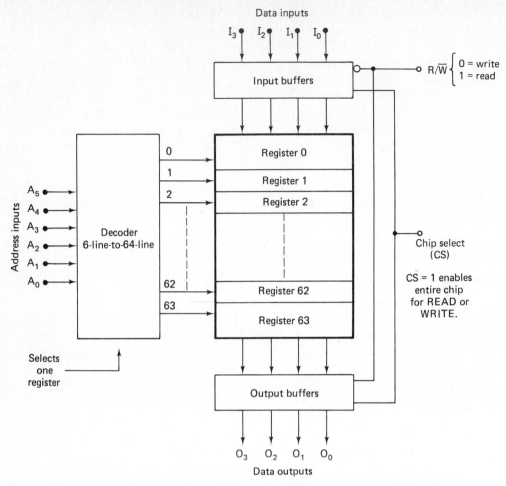

FIGURE 11.12 Internal organization of a 64 × 4 RAM.

Hi-Z state during a write operation. The write operation, of course, destroys the word that was previously stored at that address.

Chip Select Most memory chips have one or more *CS* inputs which are used to enable the entire chip or disable it completely. In the disabled mode all data inputs and data outputs are disabled (Hi-Z) so that neither a read or write operation can take place. In this mode the contents of the memory are unaffected. The reason for having *CS* inputs will become clear when we combine memory chips to obtain larger memories. Note that many manufacturers call these inputs CHIP ENABLE (*CE*) rather than *CS*.

Common Input/Output Pins In order to conserve pins on an IC package, manufacturers often combine the data input and data output functions using common input/output pins. The $R/\overline{W}$ input controls the function of these I/O pins. During a read operation, the I/O pins act as data outputs which reproduce the contents of the selected address location. During a write operation, the I/O pins act as data inputs.

We can see why this is done by considering the chip in Figure 11.12. With separate input and output pins, a total of 18 pins is required (including ground and power supply). With four common I/O pins, only 14 pins are required. The pin savings becomes even more significant for chips with larger word size.

EXAMPLE 11.7

The 2101 is an NMOS RAM that is organized as a 1K $\times$ 1 with separate data input and output, and a single active LOW chip-enable input. Draw the logic symbol for this chip, showing all pin functions.

FIGURE 11.13   (a) Logic symbol for 2101 RAM; (b) for 2114 RAM.

Solution: The logic symbol is shown in Figure 11.13(a).

EXAMPLE 11.8

The 2114 is an NMOS RAM organized as a 1K $\times$ 4 with common input/output pins and an active LOW chip enable. Draw its logic symbol.

Solution: The logic symbol is shown in Figure 11.13(b).

The architecture illustrated in Figure 11.12 for a 64 × 4 RAM will be somewhat different for larger-capacity RAMs. The registers will be arranged in a matrix such as shown for the ROM architecture in Figure 11.6. The address decoders will select the row and column of the register that is being accessed for a read or write operation. This matrix architecture reduces the size of the required decoding circuitry.

REVIEW QUESTIONS

1. Describe the input conditions needed to read a word from a specific RAM address location.
2. Why do some RAM chips have common input/output pins?
3. How many pins are required for a 256 × 8 RAM with common I/O and one *CS* input? (*Ans.* 20, including V_{cc} and ground)

11.11 STATIC RAM

The RAM operation that we have been discussing up to this point applies to a *static* RAM—one that can store data as long as power is applied to the chip. Static-RAM memory cells are essentially flip-flops that will stay in a given state (store a bit) indefinitely, provided that power to the circuit is not interrupted. Later we will describe *dynamic* RAMs, which store data as charges on capacitors. With dynamic RAMs the stored data will gradually disappear because of capacitor discharge, so it is necessary to periodically refresh the data (i.e., recharge the capacitors).

Static RAMs are available in bipolar and MOS technologies, although the vast majority of applications use NMOS or CMOS RAMs. As stated earlier, the bipolars have the advantage in speed (though NMOS is gradually closing the gap), and MOS devices have much greater capacities. Figure 11.14 shows for comparison a typical bipolar static memory cell and a typical NMOS static memory cell. The bipolar cell contains two bipolar transistors and two resistors, while the NMOS cell contains four

Bipolar cell   NMOS cell

FIGURE 11.14 Typical bipolar and NMOS static RAM cells.

N-channel MOSFETs. The bipolar cell requires more chip area than the MOS cell because a bipolar transistor is more complex than a MOSFET, and because the bipolar cell requires separate resistors while the MOS cell uses MOSFETS as resistors (Q_3 and Q_4). A CMOS memory cell would be similar to the NMOS cell except that it would use P-channel MOSFETs in place of Q_3 and Q_4. This results in lower power consumption but increases the chip complexity, so that CMOS capacities fall somewhere between bipolar and NMOS.

Table 11.1 lists some of the standard static RAMs along with their major characteristics. Note that the bipolar RAMs have the shortest access times, with the ECL device being the fastest, but note also that the 2147H-1 NMOS RAM is almost as fast as the TTL device. The CMOS RAM has the lowest power consumption while ECL has the highest. The numbers in parentheses represent the power consumption in the standby or *deselected* mode. Some RAM chips are designed to draw much less power when their chip-select inputs are inactive (i.e., the chip is deselected). This can be a very important consideration in large-capacity memories where, at any one time, only some RAM chips are being activated while others are not.

TABLE 11.1 STANDARD STATIC RAMs

Device	Technology	Capacity	Access time (ns)	Power per bit (μW)
93425A	TTL	1K × 1	30	475
10425	ECL	1K × 1	12	525
6116	CMOS	2K × 8	120	11 (.5)
2102A	NMOS	1K × 1	350	289
2114A	NMOS	1K × 4	100	50
2141-2	NMOS	4K × 1	120	85 (25)
2147H-1	NMOS	4K × 1	35	220 (35)

Static RAM Timing   RAM ICs are most often used as the internal memory of a computer. The CPU (central processing unit) continually performs read and write operations on this memory at a very fast rate determined by the limitations of the CPU. The memory chips that are interfaced to the CPU have to be fast enough to respond to the CPU read and write commands, and a computer designer has to be concerned with the RAM's various timing characteristics.

Not all RAMs have the same timing characteristics, but most of them are similar, so we will use a typical set of characteristics for illustrative purposes. The nomenclature for the different timing parameters will vary from one manufacturer to another, but the meaning of each parameter is usually easy to determine from the memory timing diagrams on the RAM data sheets. Figure 11.15 shows the timing diagrams for a complete read cycle and complete write cycle for a typical RAM chip.

The waveforms in Figure 11.15(a) show how the address inputs, chip-select input, $R/\overline{W}$ input, and data outputs behave during a read cycle. Note that the $R/\overline{W}$ line stays HIGH throughout the read cycle. In most memory systems, $R/\overline{W}$ is normally kept HIGH and is driven LOW only during a write cycle. The read cycle starts at t_0 when the address inputs change to the new address from which data are to be read; the read cycle ends at t_3 when the address inputs change to a different address to begin the next read cycle. Thus, the t_0–t_3 interval defines the read-cycle time, t_{RC}.

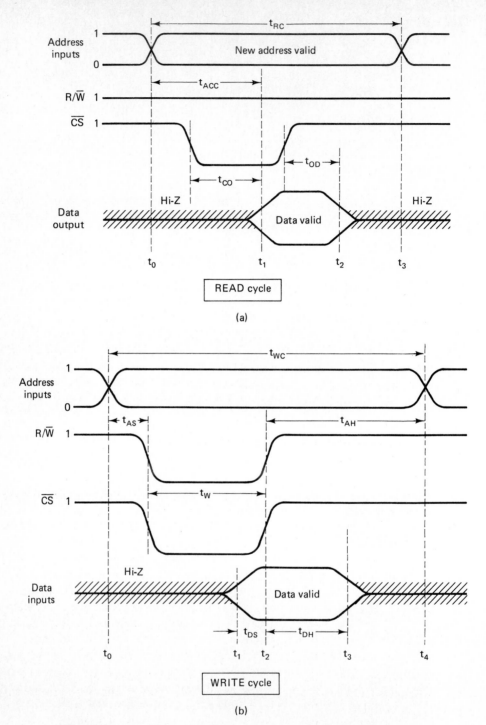

FIGURE 11.15 Typical timing for static RAM: (a) READ cycle;
(b) WRITE cycle.

The access time, t_{ACC}, occurs within the t_{RC} interval, and it represents the time required for the memory chip to produce valid data outputs. Of course, the $\overline{CS}$ input has to be driven LOW. The timing parameter t_{CO} is the time it takes for the memory outputs to go from Hi-Z to valid data after $\overline{CS}$ goes LOW. The time interval t_{OD} is the time it takes for the outputs to return to the disabled (Hi-Z) state after $\overline{CS}$ returns HIGH. Thus, the memory outputs hold valid data between t_1 and t_2, during which time the data are transferred to another circuit. Any attempt to transfer the memory data prior to t_1 will produce invalid results. In many microcomputer systems the positive-going transition of $\overline{CS}$ is used to clock the data into a CPU register.

Figure 11.15(b) shows a typical write cycle. The waveforms include the memory data inputs instead of the data outputs. The write cycle starts at t_0, when the address inputs change to the new address that is to be written into. It ends at t_4, so that the t_0–t_4 interval defines the write-cycle time, t_{WC}. The $R/\overline{W}$ line is initially HIGH and is driven LOW only after the new address has been stable for a time interval t_{AS}, called the address set-up time. This gives the RAM's address decoders time to respond to the new address.

The $\overline{CS}$ line is driven LOW at the same time as $R/\overline{W}$, since both are required for a write operation. $\overline{CS}$ and $R/\overline{W}$ must be held LOW for a time duration t_W. The data inputs that are to be written into the addressed memory location are applied at t_1. Those data have to be held stable for (at least) a time t_{OS}, data set-up time, before $R/\overline{W}$ and $\overline{CS}$ are returned HIGH. The data also have to be held stable for (at least) a time t_{OH}, data hold time, after $R/\overline{W}$ and $\overline{CS}$ return HIGH. Similarly, the address inputs have to be maintained for a time t_{AH}, address hold time. If any of these set-up or hold time requirements are not met, the write operation will not occur reliably.

REVIEW QUESTIONS

1. How does a static RAM differ from a dynamic RAM?
2. Why do NMOS memories have greater densities than bipolar memories?
3. True or false: In a read cycle, the $R/\overline{W}$ line goes from LOW to HIGH.
4. True or false: In a write cycle, the $R/\overline{W}$ and $\overline{CS}$ lines should not be activated until after the new address inputs have stabilized.

11.12 DYNAMIC RAM

A major difference between static and dynamic RAMs is that dynamic memory cells retain data only for a limited time—typically 2 ms, after which the data are lost. This requires that the data be refreshed at regular intervals in order to keep them stored in memory. The need for data refreshing is a disadvantage of dynamic RAMs. Their advantages are low power consumption and low cost—both of which are a direct result of the simplicity of the memory cell. A dynamic memory cell consists of a single MOSFET and a single MOS capacitor (typically a few picofarads) to store a single bit (Figure 11.16)

The MOSFET acts as a switch, and the capacitor is the actual storage element. The basic write operation is accomplished by applying a HIGH to the SELECT input to turn-on the MOSFET and complete the path between the I/O line and the capacitor.

FIGURE 11.16 Dynamic RAM memory cell.

During a write operation, the I/O line is used as an input line that either charges or discharges the capacitor to store either a 1 or a 0. When the write operation is complete, the SELECT input is made LOW and the MOSFET turns off, essentially opening the path to the capacitor. Ideally, the capacitor will hold its charge indefinitely, but in practice its charge will leak off after about 2 ms.

A read operation is accomplished by turning on the MOSFET with a HIGH on the SELECT input, then using the I/O line as an output sense line. Since the capacitor is now essentially connected to the I/O line, its stored charge will determine the voltage that appears on the I/O line. This output voltage is fed to a buffer amplifier that is part of the circuitry of a dynamic RAM chip.

Because of their simple cell structure, dynamic RAMs typically have four times the density of static RAMs. This allows four times as much memory capacity to be placed on a single board, or alternatively, requires one-fourth as much board space for the same amount of memory. The cost per bit of dynamic RAM storage is typically four or five times less than for static RAMs. A further cost saving is realized because the lower power requirements of a dynamic RAM, typically three to six times less than those of a static RAM, allow the use of smaller, less expensive power supplies. These advantages of dynamic RAMs are illustrated in Table 11.2, which compares two of the newer RAM chips.

TABLE 11.2

	2164 (dynamic)	2167 (static)
Capacity (bits)	64K	16K
Access time (ns)	150	70
Active power (mW)	300	625
Standby power (mW)	25	200
Cost per bit (millicents)	45	250

Dynamic RAMs do have several disadvantages. They are generally slower than static devices, as exemplified by the comparison in Table 11.2. Some of the older dynamic RAMs require more than one supply voltage, while static RAMs use only +5 V. The newer dynamic RAMs use only +5 V. Dynamic RAMs require more external support circuitry than static RAMs because of the *address multiplexing* and *refreshing* operations that must take place. These will be described shortly.

1. For how long can a dynamic RAM cell retain data?
2. What are the advantages of dynamic RAM over static RAM?
3. What are the disadvantages of dynamic RAM?

11.13 DYNAMIC RAM STRUCTURE AND OPERATION

Most dynamic RAMs (DRAMs) can be visualized as a matrix of single-bit registers (cells) as illustrated in Figure 11.17. Here there are 4096 cells arranged in a 64×64 matrix. Each cell occupies a specific row and a specific column in the matrix. Twelve address inputs are needed to select one of the cells: the lower-order address bits, A_0–A_5 select the row, and the higher-order address bits select the column. Each 12-bit address code selects a unique cell to be read from or written into. The structure in Figure 11.17 is for a $4K \times 1$ DRAM chip (i.e., it stores 4096 1-bit words). DRAM chips are currently available in capacities up to $256K \times 1$. Since most DRAMs have a 1-bit word size, several chips have to be combined to produce larger words. We will examine this later.

Address Multiplexing   In order to reduce the required number of address pins (to reduce size and cost), DRAM manufacturers use *address multiplexing*. Here the row address is first applied to the DRAM chip, and then the column address is applied. Since both parts of the address are applied at the same inputs, the number of required address input pins is cut in half. This can be illustrated for a $4K \times 1$ DRAM, such as the MOSTEK 4027, by using the simplified architecture shown in Figure 11.18(a).

FIGURE 11.17   Cell arrangement in 4K × 1 dynamic RAM.

(a)

(b)

FIGURE 11.18 (a) Simplified architecture of a 4K × 1 DRAM; (b) $\overline{RAS}/\overline{CAS}$ timing.

In this block diagram there are only six address inputs and each one has a dual function (e.g., A_0/A_6 will function as either A_0 or A_6). There is a single data input, D_{IN}, and a single data output, D_{OUT}, a $R/\overline{W}$ input, and a chip-select input, $\overline{CS}$. In addition two address *strobe* inputs are used to clock the row and column addresses into their respective latches. The *row address strobe* $\overline{RAS}$ clocks the row address latch, which is a 6-bit register inside the DRAM chip, and the *column address strobe* $\overline{CAS}$ clocks the 6-bit column address latch.

A 12-bit address is applied to the DRAM in two steps using $\overline{RAS}$ and $\overline{CAS}$. The timing is shown in Figure 11.18(b). Initially, $\overline{RAS}$ and $\overline{CAS}$ are both HIGH. At time t_0, the 6-bit row address (A_0–A_5) is applied to the address inputs. After allowing time for these address inputs to stabilize, the $\overline{RAS}$ input is driven LOW at t_1. This NGT loads the row address into the row address latch, where it is stored. At t_2 the 6-bit column address (A_6–A_{11}) is applied to the address input pins. At t_3 $\overline{CAS}$ is driven LOW and this NGT loads the column address into the column address latch.

Thus the 12-bit address is applied to the six address inputs in two steps. When both portions of the address are in their respective latches, the decoders will select the one memory cell that is being accessed, and a read or write operation takes place just as in a static RAM. The address multiplexing technique requires four fewer pins for this 4K × 1 chip than would be needed otherwise (six fewer address pins, but added $\overline{RAS}$ and $\overline{CAS}$ pins).

EXAMPLE 11.9

The Intel 2164 is a 64K × 1 DRAM that has an arrangement like that in Figure 11.18(a). How many pins are saved by using address multiplexing?

Solution: 64K = 64 × 1024 = 65,536 = 2^{16}. Thus, this memory chip requires 16 address bits to select one memory cell. Using address multiplexing, only eight address input pins are needed. This is a saving of eight address pins, which results in a net saving of *six* pins, since $\overline{RAS}$ and $\overline{CAS}$ are added.

In a typical computer system, the address inputs to the memory system come from the central processing unit (CPU). When the CPU wants to access a particular memory location, it generates the complete address and places it on address lines that make up an *address bus*. Figure 11.19(a) shows this for a memory that has a capacity of 4096 words and therefore requires a 12-line address bus going directly from the CPU to the memory.

This arrangement works for ROM or for static RAM, but it has to be modified for DRAM that uses multiplexed addressing. If the 4K memory is DRAM, it will only have six address inputs. This means that the 12 address lines from the CPU address bus have to be fed into a multiplexer circuit that will transmit 6 address bits at a time to the memory address inputs. This is shown symbolically in Figure 11.19(b). The multiplexer select input, labeled *MUX*, controls whether CPU address lines A_0–A_5 or address lines A_6–A_{11} will be present at the DRAM address inputs.

The timing of the *MUX* signal has to be sychronized with the $\overline{RAS}$ and $\overline{CAS}$ signals that strobe the addresses into the DRAM. This is shown in Figure 11.20. *MUX*

(a)

*MUX = 0 transmits CPU addresses
A_5–A_0 to DRAM. MUX = 1 transmits
A_{11}–A_6 to DRAM

(b)

FIGURE 11.19   (a) CPU address bus driving ROM or static RAM memory; (b) CPU addresses driving DRAM with multiplexer used to multiplex the CPU address lines.

MUX

$\overline{RAS}$

$\overline{CAS}$

A₀-A₅ latched into DRAM row-address register

A₆-A₁₁ latched into DRAM column-address register

FIGURE 11.20 Timing required for address multiplexing.

has to be LOW when $\overline{RAS}$ is pulsed LOW so that address lines A_0–A_5 from the CPU will reach the DRAM address inputs to be latched on the NGT of $\overline{RAS}$. Likewise, *MUX* has to be HIGH when $\overline{CAS}$ is pulsed LOW so that A_6–A_{11} from the CPU will be present at the DRAM inputs.

The actual multiplexing and timing circuitry will not be shown here but will be left as an end-of-chapter problem.

REVIEW QUESTIONS

1. Why do DRAMs use address multiplexing?
2. What are the functions of the $\overline{RAS}$ and $\overline{CAS}$ inputs?
3. What is the function of the *MUX* signal in a DRAM system?

11.14 DRAM REFRESHING

Earlier we explained that dynamic RAM cells have to be refreshed at least once every 2 milliseconds or the stored data will be lost. DRAMs are designed so that each time a read or write operation is performed on a memory cell, *all* the cells in the same row will be refreshed. This means that a complete DRAM will be refreshed provided that a read or a write operation is performed on each row every 2 msec. Unfortunately, we cannot expect that every row of a DRAM will be read from or written into every 2 msec during normal operation of a memory system, so the refresh operation has to be performed by other means.

A number of methods are used to perform the refresh operation on a DRAM, and all of them require extra circuitry external to the DRAM. Typically the external refresh circuitry contains a *refresh counter* that generates sequential row addresses for the refresh operation. The addresses from this refresh counter are used to access the various rows of the DRAM to perform a row refresh at times when the DRAM is not being used by the rest of the system. Timing, logic, and multiplexing circuitry are needed to control when and how the refresh counter outputs are steered into the DRAM address inputs to ensure that each row is refreshed within the required time limit. Special chips called *refresh controllers* have been developed by some DRAM manufacturers to provide all the necessary external refresh circuitry on a single chip.

The iRAM As stated earlier, dynamic RAMs offer high density, low power consumption, and low cost per bit. On the other hand, they require external refresh circuitry and address multiplexing circuitry. Thus, dynamic RAMs are generally used where relatively large amounts of memory are required (e.g., more than 4096 words), so that the cost of the extra external circuitry can be spread out over a large amount of memory. Conversely, static RAMs have a relatively high power consumption and cost per bit compared to DRAMs, but they need very little external control circuitry. Static RAMs usually provide a better solution for relatively small memory systems where simple system design is desired.

A new type of RAM has been developed that combines the best features of static and dynamic RAM. The *iRAM* (*integrated RAM*) is an entire dynamic memory system on a single integrated circuit; that is, it includes the memory-cell array and the refresh control logic on the same chip. This new arrangement combines the density, cost, and power advantages of DRAM with the simplicity of static RAM. iRAMs promise to have a significant impact on new large memory-system designs.

REVIEW QUESTIONS

1. What function is performed by a DRAM's external refresh circuitry?
2. What is an iRAM?

11.15 EXPANDING WORD SIZE AND CAPACITY

In most IC memory applications the required memory capacity or word size cannot be satisfied by one memory chip. Instead, several memory chips have to be combined to provide the desired capacity and word size. We will see how this is done through several examples that illustrate all the important concepts that will be needed when we interface memory chips to a microprocessor.

Expanding Word Size Suppose we need a memory that can store sixteen 8-bit words and all we have are RAM chips which are arranged as 16×4 memories with common I/O lines. We can combine two of these 16×4 chips to produce the desired memory. The configuration for doing so is shown in Figure 11.21. Examine this diagram carefully and see what you can find out from it before reading on.

Since each chip can store sixteen 4-bit words and we want to store sixteen 8-bit words, we are using each chip to store *half* of each word. In other words, RAM-0 stores the four *higher*-order bits of each of the 16 words, and RAM-1 stores the four *lower*-order bits of each of the 16 words. A full 8-bit word is available at the RAM outputs connected to the data bus.

Any one of the 16 words is selected by applying the appropriate address code to the four-line *address bus* (AB_3, AB_2, AB_1, AB_0). For now, we will not be concerned with where these address inputs come from.* Note that each address bus line is

---

*Typically, the address lines come from the CPU portion of a computer.

FIGURE 11.21 Combining two 16 × 4 RAMs for a 16 × 8 memory.

connected to the corresponding address input of each chip. This means that once an address code is placed on the address bus, this same address code is applied to both chips so that the same location in each chip is accessed at the same time.

Once the address is selected, we can read or write at this address under control of the common $R/\overline{W}$ and $\overline{CS}$ line. To read, $R/\overline{W}$ must be high and $\overline{CS}$ must be low. This causes the RAM I/O lines to act as *outputs*. RAM-0 places its selected 4-bit word on the upper four data bus lines and RAM-1 places its selected 4-bit word on the lower four data bus lines. The data bus then contains the full selected 8-bit word, which can now be transmitted to some other device (e.g., a register).

To write, $R/\overline{W} = 0$ and $\overline{CS} = 0$ causes the RAM I/O lines to act as *inputs*. The 8-bit word to be written is placed on the data bus from some external device. The upper 4 bits will be written into the selected location of RAM-0 and the lower 4 bits will be written into RAM-1.

Sec. 11.15   Expanding Word Size and Capacity

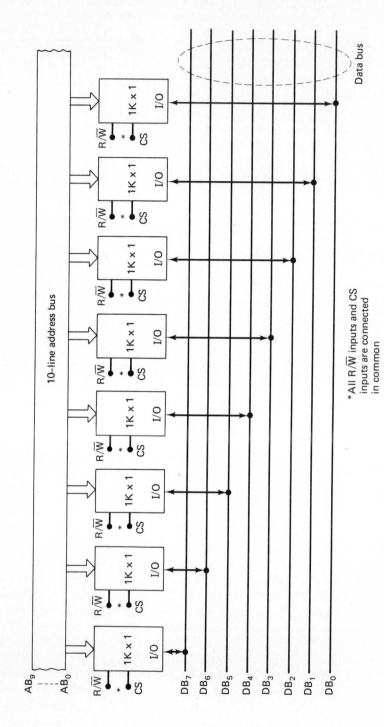

FIGURE 11.22   Eight 1K × 1 chips arranged as a 1K × 8 memory.

480

In essence, the combination of the two RAM chips acts like a single 16×8 memory chip. We would refer to this combination as a 16×8 memory module.

The same basic idea for expanding word size will work for many different situations. Read the following example and draw a rough diagram for what the system will look like before looking at the solution.

EXAMPLE 11.10

Show how to construct a 1024×8 memory module from static RAM ICs that have a capacity of $1K \times 1$.

Solution: The circuit is shown in Figure 11.22 where eight RAM chips are used. Each RAM stores one of the bits of each of the 1024 8-bit words. Note that all of the $R/\overline{W}$ and $\overline{CS}$ inputs are wired together. Also note that the 10-line address bus is connected to the address inputs of each chip. When combined in this way, the eight chips act like a $1K \times 8$ memory module.

Expanding Capacity   Suppose we need a memory that can store 32 4-bit words and all we have are the 16×4 chips. By combining two 16×4 chips as shown in Figure 11.23, we can produce the desired memory. Once again, examine this diagram and see what you can determine from it before reading on.

Word size	4 bits	
Address ranges	00000 to	01111 — words in RAM-0
	10000 to	11111 — words in RAM-1
Total →	00000 to	11111 — 32 words

FIGURE 11.23   Combining two 16×4 chips for a 32×4 memory.

Each RAM is used to store sixteen 4-bit words. The data I/O pins of each RAM are connected to a common four-line data bus. Only one of the RAM chips can be selected (enabled) at one time so that there will be no bus-contention problems. This is assured by driving the respective $\overline{CS}$ inputs from different logic signals.

Since the total capacity of this memory module is 32×4, there have to be 32 different addresses. This requires *five* address bus lines. The upper address line AB_4 is used to select one RAM or the other (via the $\overline{CS}$ inputs) as the one that will be read from or written into. The other four address lines AB_0–AB_3 are used to select the one memory location out of 16 from the selected RAM chip.

To illustrate, when $AB_4 = 0$, the $\overline{CS}$ of RAM-0 enables this chip for read or write. Then, any address location in RAM-0 can be accessed by AB_3–AB_0. The latter four address lines can range from 0000 to 1111 to select the desired location. Thus, the range of addresses representing locations in RAM-0 are

$$AB_4AB_3AB_2AB_1AB_0 = 00000 \text{ to } 01111$$

Note that when $AB_4 = 0$, the $\overline{CS}$ of RAM-1 is high, so that its I/O lines are disabled and cannot communicate (give or take data) with the data bus.

It should be clear that when $AB_4 = 1$, the roles of RAM-0 and RAM-1 are reversed. RAM-1 is now enabled and the AB_3–AB_0 lines select one of its locations. Thus, the range of addresses located in RAM-1 is

$$AB_4AB_3AB_2AB_1AB_0 = 10000 \text{ to } 11111$$

EXAMPLE 11.11

It is desired to combine several 256×8 PROMs to produce a total capacity of 1024×8. How many PROM chips are needed? How many address bus lines are required?

Solution: Four PROM chips are required, with each one storing 256 of the 1024 words. Since $1024 = 2^{10}$, the address bus must have 10 lines.

The configuration for the memory of Example 11.11 is similar to the 32×4 memory of Figure 11.23. However, it is slightly more complex, because it requires a decoder circuit for generating the $\overline{CS}$ input signals. The complete diagram for this 1024×8 memory is shown in Figure 11.24.

Since the total capacity is 1024 words, 10 address bus lines are required. The two highest-order lines, AB_9 and AB_8, are used to select *one* of the PROM chips; the other eight address bus lines go to each PROM to select the desired location within the selected PROM. The PROM selection is accomplished by feeding AB_9 and AB_8 into the decoder circuit. The four possible combinations of AB_9, AB_8 are decoded to generate active-LOW signals, which are applied to the CS inputs. For example, when $AB_9 = AB_8 = 0$, the 0 output of the decoder goes low (all others are high) and enables PROM-0. This causes the PROM-0 outputs to generate the data word stored at the address determined by AB_7–AB_0. All other PROMs are disabled.

Thus, all addresses in the following range are stored in PROM-0:

$$AB_9AB_8AB_7 \ldots AB_0 = 0000000000 \text{ to } 0011111111$$

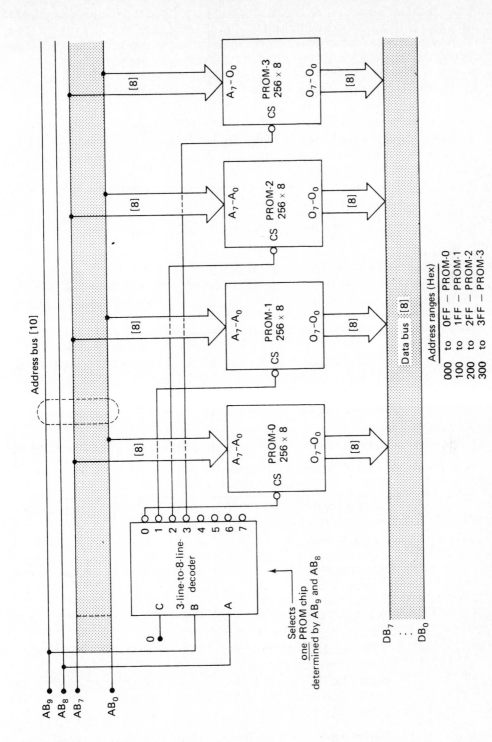

FIGURE 11.24 1K × 8 PROM memory using four 256 × 8 chips.

These are the first 256 addresses in the memory. For convenience, these addresses can be more easily written in hexidecimal code to give a range of 000_{16} to $0FF_{16}$.

Similarly, $AB_9 = 0$, $AB_8 = 1$ selects PROM-1 to give it an address range of

$$0100000000 \text{ to } 0111111111 \qquad \text{(binary)}$$

or

$$100 \text{ to } 1FF \qquad \text{(hex)}$$

The reader should verify the PROM-2 and PROM-3 address ranges given in Figure 11.24.

EXAMPLE 11.12

What size of decoder would be needed to expand the memory of Figure 11.24 to 4K $\times$ 8? How many address bus lines are needed?

Solution: A 4K capacity is $4 \times 1024 = 4096$ words, which requires $4096/256 = 16$ PROM chips. To select one of 16 chips will require a 4-line-to-16-line decoder. With a capacity of 4096 words, a total of 12 address bus lines are needed—4 go to the decoder to select the PROM and the other 8 go to each PROM to select the address in the PROM.

We have seen how to expand memory word size and memory capacity by combining memory chips. In many memory systems both types of expansion are required. We will see how this is accomplished in one of the end-of-chapter problems.

REVIEW QUESTIONS

1. The 2114 is a 1K $\times$ 4 RAM. How many of these are needed for a 1K $\times$ 16 memory system? (*Ans.* Four)
2. How many 2114s are needed for an 8K $\times$ 4 system? (*Ans.* Eight)
3. How many 2114s are required in a 16K $\times$ 8 system? (*Ans.* 32)

11.16 MAGNETIC BUBBLE MEMORIES

The newest type of solid-state memory device is the *magnetic bubble memory* (MBM). It differs from the semiconductor memories that have already been discussed in that a memory cell consists of a tiny cylindrical magnetic domain (i.e., a bubble). The presence or absence of a bubble in a specific bit position represents a logic 1 or logic 0, respectively. Continuously changing magnetic fields are used to move the bubbles around in a loop inside the magnetic material, much like a recirculating shift register. The MBM is nonvolatile in that the bubbles will stay fixed in their positions when power is removed. When power is restored to the electronic support circuitry, the bubbles resume circulating around the loop.

Nonvolatility is the principal advantage of the MBM. Products that use MBM do not require battery-powered backup memories. Unlike other nonvolatile semiconductor memories such as ROMs, EPROMs, and EEPROMs, the MBM can be written into and read from with equal ease and comparable speed. Furthermore, unlike disk or tape memories, a MBM system has no moving parts and is therefore quieter and more reliable.

MBMs are compact and dissipate very little power (typically 1 μW per bit). A typical MBM device can store a million or more bits, which is much more than any other semiconductor memory at this time. With all these advantages, the MBM does not possess the speed needed for it to function as the main internal memory of a computer; typical access times are in the 1- to 20-ms range. For this reason MBMs are better suited for the external mass storage functions that have been previously dominated by magnetic tape and disk storage. At present, MBMs are about 100 times faster than floppy disks but somewhat more costly because of the required support circuitry. As their cost comes down, we should see more and more MBM systems being used in place of the slower, less reliable floppy disk systems.

MBM Unit and Support Chips   A typical MBM unit (Figure 11.25) contains the bubble chip, two coils that generate the changing magnetic field, two permanent magnets that produce a steady bias field, and a magnetic shield that prevents disturbances from external magnetic fields. This is the actual unit that stores the binary data, and it may typically store one million bits (i.e., one megabit). This MBM unit

FIGURE 11.25   **Exploded view of an MBM unit. (Courtesy of Intel Corporation)**

requires several support chips in order to function as a complete memory system interfaced to a computer.

Figure 11.26 shows a typical bubble memory system configuration using Intel's 7110 MBM unit and support chips. The shaded portion is Intel's BPK70 Bubble Storage Subsystem. It contains a 1-megabit MBM chip and four support chips. External to the subsystem we have the bubble memory controller (BMC), which is the interface between the memory system and the equipment it serves (i.e., computer). The BMC generates all the timing signals required by the other support chips. It also performs the parallel-to-serial and serial-to-parallel conversions needed because the bubble memory subsystem uses serial data, while the computer uses parallel data. The BMC also serves as a speed-matching device between the faster computer and the slower bubble memory subsystem. The BMC can serve up to eight bubble subsystems for a total of 8 megabits.

Serial data that are to be stored in or read from the bubble memory must pass through the formatter/sense amplifier (FSA). The FSA has many functions, including shifting serial data to and from the correct magnetic bubble loops in the MBM. The coil predriver chip takes timing signals from the BMC and converts them to pulsed

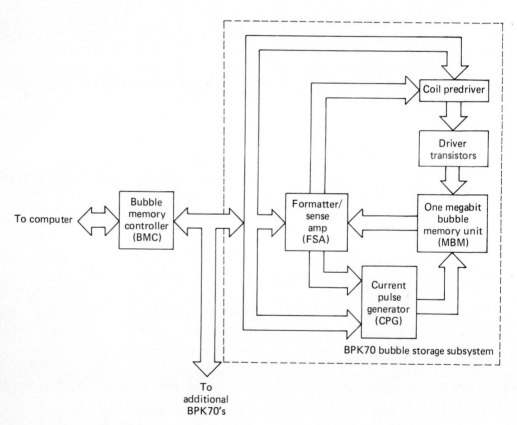

FIGURE 11.26 Intel MBM system. (Courtesy of Intel Corp.)

outputs that are amplified by the driver transistor chip to drive the coils that surround the bubble chip. The current pulse generator (CPG) generates the current pulse needed to produce the magnetic bubble when a logic 1 is to be stored.

_____ **REVIEW QUESTIONS**

1. What is the principal advantage of the MBM compared to semiconductor RAM? What is its principal disadvantage?
2. What are its advantages compared to magnetic tape and disk memory? What is its disadvantage?

11.17 CHARGE-COUPLED-DEVICE MEMORY (CCD)

Another relatively new semiconductor memory technology is the charge-coupled-device (CCD) technology. As with the dynamic RAM, data in a charge-coupled device are stored as charges on MOS capacitors. Whereas a DRAM memory cell consists of a capacitor and a MOSFET, a CCD memory cell uses only a capacitor. A CCD memory chip is made up of long rows or *channels* of these capacitors, as illustrated in Figure 11.27. One plate of each capacitor is common to all the capacitors; this is the semiconductor substrate. The other plates are separate. A bit of data is represented by a large or small packet of charge stored on each capacitor. Clock pulses are sequentially applied to cause the packets of charge to move left to right from capacitor to capacitor along the channel.

In a CCD chip the cells are arranged as a recirculating shift register that is being continuously shifted. Data are entered at one end of the shift register during a write operation, and data are read out at the other end. This makes the CCD memory a *sequential-access memory*, since the data in a particular memory cell have to be shifted to the output end before they can be accessed.

CCD memory chips are volatile, so data will be lost when power is removed. In addition, the CCD clock pulses must be continuously applied in order to maintain data. This is because the charges have to be periodically refreshed. This is accomplished by having a refresh amplifier at one end of the recirculating shift register so that the charge packet is refreshed as it recirculates around the register.

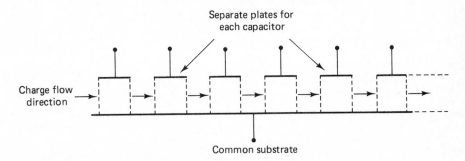

FIGURE 11.27 Charge-coupled device channel acts like a shift register.

CCD Memory Characteristics   Because of the simple cell structure, CCDs have a higher packing density than dynamic RAMS. A typical CCD chip stores 64K bits. CCDs, however, have a relatively long access time (often referred to as *latency time* for CCDs), a typical maximum value being about 4 ms. Thus, CCDs are not suited for the internal main memory of a computer. Instead, they compete with other high-density, sequential-access devices such as magnetic bubble, disk, and tape. Their small size, lack of moving parts, and fast access compete well against these other devices. Unfortunately, CCDs have a big disadvantage in that they are volatile while their competitors are not. However, their very low power requirements (1 μW per bit) permit the use of battery backup power for limited time durations.

REVIEW QUESTIONS

1.   Describe the similarities and differences between a CCD and a DRAM.
2.   What is the disadvantage of CCD versus magnetic bubble devices?

11.18   MAGNETIC CORE MEMORY

The basic memory cell in magnetic core memories is a small toroidal (ring-shaped) piece of ferromagnetic material called a *ferrite core*. These cores are typically 50 mils in diameter, although more recently 18-mil cores have been introduced. Ferrite cores have a very low reluctance, which means that they are easily magnetized, and they have a high retentivity, which means that they will stay magnetized indefinitely when not disturbed.

FIGURE 11.28   **Magnetic core.**

Figure 11.28 shows a ferrite core, many times its actual size, with a wire threaded through the core. If a current pulse is passed through the wire, magnetic flux will be produced around the wire in a direction dependent on the direction of current.* Because of the ferrite core's low magnetic reluctance, these lines of flux will enter the core and magnetize the core in either a clockwise or counterclockwise direction. When the current in the wire is reduced to zero, the retentivity of the ferrite core results in the core retaining a large portion of the flux, *provided that the wire current exceeded the threshold magnetizing current value,* I_M. The two possible states of magnetization of the ferrite core act just like the logic levels of a flip-flop and can be assigned logic values. The usual convention is:

$$CW \; flux \; = \; logical \; 1 \; state$$

$$CCW \; flux \; = \; logical \; 0 \; state$$

*The right-hand rule can be used to determine the flux direction as follows: Grasp the wire in the right hand with the thumb pointing in the direction of *conventional* current. The fingers then wrap around the wire in the direction of the flux lines.

Sensing a Core Once a core has been pulsed to a given logic state, it can retain that logical state indefinitely without the need for any electrical power. Each core is a memory cell storing 0 or 1. Since a core's logic state is magnetic rather than electrical, a somewhat unusual technique is required to convert the core's stored information to an electrical signal to determine whether the core contains a 0 or 1. This technique of *sensing* or *reading* the core requires a second wire, called a *sense line*, threaded through the core as illustrated in Figure 11.29.

To sense or read the logic state of the core, a negative current pulse is applied to the input line to bring the core to the 0 state (CCW). If the core was previously in the 0 state, there will be only a very small flux change in the core which will produce a small voltage pulse across the sense line. If the core was in the 1 state, the flux change will be much greater and will produce a large voltage pulse across the sense line.

As Figure 11.29(b) shows, the 1-state output waveform has a much larger amplitude and pulse width than the 0-state output waveform. Thus, a 1 can be detected by using an amplitude-discriminating amplifier. This amplifier will amplify the 1 output pulse to several volts so that it can drive logic circuitry, but it will not amplify the 0 pulse.

Destructive Readout The main drawback of this technique for sensing the states of magnetic cores is that all the cores end up reset to the 0 state. This means that the information previously contained in the cores is no longer there. This sensing tech-

(a)

(b)

FIGURE 11.29 Sensing the state of a core.

nique is therefore referred to as *destructive readout*. Generally speaking, it is necessary to retain the information in the core memory after it has been read. This requires that the information be written back into the memory after each read operation.

Magnetic Core Memory Characteristics Individual core memory cells are usually arranged in core planes, typically containing 4096 cores per plane. This is illustrated symbolically in Figure 11.30 on a smaller scale. Each of the four planes contains 16 cores arranged in a 4 × 4 matrix. A 4-bit data word is stored vertically with 1 bit in each plane. For example, the cores labeled *a*, *b*, *c*, *d* represent a single 4-bit word.

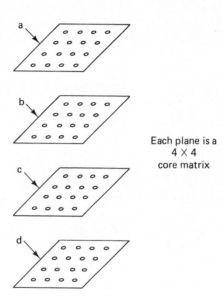

Each plane is a
4 × 4
core matrix

FIGURE 11.30 Core memory is arranged as a stack of core planes where each plane contains a matrix of cores.

A typical magnetic core memory system would contain sixteen 64 × 64 core planes to store 4096 16-bit words. Larger word sizes are obtained by increasing the number of planes; larger capacity is obtained by increasing the size of each plane.

Prior to the development of semiconductor memories, core memory systems were the dominant technology used in internal computer memories. At present, cores cannot compete with semiconductor RAM in the areas of speed, size, and cost. Typical core memory access times are in the 100–500 ns range. A core memory system requires an extensive amount of external support circuitry in addition to the cores themselves. This circuitry includes (1) address decoders to select a given data word location; (2) current generators to produce the relatively large current pulses (100 mA to 1 A) needed to magnetize the cores; (3) sense amplifiers to amplify the small sense line outputs. This circuitry makes a core memory system larger and more expensive than a typical semiconductor memory system, especially when large capacities are needed.

The major advantage of core memory is its nonvolatility. In applications where this is critical, cores offer a reasonably fast form of random-access memory.

1. How are 1s and 0s written into a magnetic core?
2. How is the logic state of a memory cell sensed?
3. What is the principal advantage of core memory over semiconductor memory?

11.19 RECORDING ON MOVING MAGNETIC SURFACES

Magnetic tape and disks are memory devices that involve recording and reading magnetic spots on a moving surface of magnetic material. For each of these devices, a thin coating of magnetic material is applied to a smooth nonmagnetic surface. Magnetic tapes, for example, consist of a layer of magnetic material deposited on plastic tape. Disks have the magnetic material coated on both sides of a flat disk that resembles a phonograph record.

Recording and reading of binary information on tapes and disks uses the same basic principles. Figure 11.31 illustrates the fundamental concept of recording on a moving magnetic surface. The read/write head is a core of high-permeability soft iron with a coil wound around it and a small air gap (typically 0.001 inch wide). In writing, a current is driven through the coil, establishing magnetic lines of flux in the core. These flux lines remain in the core until they encounter the air gap, which has a very high reluctance to magnetic flux. This causes the flux lines to deviate and travel through the magnetic coating on the moving surface. Thus, pulses of current in the coil result in spots of magnetism on the moving surface. These spots remain magnetized after they pass the read/write head.

The read operation is just the opposite of the write operation. During the read operation the coil is used as a sense line. As the magnetic surface is moved under the read/write head, the spots that have been magnetized produce a flux up through the air gap and into the core. This change in the core's flux induces a voltage signal in the sense winding which is then amplified and interpreted.

FIGURE 11.31 Fundamental parts for recording on moving magnetic surface.

The coil around the core usually has a center tap, so that half the coil is used as a sense winding and the other half is heavier-gauge wire used for supplying current in the write operation.

11.20 MAGNETIC DISKS

Magnetic-disk memories provide a very large storage capacity at moderate operating speeds. The magnetic disk memory costs somewhat more than magnetic tape. In this method of storage, data are recorded on a layer of magnetic material that is coated on hard flat disks resembling phonograph records. A number of rotating disks are stacked with space between each disk (refer to Figure 11.32). Information is recorded *serially* on concentric bands or tracks on both surfaces of each disk by movable read/write heads. The packing density along a given track can be greater than 5000 bits per inch. Typical rotation speeds can range up to 6000 rpm. A single disk surface usually has from 200 to 500 tracks and can store up to 250 million bits.

The read/write heads move radially in and out between disks in the space provided. The access time depends on rotation speed and disk diameter. For faster access time, additional heads are used to serve the same disks, some serving the inner tracks and others the outer tracks. The heads do not contact the disk surface, because this would cause excessive wear. Instead, most manufacturers use a *flying head*, which is shaped to ride along on a thin layer of air that adheres to the rotating disk. In most fast-acting mechanisms, a compressed-air piston is used to force the head toward the disk into or out of the air cushion.

Hard disk memories have typical access times of 10–40 ms; basically, this is the time required for the read/write head to become positioned on the desired track. Once the head is in position, data can be transferred to or from the disk at a typical rate of 2.5 megabits per second.

FIGURE 11.32 **Typical hard disk arrangement.**

In a hard disk system, the flying head is only a few millionths of an inch above the disk surface; this is smaller than even a particle of smoke. Since the air is full of particles that could interfere with the operation of the read/write head, a hard disk unit requires that the head and disk be situated in an environmentally protected area.

A *Winchester* disk system houses the disk and head in an airtight container that is filled with filtered air or an inert gas. This provides excellent protection against environmental contaminants.

Floppy Disks   A relatively recent innovation in disk storage, which was originally developed at IBM, uses a small flexible disk with a plastic base in place of the more conventional metal disk. A floppy disk (also called a *diskette*) is a flexible disk that looks like a 45-rpm phonograph record without grooves. Because they can be bent easily, floppy disks are enclosed in a protective cardboard envelope. This keeps them rigid as they are spun around by the disk drive unit in much the same way as phonograph records.

Figure 11.33 shows the standard floppy disk, which has a diameter slightly less than 8 inches.* The protective envelope has cutouts for the drive spindle, the read/record head, and an index position hole. The access slot allows the read/record head to make contact with the surface of the disk as it spins inside the envelope. The index hole allows a photoelectric sensor to be used to determine a reference point for all the tracks on the disk.

Floppy disks are normally rotated at a speed of 360 rpm, which corresponds to one rotation in about 167 ms. As the disk rotates, the read/record head makes contact with it through the access slot so that it can write or sense magnetic pulses on the disk

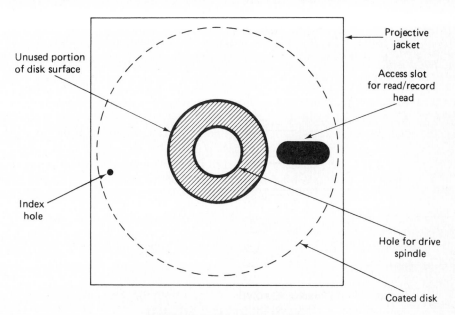

FIGURE 11.33   Floppy disk.

*A "mini-floppy" has a diameter of $5\frac{1}{4}$ inches. This is the most popular size for personal computers.

surface. The standard 8-inch disk surface is divided into 77 concentric tracks numbered 0 to 76 [see Figure 11.34(a)], with track number 0 being the outside track. These tracks are normally spaced about 0.02 inch apart, with track 0 having a total length of about 25 inches and track 77 a total length of 15 inches. This means that the longer outside tracks have space for more information than do the shorter inside tracks. For convenience in accessing a given point on the disk, the disk circumference is divided into 26 equal-size *sectors* [see Figure 11.34(b)] in which information can be stored. This means that the same amount of storage space is used on each track irrespective of the length of the track; thus, more space is wasted (not used) on the longer tracks.

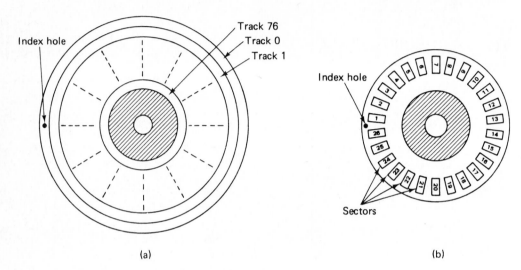

(a)   (b)

FIGURE 11.34   (a) Disk is divided into 77 concentric tracks; (b) each track is divided into 26 equal sectors.

The index hole is placed between the last sector (26) and the first sector (1). Using a light source/photodetector combination, the disk-drive circuitry will know when the disk is in its reference position with sector 1 next to the index hole. By counting each sector as it passes under the read/record head, the drive circuitry will always know which sector is passing under the read/record head.

Each sector usually stores 128 bytes of information, so that the total storage capacity of a single floppy disk surface will be

$$128\frac{\text{bytes}}{\text{sector}} \times 26\frac{\text{sectors}}{\text{track}} \times 77\frac{\text{tracks}}{\text{disk}} \approx 256{,}000\frac{\text{bytes}}{\text{disk}} \approx 2 \times 10^6\frac{\text{bits}}{\text{disk}}$$

Some manufacturers use "double-density" floppy disks, which store 256 bytes into each sector.

Access Time   Data are recorded or read from the disk surface one bit at a time at a nominal rate of 1 bit every 4 μs* or 1 byte every 32 μs. This rate, however, occurs

*This is a data transfer rate of 250 kbits/sec, about 10 times slower than conventional hard disk systems.

after the read/record head has been positioned on the desired sector of the desired track. It takes approximately 6 ms for the head-positioning mechanism to move the head from one track to the next and 16 ms more to lower the head onto the track. With 77 tracks, it will take about $\frac{1}{2}$ s in the worst case for the head to move from its present track to the track that is to be accessed for reading or writing. Once the head is on the desired track, it has to find the desired sector on that track. Since the disk makes one revolution in 167 ms, it will take 167 ms to find the right sector in the worst case. Thus, in the worst case it takes a little over 0.6 s to find the desired record on the disk. On the average, this access time will be more like 0.3 s. This is much faster than the time required to randomly access a record on even the fastest magnetic tape system.

11.21   MAGNETIC TAPE

Currently, magnetic tape is the most popular medium for storing very large quantities of digital information. Most medium-size and large-size computers have one or more magnetic tape units. Magnetic tape represents the least expensive type of storage; the cost averages about 0.0001 cent per bit. Tapes are usually $\frac{1}{2}$–1 inch wide and have 7–14 tracks across the width of the tape. Packing density usually ranges from 200 to 1600 bits per inch along each track. A typical tape reel is about 2400 ft and can store more than 10^8 bits of information.

A major difference between tape and drums or disks is that the tape is not kept in continuous motion but is started and stopped each time it is used. The tape transports have the ability to start and stop very quickly. The necessity to accelerate and decelerate the tape quickly is due to two reasons. First, since the reading and writing processes cannot begin until the tape has reached sufficient speed, a time delay occurs as the tape comes up to speed. Second, the tape which passes under the heads during the acceleration and deceleration processes is wasted, so that fast starting and stopping conserves tape space. Typical start and stop times are of the order of 1 msec.

Most modern tape units use one READ/WRITE head per track and usually these heads have *two* air gaps, one for writing and one for reading [refer to Figure 11.35(a)]. The read gap is positioned after the write gap so that the information written on the tape can be immediately read out and compared with the input information (usually in a register) to see that it is correct.

These are several formats for recording data on magnetic tape. Typically, one character (number, letter, symbol) is recorded on a given row along the tape. Figure 11.35(b) shows a seven-track format where one of the tracks is for a parity bit. Data are recorded on tape in blocks called *records* with spaces between records to allow for starting and stopping of the tape.

Magnetic tape has become a very reliable storage medium, and its basic limitation continues to be its relatively long access time. However, despite this limitation, it continues to be an economical and convenient medium for bulk storage of large quantities of information.

Tape Cassette Storage   Microcomputers often use a low-cost type of magnetic tape storage utilizing standard audio cassettes. A typical cassette storage system records different audio tones to represent digital data. These systems are relatively slow,

(a)

(b)

FIGURE 11.35 Two-gap READ/WRITE head; (b) seven-track tape format.

having access times that can be several minutes, and data transfer rates of 300–1200 bits per second. Their advantages are simplicity and low cost.

REVIEW QUESTIONS

1. True or false: Floppy disks have a greater data transfer rate than hard disks.
2. True or false: Magnetic tape is slower than disk.
3. How does a Winchester disk unit differ from a standard hard disk?

PROBLEMS

11.1 A certain memory has a capacity of 16K × 32. How many words does it store? What is the number of bits per word? How many memory cells does it contain?

11.2 How many different addresses are required by the memory of Problem 11.1?

11.3 What is the capacity of a memory that has 16 address inputs, four data inputs, and four data outputs?

11.4 Refer to Figure 11.6. Determine the data outputs for each of the following input conditions:
  (a) $[A] = 1011; CS = 0$
  (b) $[A] = 0111; CS = 1$

11.5 Refer to Figure 11.7.
  (a) What register is enabled by input address 1011?
  (b) What input address code selects register 4?

11.6 A certain ROM has a capacity of $16K \times 4$ and an internal structure like that shown in Figure 11.7.
  (a) How many registers are in the array?
  (b) How many bits are there per register?
  (c) What size decoders does it require?

11.7 Figure 11.36 shows how data from a ROM can be transferred to an external register. The ROM has the following timing parameters: $t_{ACC} = 250$ ns and $t_{OE} = 120$ ns. Assume that the new address inputs have been applied to the ROM 500 ns before the occurrence of the TRANSFER pulse. Determine the minimum duration of the TRANS-FER pulse for reliable transfer of data.

FIGURE 11.36

11.8 Repeat Problem 11.7 if the address inputs are changed 70 ns prior to the TRANSFER pulse.

11.9 Refer to Figure 11.9 and determine the data output levels when ENABLE1 = ENABLE2 = LOW, and ENABLE3 = HIGH.

11.10 Change the MROM of Figure 11.9 so that the following words are stored from top to bottom: 1011, 0100, 1010.

11.11 Figure 11.37 shows a simple circuit for manually programming a 2716 EPROM. Each EPROM data pin is connected to a switch that can be set at a 1 or 0 level. The address inputs are driven by a 12-bit counter. The 50-ms programming pulse comes from a one-shot each time the PROGRAM pushbutton is actuated.

FIGURE 11.37

(a) Explain how this circuit can be used to sequentially program the EPROM memory locations with the desired data.

(b) Show how 7493s (Figure 7.7) can be used for the counter, and a 74121 (Figure 5.69) for the one-shot.

(c) Should switch bounce have any effect on the circuit operation?

11.12 Another ROM application is the generation of timing and control signals. Figure 11.38 shows a 16 × 8 ROM with its address inputs driven by a MOD-16 counter so that the ROM addresses are incremented with each input pulse. Assume the ROM is programmed as in Figure 11.6, and sketch the waveforms at each ROM output as the pulses are applied. Ignore ROM delay times.

FIGURE 11.38

D_7

10 μs

FIGURE 11.39

11.13 Change the program stored in the ROM of Problem 11.12 to generate the D_7 waveform of Figure 11.39.

11.14 Draw the logic symbol for a 256 $\times$ 4 RAM with common I/O pins and two active LOW chip-enable inputs.

11.15 A certain static RAM has the following timing parameters (in nanoseconds):

$$t_{RC} = 100 \qquad t_{AS} = 20$$

$$t_{ACC} = 100 \qquad t_{AH} = \text{not given}$$

$$t_{CO} = 70 \qquad t_W = 40$$

$$t_{OD} = 30 \qquad t_{DS} = 10$$

$$t_{WC} = 100 \qquad t_{DH} = 20$$

(a) How long after the address lines stabilize will valid data appear at the outputs during a read cycle?

(b) How long will output data remain valid after $\overline{CS}$ returns HIGH?

(c) How many read operations can be performed per second?

(d) How long should $R/\overline{W}$ and $\overline{CS}$ be kept HIGH after the new address stabilizes during a write cycle?

(e) What is the minimum time that input data have to remain valid for a reliable write operation to occur?

(f) How long must the address inputs remain stable after $R/\overline{W}$ and $\overline{CS}$ return HIGH?

(g) How many write operations can be performed per second?

11.16 Draw the logic diagram for a 16K $\times$ 1 DRAM with an active LOW chip-select input.

11.17 Figure 11.40(a) shows a circuit that generates the $\overline{RAS}$, $\overline{CAS}$, and MUX signals needed for proper operation of the circuit of Figure 11.19(b). The 10-MHz master clock signal provides the basic timing for the computer. The memory request signal ($MEMR$) is generated by the CPU in synchronism with the master clock as shown in part (b) of the figure. $MEMR$ is normally LOW and is driven HIGH whenever the CPU wants to access memory for a read or write operation. Determine the waveforms at Q_0, $\overline{Q}_1$, and Q_2 and compare them to the desired waveforms of Figure 11.20.

11.18 Show how to connect two 74157 multiplexers (Figure 9.23) to provide the multiplexing function required in Figure 11.19(b).

11.19 Show how to combine two 2114 RAM chips (Figure 11.13) to produce a 1K $\times$ 8 module.

11.20 Compare your circuit from Problem 11.19 with the 1K $\times$ 8 module of Figure 11.22. In addition to the obvious difference in the required number of chips, there is a significant difference in the amount of loading which each address bus line will experience. For each of these modules determine the number of loads that each address bus line has to drive.

MEMR
(from CPU)

10-MHz
master
CLOCK

(a)

Master
CLOCK

MEMR

(b)

FIGURE 11.40

11.21 Refer to the ROM chip represented in Figure 11.41. Show how to combine two of these chips to produce a 2K × 8 memory module. The circuit should use no other logic devices.

11.22 Modify the memory circuit of Figure 11.24 so that it has a capacity of 2K × 8. Then determine the range of addresses for each ROM.

11.23 Examine the memory circuit of Figure 11.42.
(a) Determine the total capacity and word size.

FIGURE 11.41

500 Chap. 11 **Memory Devices**

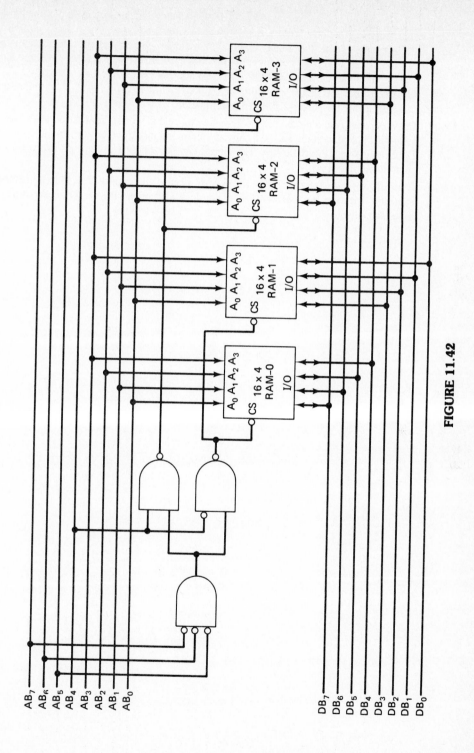

FIGURE 11.42

(b) Note that there are more address bus lines than are necessary to select one of the memory locations. This is not an unusual situation, especially in small computer systems, where the actual amount of memory circuitry is much less than the maximum which the computer address bus can handle. Which RAMs will put data on the data bus when $R/\overline{W} = 1$ and the address bus is at 00010110?

(c) Determine the range of addresses stored in the RAM-0/RAM-1 combination. Repeat for the RAM-2/RAM-3 combination.

11.24 Draw the complete diagram for a 4K × 4 memory that uses static RAM chips with the following specifications: 1K × 1 capacity, common input/output line, and two active-LOW chip-select inputs. [*Hint:* The circuit can be designed using only two inverters (plus memory chips).]

11.25 Compare the number of 2114 static RAMs needed to produce a 32K × 8 memory capacity, with the number of 2118 dynamic RAM chips (16K × 1) needed for the same capacity.

12

INTRODUCTION
to the MICROPROCESSOR
and MICROCOMPUTER

It is no exaggeration to say that the microprocessor and microcomputer have revolutionized the electronics industry and have had a tremendous impact on many aspects of our lives. Large-scale integration (LSI) has so sharply reduced the size and cost of computers that designers routinely consider using the power and versatility of the microprocessor and microcomputer in a wide variety of applications.

In this chapter we will study the basic principles of operation of microcomputers. Since complete textbooks are devoted to the study of computers, our main objective here is only to provide a firm foundation for further study. Although we will concentrate on the microcomputer, most of the concepts and principles apply equally to computers of all sizes.

12.1 WHAT IS A DIGITAL COMPUTER?

A digital computer is a combination of digital devices and circuits that can perform a *programmed* sequence of operations with a minimum of human intervention. The sequence of operations is called a *program*. The program is a set of coded instructions that is stored in the computer's internal memory along with all the data that the program requires. When the computer is commanded to execute the program, it performs the instructions in the order that they are stored in memory until the program is completed. It does this at extremely high speeds without making any errors.

12.2 HOW DO COMPUTERS THINK?

Computers do not think! The computer *programmer* provides a *program* of instructions and data which specifies every detail of what to do, what to do it to, and when to do it. The computer is simply a high-speed machine that can manipulate data, solve problems, and make decisions, all under the control of the program. If the programmer makes a mistake in the program or puts in the wrong data, the computer will produce wrong results. A popular saying in the computer field is "garbage in gives you garbage out."

Perhaps a better question to ask at this point is: How does a computer go about executing a program of instructions? Typically, this question is answered by showing a diagram of a computer's architecture (arrangement of its various elements) and then going through the step-by-step process which the computer follows in executing the program. We will do this—but not yet. First, we will look at a somewhat far-fetched analogy that contains many of the concepts involved in computer operation.

12.3 SECRET AGENT 89

Secret Agent 89 is trying to find out how many days before a certain world leader is to be assassinated. His contact tells him that this information is located in a series of post office boxes. To ensure that no one else gets the information, it is spread through 10 different boxes. His contact gives him 10 keys along with the following instructions:

1. The information in each box is written in code.
2. Open box 1 first and execute the instruction located there.
3. Continue through the rest of the boxes in sequence unless instructed to do otherwise.
4. One of the boxes is wired to explode upon opening.

Agent 89 takes the 10 keys and proceeds to the post office, code book in hand.

Figure 12.1 shows the contents of the 10 post office boxes after having been decoded. Assume that you are Agent 89; begin at box 1 and go through the sequence of operations to find the number of days before the assassination attempt. Of course, it should not be as much work for you as it was for Agent 89 because you don't have to decode the messages. The answer is given in the next paragraph.

① Add the number stored in box ⑨ to your secret agent code number.	② Divide the previous result by the number stored in box ⑩.
③ Subtract the number stored in box ⑧	④ If the previous result is not equal to 30, go to box ⑦. Otherwise continue to next box.
⑤ Subtract 13 from the previous result.	⑥ HALT. You now have the answer
⑦ BOMB! (too bad)	⑧ 20
⑨ 11	⑩ 2

FIGURE 12.1   Ten post office boxes with coded message for Agent 89.

If you have proceeded correctly, you should have ended up at box 6 with an answer of 17. If you made a mistake, you might have opened box 7, in which case you are no longer with us. As you went through the sequence of operations, you essentially duplicated the types of operations and encountered many of the concepts that are part of a computer. We will now discuss these operations and concepts in the context of the secret-agent analogy and see how they are related to actual computers.

In case you have not already guessed, the post office boxes are like the *memory* in a computer, where *instructions* and *data* are stored. Post office boxes 1–6 contain instructions to be executed by the secret agent and boxes 8–10 contain the data called for by the instructions. (The contents of box 7, to our knowledge, has no counterpart in computers.) The numbers on each box are like the *addresses* of the locations in memory.

Three different classes of instructions are present in boxes 1–6. Boxes 1, 2, 3, and 5 are instructions that call for *arithmetic operations*. Box 4 contains a *decision-making* instruction called a *conditional jump* or *conditional branch*. This instruction calls for the agent (or computer) to decide whether to jump to address 7 or to continue to address 5, depending on the result of the previous arithmetic operation. Box 6 contains a simple control instruction that requires no data or refers to no other address (box number). This *halt* instruction tells the agent that the problem is finished (program is completed) and to go no further.

Each of the arithmetic and conditional jump instructions consists of two parts— an *operation* and an *address*. For example, the first part of the first instruction specifies the operation of addition. The second part gives the address (box 9) of the data to be used in the addition. These data are usually called the *operand* and their address the *operand address*. The instruction in box 5 is a special case in which no operand address is specified. Instead, the operand (data) to be used in the subtraction operation is included as part of the instruction.

A computer, like the secret agent, decodes and then executes the instructions stored in memory *sequentially,* beginning with the first location. The instructions are executed in order unless some type of *branch* instruction (such as box 4) causes the operation to branch or jump to a new address location to obtain the next instruction. Once the branching occurs, instructions are executed sequentially beginning at the new address.

This is about as much information as we can extract from the secret-agent analogy. Each of the concepts we encountered will be encountered again in subsequent material. Hopefully, the analogy has furnished insights that should prove useful as we begin a more technical study of computers.

12.4 BASIC COMPUTER SYSTEM ORGANIZATION

Every computer contains five essential elements or units: the *arithmetic logic unit* (ALU), the *memory unit,* the *control unit,* the *input unit,* and the *output unit.* The basic interconnection of these units is shown in Figure 12.2. The arrows in this diagram indicate the direction in which data, information, or control signals are flowing. Two different-size arrows are used; the larger arrows represent data or information that actually consists of a relatively large number of parallel lines, and the smaller arrows represent control signals that are normally only one or a few lines. The various arrows are also numbered to allow easy reference to them in the following descriptions.

Arithmetic/Logic Unit The ALU is the area of the computer in which arithmetic and logic operations are performed on data. The type of operation that is to be performed is determined by signals from the control unit (arrow 1). The data that are to be operated on by the ALU can come from either the memory unit (arrow 2) or the input unit (arrow 3). Results of operations performed in the ALU can be transferred to either the memory unit for storage (arrow 4) or to the output unit (arrow 5).

FIGURE 12.2   Basic computer organization.

Memory Unit   The memory stores groups of binary digits (words) that can represent instructions (program) that the computer is to perform and the data that are to be operated on by the program. The memory also serves as storage for intermediate and final results of arithmetic operations (arrow 4). Operation of the memory is controlled by the control unit (arrow 6), which signals for either a read or a write operation. A given location in memory is accessed by the control unit, providing the appropriate address code (arrow 7). Information can be written into the memory from the ALU or the input unit (arrow 8), again under control of the control unit. Information can be read from memory into the ALU (arrow 2) or into the output unit (arrow 9).

Input Unit   This unit consists of all of the devices used to take information and data that are external to the computer and put them into the memory unit (arrow 8) or the ALU (arrow 3). The control unit determines where the input information is sent (arrow 10). The input unit is used to enter the program and data into the memory unit prior to starting the computer. This unit is also used to enter data into the ALU from an external device during the execution of a program. Some of the common input devices are keyboards, toggle switches, teletypewriters, punched-card readers, magnetic disk units, magnetic tape units, and analog-to-digital converters (ADC).

Output Unit   This unit consists of the devices used to transfer data and information from the computer to the "outside world." The output devices are directed by the control unit (arrow 12) and can receive data from memory (arrow 9) or the ALU (arrow 5), which is then put into appropriate form for external use. Examples of common output devices are LED readouts, indicator lights, printers, disk or tape units, cathode-ray-tube displays, and digital-to-analog converters (DAC).

As the computer executes its program, it usually has results or control signals that it must present to the external world. For example, a large computer system might have a line printer as an output device. Here, the computer sends out signals to print out the results on paper. A microcomputer might display its results on indicator lights or on LED displays.

Interfacing   The most important aspect of the I/O units involves *interfacing,* which can be defined as the joining of dissimilar devices in such a way that they are able to function in a compatible and coordinated manner. *Computer interfacing* is more specifically defined as the synchronization of digital information transmission between the computer and external input/output devices.

Many input/output devices are not directly compatible with the computer because of differences in such characteristics as operating speed, data format (e.g., hex, ASCII, binary), data transmission mode (e.g., serial, parallel), and logic signal level. Such I/O devices require special interface circuits which allow them to communicate with the CONTROL, MEMORY, and ALU portions of the computer system. A common example is the popular teletypewriter (abbreviated TTY), which can operate both as an input and an output device. The TTY transmits and receives data serially (one bit at a time) while most computers handle data in parallel form. Thus, a TTY requires interface circuitry in order to send data to or receive data from a computer.

Control Unit   The function of the control unit should now be obvious. It directs the operation of all the other units by providing timing and control signals. In a sense, the control unit is like the conductor of an orchestra, who is responsible for keeping each of the orchestra members in proper synchronization. This unit contains logic and timing circuits that generate the proper signals necessary to execute each instruction in a program.

The control unit *fetches* an instruction from memory by sending an address (arrow 7) and a read command (arrow 6) to the memory unit. The instruction word stored at the memory location is then transferred to the control unit (arrow 11). This instruction word, which is in some form of binary code, is then decoded by logic circuitry in the control unit to determine which instruction is being called for. The control unit uses this information to generate the necessary signals for *executing* the instruction.

Central Processing Unit (CPU)   In Figure 12.2, the ALU and control units are shown combined into one unit called the central processing unit (CPU). This is commonly done to separate the actual "brains" of the computer from the other units. We will use the CPU designation often in our work on microcomputers because, as we shall see, in microcomputers the CPU is often contained in a single LSI chip called the *microprocessor* chip.

1. Name the five basic units of a computer and describe the major functions of each.
2. What is the CPU?
3. What is meant by interfacing in a computer system?

12.5 BASIC μC ELEMENTS

It is important that we understand the distinction between the microcomputer (μC) and the microprocessor (μP). A μC contains many elements, one of which is the μP. The μP is the central processing unit (CPU) portion of the μC. This is illustrated in Figure 12.3, where the basic elements of a μC are shown. The μP is typically a single LSI chip that contains all the control and arithmetic circuits of the μC. The μP may consist of more than one chip. This is true, for example, of bipolar μPs (TTL, Shottky TTL, ECL) which do not have the high packing densities of the MOS devices, and so require two or more chips to produce a μP with appropriate word size.

The memory unit shows both RAM and ROM devices, typical of most μCs, although one or the other might not be present in certain applications. The RAM section consists of one or more LSI chips arranged to provide the designed memory capacity. This section of memory is used to store programs and data, which will change often during the course of operation. It is also used as storage for intermediate and final results of operations performed during execution of a program.

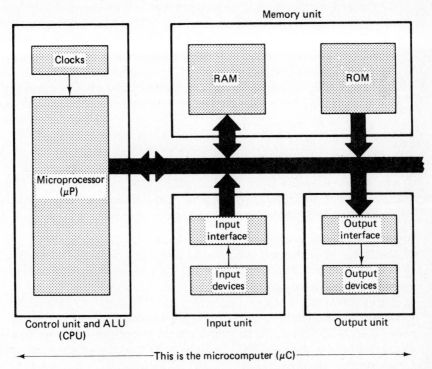

FIGURE 12.3 Basic elements of a μC.

The ROM section consists of one or more LSI chips to store instructions and data that do not change. For example, it might store a program that causes the μC to continually monitor a keyboard, or it might store a table of ASCII codes needed for outputting information to a teletype unit.

The input and output sections contain the interface circuits needed to allow the I/O devices to properly communicate with the rest of the computer. In some cases these interface circuits are LSI chips designed by the μP manufacturer to interface his μP to a variety of I/O devices. In other cases the interface circuits may be as simple as a buffer register.

REVIEW QUESTIONS

1. Explain the difference between a microprocessor and a microcomputer.
2. Name the basic elements of a microcomputer.

12.6 COMPUTER WORDS

The preceding description of how the various units in a computer interact has been, by necessity, somewhat oversimplified. To proceed in more detail, we must define the various forms of information that are continually being transferred and manipulated within the computer.

In a computer, the most elementary unit of information is the binary digit (bit). A single bit, however, can impart very little information. For this reason, the primary unit of information in a computer is a group of bits referred to as the *computer word*. Word size is so important that it is often used in describing a computer. For example, a 16-bit computer is a computer in which data and instructions are processed in 16-bit units. Of course, the word size also indicates the word size of the memory unit. Thus, a 16-bit computer has a memory unit that stores a certain number of 16-bit words.

A large variety of word sizes have been used by computer manufacturers. The larger (maxi) computers have word sizes that range from 16 to 64 bits, with 32 bits being the most common. Minicomputer word sizes run from 8 to 32, with 16 bits representing the overwhelming majority. Most microcomputers use an 8-bit word size. There are several 4-bit microcomputers which are designed for replacing digital logic circuits, some 16-bit and a few 32-bit microcomputers which are aimed at competing with minicomputers.

The Byte   A group of 8 bits is called a *byte* and represents a universally used unit in the computer industry. For example, a microcomputer with an 8-bit word size is said to have a word size of one byte. A 16-bit computer can be said to have a word size of two bytes. When we deal with microcomputers that have an 8-bit word size, we will use the terms "word" and "byte" interchangeably.

The 4-bit microcomputers have a word size of one-half byte. This is commonly referred to as a "nibble." Thus, each word in a 4-bit microcomputer is a nibble and two nibbles constitute a byte.

Types of Computer Words   A word stored in a computer's memory unit can contain several different types of information, depending on what the programmer intended for that particular word. We can classify computer words into three categories: (1) pure binary numerical data, (2) coded data, and (3) instructions. These will now be examined in detail.

12.7   BINARY DATA WORDS

These are words that simply represent a numerical quantity in the binary number system. For example, a certain location in the memory of an 8-bit process control microcomputer might contain the word 01110011, representing the desired process temperature in Fahrenheit degrees. This binary number 01110011 is equivalent to 115_{10}.

Here is an example of a 16-bit data word:

$$1010000101001001$$

which is equivalent to $41,289_{10}$.

Obviously, a wider range of numerical data can be represented with a larger word size. With an 8-bit word size, the largest data word (11111111_2) is equivalent to $2^8 - 1 = 255_{10}$. With a 16-bit word size, the largest data word is equivalent to $2^{16} - 1 = 65,535_{10}$. With 32 bits (four bytes) we can represent numbers greater than 4 billion.

Signed Data Words   A computer would not be too useful if it could only handle positive numbers. For this reason, most computers use the signed 2's-complement system. Recall that the most significant bit (MSB) is used as the *sign* bit (0 is positive and 1 is negative). Here is how the values $+9$ and -9 would be represented in an 8-bit computer:

$$+9 \longrightarrow \underbrace{0\overbrace{0001001}}_{+}$$
binary for 9_{10}

$$-9 \longrightarrow \underbrace{1\overbrace{1110111}}_{-}$$
2's complement of 0001001

Here, of course, only 7 bits are reserved for the magnitude of the number. Thus, in the signed 2's-complement system, we can only represent numbers from -128_{10} to $+127_{10}$. Similarly, with 16-bit words, we can have a range from $-32,768_{10}$ to $+32,767_{10}$.

Multiword Data Units   Very often a computer needs to process data that extend beyond the range possible with a single word. For such cases, two or more memory words can be used to store the data in parts. For example, the 16-bit data word 1010101100101001_2 can be stored in two consecutive 8-bit memory locations, as shown below:

Memory Address (hex)	Memory Contents	
0030	10101011 →	8 high-order bits of 16-bit number
0031	00101001 →	8 low-order bits of 16-bit number

Here, address location 0030_{16} stores the 8 higher-order bits of the 16-bit data word. This is also called the *high-order byte*. Similarly, address 0031_{16} stores the *low-order byte*. The two bytes combined make up the full data word.

There is no actual limit to the number of memory words that can be combined to store large numbers.

Octal and Hex Data Representation For purposes of convenience in writing and displaying data words, they can be represented in either octal or hexadecimal codes. For example, the number $+116_{10}$ can be represented in a single byte as 01110100_2. The hex and octal representations are:

$$01110100_2 = 74_{16}$$

$$01110100_2 = 164_8$$

It is important to realize that the use of hex or octal representations is solely for convenience of the computer user; the computer memory still stores the binary numbers (0s and 1s), and these are what the computer processes.

12.8 CODED DATA WORDS

Data processed by a computer do not have to be pure binary numbers. One of the other common data forms uses the BCD code, where each group of 4 bits can represent a single decimal digit. Thus, an 8-bit word can represent two decimal digits, a 16-bit word can represent four decimal digits, and so on. Many computers can perform arithmetic operations on BCD-coded numbers as part of their normal instruction repertoire; others, especially some microcomputers, require special effort on the part of the programmer in order to do BCD arithmetic.

Data words are not restricted to representing only numbers. They are often used to represent alphabetic characters and other special characters or symbols using codes such as the 7-bit ASCII code (Chapter 2). The ASCII code is used by all minicomputer and microcomputer manufacturers. Although the basic ASCII code uses 7 bits, an extra parity bit (Chapter 2) is added to each code word, producing a one-byte ASCII code. The example below shows how a message might be stored in a sequence of memory locations using ASCII code with an *even* parity bit. The contents of each location are also given in hex code. Use Table 2.5 to determine the message. Note that the leftmost bit is the parity bit and the first character is stored in location $012A_{16}$. The decoded message is the familiar electrical Ohm's law, $I = V/R$.

Address Location	Contents	
	Binary	Hex
012A	11001001	C9
012B	10111101	BD
012C	01010110	56
012D	10101111	AF
012E	11010010	D2
	ASCII	

Interpretation of Data Words   Suppose you are told that a particular data word in a microcomputer's memory is 01010110. This word can be interpreted in several ways. It could be the binary representation of 86_{10}; it could be the BCD representation of 56_{10}; or it could be the ASCII code for the character V. How should this data word be interpreted? It is up to the programmer, since he or she is the one who places the data in memory along with instructions that make up the program. The programmer knows what type of data word he or she is using and must make sure that the program of instructions executed by the computer interprets the data properly.

12.9   INSTRUCTION WORDS

The format used for *data* words varies only slightly among different computers, especially those with the same word size. This is not true, however, of the format for *instruction* words. These words contain the information necessary for a computer to execute its various operations, and the format and codes for these can vary widely from computer to computer. Depending on the computer, the information contained in an instruction word can be different. But, for most computers, the instruction words carry two basic units of information: the *operation* to be performed and the *address* of the *operand* (data) that is to be operated upon.

Figure 12.4 shows an example of a *single-address instruction word* for a hypothetical 20-bit computer. The 20 bits of the instruction word are divided into two parts. The first part of the word (bits 16–19) contains the *operation code* (*op code*, for short). The 4-bit op code represents the operation that the computer is being instructed to perform, such as addition or subtraction. The second part (bits 0–15) is the *operand address* which represents the location in memory where the operand is stored.

With 4 bits used for the op code, there are $2^4 = 16$ different possible op codes, with each one indicating a different instruction. This means that a computer using this instruction-word format is limited to 16 different possible instructions which it can

FIGURE 12.4   Typical single-address instruction word.

perform. A more versatile computer would have a greater number of instructions and would therefore require more bits in its op code. In any case, each instruction that a computer can perform has a specific op code which the computer (control unit) must interpret (decode).

The instruction word of Figure 12.4 has 16 bits reserved for the operand address code. With 16 bits, there are $2^{16} = 65,536$ different possible addresses. Thus, this instruction word can specify 16 different instructions and 65,536 operand addresses. As an example, a 20-bit instruction word might be

The op code 0100 represents one of 16 possible operations; let's assume that it is the code for *addition* (ADD). The address code is 0101101001110010 or, more conveniently, 5A72 in hexadecimal. In fact, this complete instruction word can be expressed in hexadecimal as

$$\underbrace{4}_{\substack{\text{Op}\\\text{code}}} \quad \underbrace{5 \quad A \quad 7 \quad 2}_{\text{Address}}$$

This complete instruction word, then, tells the computer to do the following:

> Fetch the data word stored in address location 5A72, send it to the ALU and *add* it to the number in the accumulator register. The sum will then be stored in the accumulator. (Previous contents of accumulator is lost.)

We will examine this and other instructions more thoroughly later.

Multiple-Address Instructions   The single-address instruction described above is the basic type used in small computers and was once the principal type used in larger computers. The larger computers, however, have begun to use several other instruction formats which provide more information per instruction word.

Figure 12.5 shows two instruction-word formats that contain more than one address. The two-address instruction has the op code plus the addresses of *both* operands which are to take part in the specified operation. The three-address instruction has the addresses of both operands plus the address in memory where the result is to be stored.

These multiple-address instruction words have the obvious advantage that they contain more information than does a single-address instruction. This means that a computer using multiple-address instructions will require fewer instructions to execute a particular program. Of course, the longer instruction words require a memory unit with a larger word size. We will not concern ourselves further with multiple-address instructions since they are not used in microcomputers.

	Op code	Address of operand 1	Address of operand 2

Two-address instruction

(a)

	Op code	Address of operand 1	Address of operand 2	Address of where to store result

Three-address instruction

(b)

FIGURE 12.5 Multiple-address instruction formats.

Multibyte Instructions We have looked at instruction-word formats that contain op code and operand address information in a *single* word. In other words, a complete instruction such as those in Figure 12.4 or Figure 12.5 is stored in a *single* memory location. This is typical of computers with relatively large word sizes. For most microcomputers and many minicomputers, the smaller word size makes it impossible to provide the op code and operand address in a single word.

Since the vast majority of microcomputers use an 8-bit (one-byte) word length, we will describe the instruction formats used in 8-bit computers. With a one-byte word size there are *three* basic instruction formats: single-byte, two-byte, and three-byte instructions. These are illustrated in Figure 12.6.

The single-byte instruction contains only an 8-bit op code, with no address portion. Clearly, this type of instruction does not specify any data from memory to be operated on. As such, single-byte instructions are used for operations that do not require memory data. An example would be the instruction *clear the accumulator register to zero* (CLA), which instructs the computer to clear all the FFs in the ALU's accumulator.

The first byte of the two-byte instruction is an op code and the second byte is an 8-bit address code specifying the memory location of the operand. In the three-byte instruction, the second and third bytes form a 16-bit operand address. For these multibyte instructions, the two or three bytes making up the complete instruction have to be stored in successive memory locations. This is illustrated below for a three-byte

Memory Address (hex)	Memory Word		Description
	Binary	Hex	
0020	01001010	4A	Op code for ADD
0021	00110101	F6	Low-order address bits (LO)
0022	11110110	35	High-order address bits (HI)
.	.	.	
.	.	.	
.	.	.	
.	.	.	
.	.	.	
35F6	01111100	7C	Operand

Sec. 12.9   Instruction Words

FIGURE 12.6   Instruction formats used in 8-bit microcomputers.

instruction. The left-hand column lists the address locations in memory where each byte (word) is stored. These addresses are given in hexadecimal code. The second column gives the binary word as it is actually stored in memory; the third column is the hex equivalent of this word. Examine this complete example before reading further and try to figure out what it represents.

The three bytes stored in locations 0020, 0021, and 0022 constitute the complete instruction for adding the data word stored in address location 35F6 to the accumulator. The second and third bytes hold the 8 low-order bits (LO), and 8 high-order bits (HI), respectively, of the operand address. Some microcomputers use the reverse order, with HI stored in the second byte and LO in the third byte of the instruction sequence. Memory location 35F6 is also shown; its contents is the data word (7C) which the control unit of the computer will fetch and send to the ALU for addition.

1. Explain the difference between a data word and an instruction word.
2. What information is contained in a single-address instruction word?
3. How is this information represented in a three-byte instruction?

12.10 EXECUTING A MACHINE-LANGUAGE PROGRAM

The instruction words we have been describing are called *machine-language* instructions because they are represented by 1s and 0s, the only language the machine (computer) understands. Many other languages can be used to program a computer, and you may be familiar with some of them, such as BASIC or FORTRAN. These *high-level languages* are designed to make it easy to write a program. It is important to understand that these high-level programs have to be converted to machine-language instructions and placed in the computer's internal memory before the computer can execute them. In a typical microcomputer, the conversion from BASIC to machine language is accomplished through a special machine-language program stored in ROM.

In order to illustrate how a microcomputer executes a machine-language program, we will use the instructions described in Table 12.1. These instructions do not pertain to any particular microcomputer but they are typical of the kinds of instructions

TABLE 12.1

Mnemonic	Op code, binary/hex	Description of operation
LDA	01011011/5B	*Load accumulator:* the data stored at the operand address are loaded into the accumulator register.
ADD	01101101/6D	*Add:* the operand is added to the number stored in the accumulator and the resultant sum is stored in the accumulator.
SUB	10100010/A2	*Subtract:* the operand is subtracted from the contents of the accumulator and the result is stored in the accumulator.
STA	10001010/8A	*Store accumulator:* the contents of the accumulator is stored in memory at the location specified by the operand address.
JMP	01001100/4C	*Jump (unconditionally):* the next instruction is taken from the location specified by the operand address instead of in sequence.
JPZ	01110110/76	*Jump on zero:* the next instruction is taken from operand address *if* the accumulator contents is *zero.* Otherwise, the next instruction is taken in sequence.
HLT	00111111/3F	*Halt:* the computer operation is halted. No further instructions are executed.

that most microcomputers can execute. Each instruction is accompanied by a *mnemonic* or abbreviation that is easier to remember than the op code. Read the description of each instruction carefully and keep in mind that the operand that is being referred to is the data word stored at the operand address location.

We will use some of the instructions listed in Table 12.1 to write a machine-language program that starts at 0000 and does the following:

1. Subtracts an 8-bit data word (*X*) from another 8-bit data word (*Y*). The values of *X* and *Y* are stored in hex address locations 0300 and 0301, respectively.
2. Stores the difference (DIF) in memory location 0302.
3. If the value of DIF is zero, the program jumps to address 0400 for its next instruction. Otherwise, the program halts.

Table 12.2 shows the complete program as it would be entered into the computer's memory. Actually, the first two columns are the machine-language program; the other information is included for descriptive purposes. The first column lists the hex address of each memory location being used by the program. The second column gives the hex equivalent of the word stored in each memory location. Remember, these hex values represent the actual binary addresses and instruction codes that the computer understands.

TABLE 12.2

Memory address (hex)	Memory word (hex)	Mnemonic	Description
0000	5B	LDA$0300	Load data (*Y*) from address
0001	00 ⎫		0300 into accumulator
0002	03 ⎭		
0003	A2	SUB$0301	Subtract data (*X*) at address
0004	01 ⎫		0301 from accumulator
0005	03 ⎭		
0006	8A	STA$0302	Store difference in address
0007	02 ⎫		0302
0008	03 ⎭		
0009	76	JPZ$0400	If difference (DIF) is zero,
000A	00 ⎫		jump to 0400 for next
000B	04 ⎭		instruction
000C	3F	HLT	If DIF is *not* zero, halt program.
.	.	.	.
.	.	.	.
.	.	.	.
0300	??	*Y*	Address of data *Y*
0301	??	*X*	Address of data *X*
0302	??	DIF	Address where difference is to be stored

The third column gives the mnemonic and operand address (if any) associated with each instruction. The last column describes the operation performed by each instruction. For example, the first instruction in the program is a three-byte instruction. The first byte stored in memory address 0000 is the op code 5B. The second and third bytes stored in addresses 0001 and 0002 represent the operand address 0300. The mnemonic for this instruction is LDA$0300. The LDA is the abbreviation for the load accumulator operation, and the $0300 is the operand address. The $ is often used to indicate that the address is represented in hex. When the computer executes this instruction, it will read the data Y that are stored in location 0300 and will load them into the accumulator.

The program starts at address 0000 and ends at 000C. Address locations 0300–0302 are used for data storage. The contents of these latter locations are not given because they will generally be variables. Address locations between 000C and 0300 are not being used.

Program Execution   We will now proceed through the complete execution of this program and describe what the computer does at each step. Our intent here is to outline only the main operations without getting bogged down in the details of all the activity taking place inside the computer. In particular, we will see that the computer is always in one of two kinds of operating cycles: (1) a *fetch cycle* during which the CONTROL unit fetches the instruction codes (op code and operand address) from memory and (2) an *execute cycle* during which the CONTROL unit performs the operation called for by the op code.

The operation starts when the operator activates a START or RUN switch. This will initialize a PROGRAM COUNTER (PC) to a starting count of 0000. The PC is a counter within the CONTROL unit that keeps track of the program addresses as the computer sequences through them.

1. The CONTROL unit fetches the first byte from address 0000 as determined by the PC. This byte is 5B, which is the op code for the first instruction. Circuitry within the CONTROL unit determines that this op code calls for the LDA (load accumulator) operation and that an operand address follows the op code.

2. The CONTROL unit increments the PC to 0001 and then fetches the byte stored at this address. This byte is 00 and represents the low-order byte of the operand address. The CONTROL unit then increments PC again to 0002 and fetches the byte that is stored at this address. This byte is 03 and represents the high-order byte of the operand address.

3. This completes the first fetch cycle, and the CONTROL unit now holds the op code and operand address needed to execute the LDA instruction.

4. The CONTROL unit executes this instruction by reading the data word from address 0300 and loading it into the accumulator register in the ALU. This completes the execute cycle.

5. The PC is incremented to 0003 and the CONTROL unit begins a new fetch cycle by fetching the op code stored at this address. The CONTROL unit recognizes the op code A2 as calling for the SUB operation which requires an operand address that is stored in the next two bytes of the program.

6. The CONTROL unit fetches the operand address 0301 in the same manner as in step 2 above. The PC is now at 0005.

7. This completes the second fetch cycle, and the CONTROL unit now holds the op code and operand address needed to perform the SUB instruction. The CONTROL unit executes this instruction by reading the data word from address 0301 and sending it to the ALU to be subtracted from the contents of the accumulator.

8. The PC is incremented to 0006 to initiate a new fetch cycle. The CONTROL unit fetches the op code (8A) stored at this address, recognizes it as calling for the STA operation, and knows that the operand address is stored in the next two bytes of the program.

9. The CONTROL unit fetches the operand address 0302. The PC is now at 0008. The contents of the accumulator is stored in 0302.

10. The PC is incremented to 0009 to initiate a new fetch cycle, and the CONTROL unit fetches the op code (76) stored at this address. The CONTROL unit recognizes the JPZ instruction and knows that it has to make a decision based on the current contents of the accumulator.

11. If the data in the accumulator are exactly equal to zero, the CONTROL unit fetches the operand address 0400 from the next two memory addresses (000A and 000B). This operand address is loaded into the PC so that the CONTROL unit will take its next instruction from address 0400 and will continue executing instructions from that point.

12. If the data in the accumulator are *not* equal to zero, the CONTROL unit increments the PC three times so that PC = 000C. The next instruction is fetched from this address, and it is recognized as the HLT instruction which causes the CONTROL unit to stop fetching and executing instructions.

This simple program illustrates many of the kinds of things that take place as a computer executes a machine-language program, but it does not even begin to show the capabilities and versatility of a computer. It is important to understand that the program execution is performed in a step-by-step sequential manner starting at the first address in the program (0000 in our example). The program's address sequence can change when either a JMP or JPZ instruction is encountered (e.g., steps 10–11 above). Of course, even though the computer performs one step at a time, each step can be done so rapidly that the overall execution time for a program can be very short. For example, this short program would take less than 20 μs on even the slowest microcomputer.

REVIEW QUESTIONS

1. What is the difference between a machine-language program and a high-level-language program?
2. What is an instruction mnemonic?
3. What generally takes place during a fetch cycle? During an execute cycle?
4. What is the function of the PROGRAM COUNTER?

12.11 TYPICAL μC STRUCTURE

We are now prepared to take a more detailed look at μC organization. The many possible μC structures are essentially the same in principle, although they vary as to the size of the data and address buses, and the types of control signals they use. In order to provide the clearest means for learning the principles of μC operation, it is necessary to choose a single type of μC structure and study it in detail. Once a solid understanding of this typical μC is obtained, it will be relatively easy to learn about any other type. The μC structure we have chosen to present here represents a common one in use today and is shown in Figure 12.7.

This diagram shows the basic elements of an 8-bit microcomputer system and the various buses that connect them together. Although this diagram looks somewhat complex, it still does not show all the details of the μC system. For our purposes, however, it will be sufficient.

The Bus System The μC has three buses which carry all the information and signals involved in the system operation. These buses connect the microprocessor (CPU) to each of the memory and I/O elements so that data and information can flow between the μP and any of these other elements. In other words, the CPU is continually involved in sending or receiving information to or from a location in memory, an input device, or an output device.*

In the μC, all information transfers are referenced to the CPU. When the CPU is sending data to another computer element, it is called a WRITE operation and the CPU is WRITING into the selected element. When the CPU is receiving data from another element, it is called a READ operation and the CPU is READING from the selected element. It is very important to realize that the terms READ and WRITE always refer to operations performed by the CPU.

The buses involved in all the data transfers have functions that are described as follows:

Address bus: This is a *unidirectional* bus, because information flows over it in only one direction, from the CPU to the memory or I/O elements. The CPU alone can place logic levels on the 16 lines of the address bus, thereby generating $2^{16} = 65,536$ different possible addresses. Each of these addresses corresponds to one memory location or one I/O element. For example, address $20A0_{16}$ might be a location in RAM or ROM where an 8-bit word is stored, or it might be an 8-bit buffer register that is part of the interface circuitry for a keyboard input device.

When the CPU wants to communicate (READ or WRITE) with a certain memory location or I/O device, it places the appropriate 16-bit address code on its 16 address pin outputs, A_0–A_{15}, and onto the address bus. These address bits are then *decoded* to select the desired memory location or I/O device. This decoding process usually requires decoder circuitry not shown on this diagram but which will be added later.

Data bus: This is a *bidirectional* bus, because data can flow to or from the CPU. The CPU's eight data pins, D_0–D_7, can be either inputs or outputs, depending on whether the CPU is performing a READ or a WRITE operation. During a READ operation they act as inputs and receive data that have been placed on the data bus by the memory or I/O element selected by the address code on the address bus. During a WRITE operation the CPU's data pins act as outputs and place data on the data bus, which are then sent to the selected memory or I/O

*In some μC systems, it is possible for I/O devices to send data directly to or receive data directly from memory without the CPU being involved. This type of operation is called *direct memory access* (DMA).

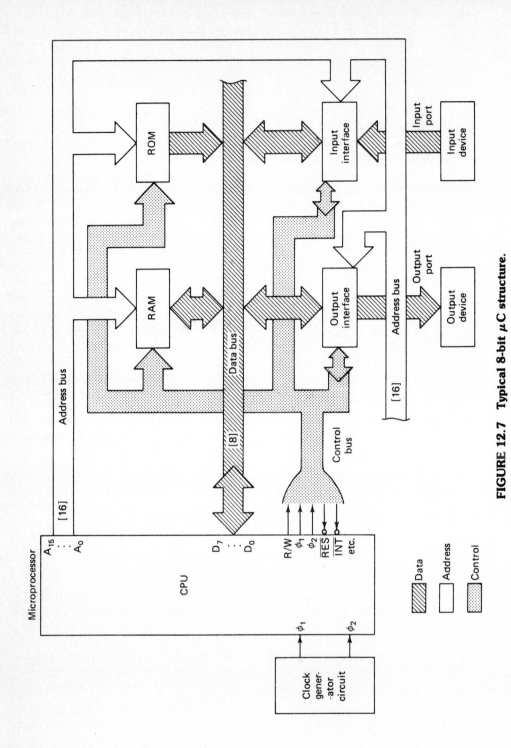

FIGURE 12.7 Typical 8-bit μC structure.

element. In all cases, the transmitted data words are 8 bits long becauase the CPU handles 8-bit data words, making this an 8-bit μC.

In some microprocessors, the data pins are used to transmit other information in addition to data (e.g., address bits or CPU status information). That is, the data pins are time-shared or *multiplexed*, which means that special control signals must be generated by the CPU to tell the other elements exactly what is on the data bus at a particular time. We will not concern ourselves with this type of operation.

Control bus: This is the set of signals that is used to synchronize the activities of the separate μC elements. Some of these control signals, such as $R/\overline{W}$, are signals the CPU sends to the other elements to tell them what type of operation is currently in progress. The I/O elements can send control signals to the CPU. An example is the reset input ($\overline{RES}$) of the CPU, which, when driven LOW, causes the CPU to reset to a particular starting state. Another example is the CPU's interrupt input ($\overline{INT}$), used by I/O devices to get the attention of the CPU when it is performing other tasks.

The control bus signals will vary widely from one μC to another. There are certain control signals that all μCs use, but there are also many control signals that are peculiar to the μP upon which the μC is based. We will include only the essential control signals in this discussion.

Timing Signals   The most important signals on the control bus are the system clock signals that generate the time intervals during which all system operations take place. Different μCs use different kinds of clock signals, depending on the type of μP being used. Some μPs, like the 8085, the 6502, and the Z-80, do not require an external clock-generating circuit. A crystal or RC network connected to the appropriate μP pins sets the operating frequency for the clock signals which are generated on the μP chip. Other μPs, such as the 8080A and 6800, require an external circuit to generate the clock signals needed by the CPU and the other μC elements. These manufacturers often provide a special clock-generator chip designed to be used with their μP.

Many of the currently popular μPs (8085, 6800, 6502) use a two-phase clock system with nonoverlapping pulses such as those shown in Figure 12.8. Other widely used μPs (Z-80, RCA 1802) operate from a single clock signal. In our subsequent discussions we will use the two-phase clock system. The two clock phases, ϕ_1 and ϕ_2, are always part of the control bus. Other timing signals, derived from ϕ_1 and ϕ_2, are sometimes generated by the CPU and become part of the control bus.

FIGURE 12.8   Two phase clock system.

I/O Ports   During the execution of a program, the CPU is constantly READING or WRITING into memory. The program may also call on the CPU to READ from one of the input devices or WRITE into one of the output devices. Although the diagram of the 8-bit μC (repeated in Figure 12.9) only shows one input and one output device, there can be any number of each tied to the μC bus system. Each I/O device is normally connected to the μC bus system through some type of interface circuit. The function of the interface is to make the μC and the device compatible so that data can be easily passed between them. The interface is needed whenever the I/O device uses different signal levels, signal timing, or signal format than the μC.

For example, a typical I/O device is the standard teletype unit (abbreviated TTY), which sends ASCII-coded information to the computer in *serial* fashion (1 bit at a time over a single line). The μC, however, accepts data from the data bus as 8 *parallel* bits. Thus, an interface circuit is used to convert the TTY's serial signal to an 8-bit parallel data word, and another to convert the μC's parallel output data to a serial signal for the TTY.

It was mentioned during the discussion of the address bus that the CPU places a 16-bit address on this bus to select a certain memory location or a certain I/O device. This means that each I/O device has a specific address just like any location in memory. In many μCs, the CPU does not distinguish between memory and I/O, and it communicates with both in the same way using the same control signals. This method is called *memory-mapped I/O*. Other μCs use separate control signals and separate address decoders for I/O. This is called *isolated I/O*. We will concentrate mainly on the memory-mapped I/O technique, since it is the most common and has several advantages over the isolated I/O technique.

Although I/O devices are treated like memory locations, they are significantly different from memory in some respects. One big difference is that I/O devices can have the capability to *interrupt* the μC while it is executing a program. What this means is that an I/O device can send a signal to the μP chip's interrupt ($\overline{INT}$) input to tell the CPU that it wishes to communicate with it. The CPU will then suspend execution of the program it is currently working on and will perform the appropriate operation with the interrupting I/O device. RAM and ROM do not have interrupting capability.

REVIEW QUESTIONS

1. Describe the functions of the three buses that are part of a typical microcomputer system.
2. Which bus is unidirectional?
3. How does memory-mapped I/O differ from isolated I/O?

12.12   READ AND WRITE OPERATIONS

We are now ready to take a more detailed look at how the μP communicates with the other μC elements. Remember, the **μP is the CPU** and contains all the control and arithmetic/logic circuitry needed to execute a program of instructions stored in RAM or ROM. The CPU is continually performing READ and WRITE operations as it executes a program. It fetches each instruction from memory with a READ operation. After interpreting the instruction, it may have to perform a READ operation to obtain

FIGURE 12.9 Typical 8-bit μC structure.

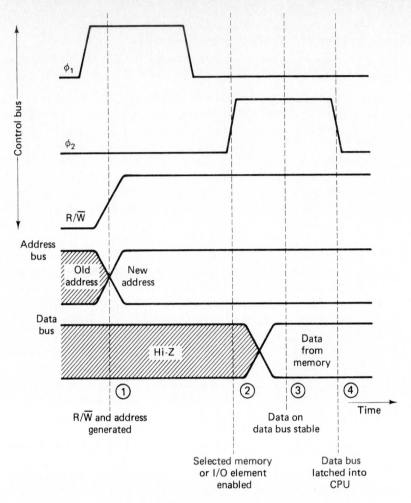

FIGURE 12.10 Typical μC timing for a READ operation.

the operand from memory, or it may have to WRITE data into memory. In some cases, the instruction may call for the CPU to READ data from an input device (such as a keyboard or TTY) or to WRITE data into an output device (like an LED display or a magnetic tape cassette).

The READ Operation The following steps take place during a READ operation:

1. The CPU generates the proper logic level on its $R/\overline{W}$ line for initiating a READ operation. Normally, $R/\overline{W} = 1$ for READ. The $R/\overline{W}$ line is part of the control bus and goes to all the memory and I/O elements.

2. Simultaneously, the CPU places the 16-bit address code onto the address bus to select the particular memory location or I/O device from which the CPU wants to receive data.

3. The selected memory or I/O element places an 8-bit word on the data bus. All

nonselected memory and I/O elements will not affect the data bus because their *tristate* outputs will be in the disabled (Hi-Z) condition.

4. The CPU receives the 8-bit word from the data bus on its data pins, D_0–D_7. These data pins act as inputs whenever $R/\overline{W} = 1$. This 8-bit word is then latched into one of the CPU's internal registers, such as the accumulator.

This sequence can be better understood with the help of a timing diagram showing the interrelationship between the signals on the various buses (see Figure 12.10). Everything is referenced to the ϕ_1 and ϕ_2 clock signals. The complete READ operation occurs in one clock cycle. This is typically 1 μs for MOS microprocessors. The leading edge of ϕ_1 initiates the CPU's generation of the proper $R/\overline{W}$ and address signals. After a short delay, typically 100 ns for a MOS μP, the $R/\overline{W}$ line goes HIGH and the address bus holds the new address code (point 1 on timing diagram). Note that the address-bus waveform shows both possible transitions (LOW-to-HIGH and HIGH-to-LOW), because some of the 16 address lines will be changing in one direction while others will be changing in the opposite direction.

During the ϕ_2 pulse, the selected memory or I/O device is enabled (point 2) and it proceeds to put its data word on the data bus. Prior to this, the data bus is in its Hi-Z state, since no device connected to it has been enabled. At some point during the ϕ_2 pulse, the data on the data bus become stable (point 3). Again, both possible data-line transitions are shown on the diagram. The delay between the start of the ϕ_2 pulse and the data-bus stabilizing depends on the speed of the memory and I/O elements. For memory this delay would be its *access time*. On the falling edge of ϕ_2, the data on the data bus are latched into the CPU (point 4). Clearly, then, the memory and I/O devices must be capable of putting data on the bus prior to the falling edge of ϕ_2, or proper transfer to the CPU will not occur. Thus, it is necessary to ensure that these devices have a speed compatible with the μC clock frequency.

EXAMPLE 12.1

A certain type of PROM has an access time specified as 750 ns (typical) and 1 μs (maximum). Can it be used with a μC that has a clock frequency of 1 MHz?

Solution: No, with a clock frequency of 1 MHz, the ϕ_2 pulse duration will be less than 500 ns. Thus, the PROM would have to have an access time of less than 500 ns for proper data transfer to the CPU.

The WRITE Operation   The following steps occur during a WRITE operation:

1. The CPU generates the proper logic level on the $R/\overline{W}$ line for initiating a WRITE operation. Normally, $R/\overline{W} = 0$ for WRITE.
2. Simultaneously, the CPU places the 16-bit address code onto the address bus.
3. The CPU then places an 8-bit word on the data bus via its data pins D_0–D_7, which are not acting as outputs. This 8-bit word typically comes from an internal CPU register, such as the accumulator. All other devices connected to the data bus have their outputs disabled.
4. The selected memory or I/O element takes the data from the data bus. All non-selected memory and I/O elements will not have their inputs enabled.

Sec. 12.12   READ and WRITE Operations   **527**

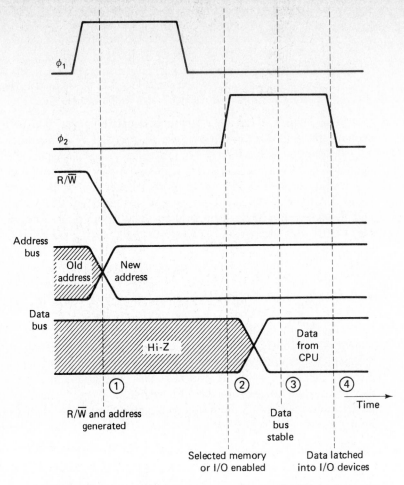

FIGURE 12.11 Typical μC timing for a WRITE operation.

This sequence has the timing diagram shown in Figure 12.11. Once again, the leading edge of ϕ_1 initiates the $R/\overline{W}$ and address-bus signals (point 1). During the ϕ_2 pulse, the selected memory or I/O device is enabled (point 2) and the CPU places its data on the data bus. The data-bus levels become stabilized a short time into the ϕ_2 pulse (typically 100 ns). These data are then written into the selected memory location while ϕ_2 is high. If an I/O device has been selected, it usually latches the data from the data bus on the falling edge of ϕ_2 (point 4).

The READ and WRITE operations encompass most of the μC activity that takes place outside the CPU. The following example illustrates.

EXAMPLE 12.2

Below is a short program that is stored in memory locations 0020_{16}–0029_{16} of an 8-bit μC. Note that since each word is one byte (8 bits), it takes two successive bytes to represent a 16-bit operand address code. Determine the total number of READ and WRITE operations that the μC will perform as it executes this program.

Memory Address (hex)	Memory Word (hex)	Mnemonic	Description
0020	5B	LDA$5001	Load accumulator (ACC) with X.
0021	01 ⎫		⎡ Address of ⎤
0022	50 ⎭		⎣ operand X ⎦
0023	6D	ADD$5002	Add Y to contents of ACC.
0024	02 ⎫		⎡ Address of ⎤
0025	50 ⎭		⎣ operand Y ⎦
0026	8A	STA$5003	Store ACC contents.
0027	03 ⎫		⎡ Address where ACC ⎤
0028	50 ⎭		⎣ will be stored ⎦
0029	3F	HLT	Halt operation.

Solution: The CPU begins executing the program by READING the contents of memory location 0020. The word stored there (5B) is taken into the CPU and is interpreted as an instruction op code. (In other words, the CPU *always* interprets the first word of a program as an op code and the programmer must *always* adhere to this format.) The CPU decodes this op code to determine the operation to be performed and to determine if an operand address follows the op code. In this case, an operand address is required and is stored in the next two successive bytes in the program. Thus, the CPU must READ locations 0021 and 0022 to obtain the address of data X. Once this address (5001) has been read into the CPU, the CPU proceeds to execute the LDA instruction by READING memory location 5001 and putting its contents into the accumulator. Thus, the complete execution of this first instruction requires *four* separate READ operations; one for the op code, two for the address, and one for the LDA operation.

Similarly, *four* READ operations are needed for the second instruction, which begins at 0023. This instruction also uses a two-byte operand address and requires reading the contents of this address for transfer to the CPU's arithmetic unit.

The third instruction begins at 0026 and uses a two-byte operand address. The CPU uses *three* READ operations to obtain these. However, the instruction STA calls for a WRITE operation whereby the CPU transfers the contents of the accumulator to memory location 5003.

The final instruction at 0029 is simply an op code with no operand address. The CPU READS this op code (3F), decodes it, and then halts further operations.

The total number of READ operations, then, is *twelve* and the total number of WRITE operations is *one*.

REVIEW QUESTIONS

1. Describe the steps that take place when the CPU performs a READ operation.
2. Repeat for a WRITE operation.
3. What device puts data on the data bus during a WRITE operation?

12.13 THE MICROPROCESSOR (μP)

All microcomputers, although they vary in their architecture, have one element in common—the microprocessor chip. As we know, the μP functions as the central processing unit of the μC. In essence, the μP is the heart of the μC because its capabilities determine the capabilities of the μC. Its speed determines the maximum speed of the μC, its address and data pins determine the μC's memory capacity and word size, and its control pins determine the type of I/O interfacing that must be used.

The μP performs a large number of functions, including:

1. Providing timing and control signals for all elements of the μC.
2. Fetching instructions and data from memory.
3. Transferring data to and from I/O devices.
4. Decoding instructions.
5. Performing arithmetic and logic operations called for by instructions.
6. Responding to I/O-generated control signals such as RESET and INTERRUPT.

The μP contains all the logic circuitry for performing these functions, but it should be kept in mind that a great deal of the μP's internal logic is not externally accessible. For example, we cannot apply an external signal to the μP chip to increment the program counter (PC). Instead, the μP elements are *software*-accessible. This means that we can affect the internal μP circuitry only by the *program* we put in memory for the μP to execute. This is what makes the μP so versatile and flexible—when we want to change the μP's operation, we simply change the program (e.g., by changing the ROMs that store the program). This is generally easier than rewiring hardware.

Since we must construct the programs that tell the internal μP logic what to do, we need to become familiar with the internal μP structure, its characteristics, and its capabilities. The μP logic is extremely complex, but we can divide it into three areas: the register section, the control and timing section, and the ALU (see Figure 12.12). Although there is some overlap among these three areas, for clarity we will consider them separately.

FIGURE 12.12 Major functional areas of a μP chip.

12.14 TIMING AND CONTROL SECTION

We will not discuss this part of the μP in too much detail for two reasons. First, it is the one area on the μP chip over which we have very little control. Second, we do not have to know the detailed structure of the timing and control section to be able to develop useful programs for the μP.

The major function of this μP section is to fetch and decode instructions from program memory, and then to generate the necessary control signals required by the ALU and register section for executing these instructions. The fetching and decoding functions correspond to the *fetch cycle*, and the control-signal generation takes place during the *execute cycle*. Both of these cycles were discussed earlier and we need not elaborate on them further.

Control-Bus Signals   The control section also generates *external* control signals that are sent to other μC elements as part of the system's control bus. We used some of these control-bus signals in our description of the μC, namely $R/\overline{W}$ and ϕ_2. In addition to generating output signals for the control bus, the μP control section also responds to control-bus signals that are sent from other μC elements to the μP chip. The RESET and INT (interrupt) mentioned earlier are examples.

Each μP has its own unique set of input and output control signals that are described in detail in the manufacturers' operation manuals. We will not attempt to define all of these here. Instead, we will describe some of the control-bus signals that are common to several different microprocessors.

RESET: All μPs have this input. When this input is activated, most of the μP's internal registers are reset to 0. In many μPs, the program counter (PC) is reset to 0 so that the instruction stored at memory location 0000_{16} is the first to be executed. In some μPs, activating the RESET input does not clear the PC. Instead, the PC is loaded from two specific memory locations (such as FFFE and FFFF). In other words, upon RESET the address of the first instruction is taken from these memory locations (each one stores one byte of the 16-bit address). Usually, this starting address is stored in ROM and is often referred to as an *address vector*.

$R/\overline{W}$:   This μP output line informs the rest of the μC as to whether the μP is in a READ or WRITE operation. Some μPs use separate control lines, *RD* to indicate a READ operation and *WR* to indicate a WRITE operation.

MREQ (memory request):   This μP output indicates that a memory access is in progress.

IORQ (I/O request):   A μP output which indicates that an I/O device is being accessed. Some μPs use this signal along with MREQ to distinguish between memory and I/O operations. This allows memory and I/O to use the same addresses because the IORQ and MREQ signals determine which one (I/O or memory) is enabled. This technique of treating I/O separately from memory is called *isolated I/O*.

READY: This μP input is used by slow memory or I/O devices which cannot respond to a μP access request within one μP clock cycle. When the slow device is selected by the address decoding circuitry, it immediately sends a READY signal to the μP. In response, the μP suspends all its internal operations and enters what is called a WAIT state. It remains there until the device is ready to send or receive data, which the device indicates by removing the READY signal.

INT or IRQ (interrupt request):   This is a μP input used by I/O devices to interrupt the execution of the current program and cause the μP to jump to a special program, called the *interrupt service routine*. The μP executes this special program, which normally involves servicing the interrupting device. When this execution is complete, the μP resumes execution of the program it had been working on when it was interrupted.

INTE (interrupt enable):   This is a μP output that indicates to external devices whether or not the internal μP interrupt logic is enabled or disabled. If enabled (*INTE* = 1), the μP can be interrupted as described above. If disabled (*INTE* = 0), the μP will not respond to the *INT* or *IRQ* inputs. The state of *INTE* can be software-controlled. For example, a program can contain an instruction that makes *INTE* = 0 so that the interrupt operation is disabled.

NMI (nonmaskable interrupt):   This is another μP interrupt input, but it differs from *INT* or *IRQ* in that its effect cannot be disabled. In other words, the proper signal on *NMI* will always interrupt the μP, regardless of the interrupt enable status.

REVIEW QUESTIONS

1.   What are the major sections of a microprocessor?
2.   What are the main functions of the timing and control section?

12.15   THE REGISTER SECTION

The most common operation that takes place *inside* the μP chip is the transfer of binary information from one register to another. The number and types of registers that a μP contains is a key part of its architecture, and it has a major effect on the programming effort required in a given application. The register structure of different μPs varies considerably from manufacturer to manufacturer. However, the basic functions performed by the various registers is essentially the same in all μPs. They are used to store data, addresses, instruction codes, and information on the status of various μP operations. Some are used as counters which can be controlled by software (program instructions) to keep track of such things as the number of times a particular sequence of instructions has been executed or sequential memory locations from which data are to be taken.

We will briefly describe the most common types of registers and their functions.

The Instruction Register (IR)   This register is used to store the op code of the current instruction that is being fetched and executed. When the μP fetches the op code from memory, it stores it in the IR while the decoding circuitry determines which operation is to be performed. The IR is automatically used by the CPU during each instruction cycle and the programmer never needs to access this register. The size of the IR will be the same as the word size. For an 8-bit μP, the IR is 8 bits long.

The Program Counter (PC)   This register has also been discussed previously. The PC always contains the address in memory of the next instruction (or portion of an instruction) that the CPU is to fetch. When the μP RESET input is activated, the PC is set to the address of the first instruction to be executed. The μP places the contents of the PC on the address bus and fetches the first byte of the instruction from that memory location. (Recall that in 8-bit μCs one byte is used for the op code and the following one or two bytes for the operand address.) The μC automatically increments the PC after each use and in this way executes the stored program sequentially unless the program contains an instruction which alters the sequence (e.g., a JUMP instruction).

The size of the PC depends on the number of address bits the μP can handle. Most of the more common μPs use 16-bit addresses, but some use 12 bits. In either case, the PC will have the same number of bits as the address bus. In many μPs, the PC is divided into two smaller registers, PCH and PCL, each of which holds half an address. This is illustrated in Figure 12.13. PCH holds the 8 high-order bits of the 16-bit address and PCL holds the 8 low-order bits. The reason for using PCH and PCL is that it is often necessary to store the contents of PC in memory. Since the memory stores 8-bit words, the 16-bit PC must be broken into two halves and stored in two successive memory locations.

FIGURE 12.13 A 16-bit PC broken into two 8-bit units.

The programmer has no direct access to the PC. That is, there are no direct instructions to load the PC from memory or to store the PC contents in memory. However, there are many instructions that cause the PC to take on a value other than its normal sequential value. JMP (jump) and JPZ (jump on zero) are examples of instructions that change the PC so that the CPU executes instructions out of their normal sequential order.

Memory Address Register (MAR)   This is sometimes called a *storage address register* or an *address latching register*. It is used to hold the address of *data* the CPU is reading from or writing into memory. For example, when executing an ADD instruction, the CPU places the operand address portion of the ADD instruction into the MAR. The contents of the MAR are then placed on the address bus so that the CPU can fetch the data during the next clock cycle.

It should now be apparent that there are two sources of addresses for the μP's address bus, the PC and the MAR. The PC is used for *instruction* addresses and the MAR is used for *data* addresses. A multiplixer is used to switch either the PC or MAR onto the address bus, depending on whether the CPU is an instruction cycle or an execution cycle. This is illustrated in Figure 12.14. Here the PC and MAR outputs are fed to a two-port multiplexer. The multiplexer's single output port is connected to the μP's address bus via the address bus buffers. The multiplexer's select input S is controlled by a signal from the μP's control section. One value of S selects the PC to be placed onto the address bus, and the other value of S selects the MAR.

Accumulator   The accumulator is a register that takes part in most of the operations performed by the ALU. It is also the register in which the results are placed after most

FIGURE 12.14 Portion of the μP showing how address source is selected for the address bus.

ALU operations. In many ALU instructions, the accumulator is the source of one of the operands and the destination of the result.

In addition to its use in ALU instructions, the accumulator has other uses; for instance, it can be used as a storage register for data that are being sent to an output device or as a receiving register for data that are being read from an input device. Some μPs have more than one accumulator. In these μPs, the instruction op code specifies which accumulator is to be used.

The accumulator generally has the same number of bits as the μP's word size. Thus, an 8-bit μP will have an 8-bit accumulator. Some μPs also have an *extension register,* which is used in conjunction with the accumulator for handling binary numbers with more than 8 bits. For example, the μP has an arithmetic mode called "double-precision arithmetic," where each number is 16 bits long and is stored in two words of memory. When these 16-bit data words are sent to the CPU, the extension register stores the 8 least significant bits and the accumulator stores the 8 most significant bits. These two registers can be considered to be one 16-bit register.

General-Purpose Registers   These registers are used for many of the temporary storage functions required inside the CPU. They can be used to store data that are used frequently during a program, thereby speeding up the program execution, since the CPU does not have to perform a memory READ operation each time the data are needed. These registers are often used as counters which are used to keep track of the number of times a particular instruction sequence in a program has been executed. In some μPs, the general-purpose registers can also be used as index registers (described below) or as accumulators. The number of general-purpose registers will vary from μP to μP. Some μPs have none, while others may have 12 or more.

Index Registers   An index register, like general-purpose registers, can be used for general CPU storage functions and as a counter. In addition, it has a special function which is of great use in a program where tables or arrays of data must be handled. In this function, the index register takes part in determining the addresses of data that the CPU is accessing. This operation is called *indexed addressing* and is a special form of memory addressing available to the programmer. There are several different forms of indexed addressing, but the basic idea of indexed addressing is that the actual or *effective* operand address called for by an instruction is the sum of the operand address portion of the instruction *plus* the contents of the index register.

The Status Register   Also referred to as a *process status register* or *condition register*, the status register consists of individual bits with different meanings assigned by the μP manufacturer. These bits are called *flags*, and each flag is used to indicate the status of a particular μP condition. The value of some of the flags can be examined under program control to determine what sequence of instructions to follow. Two of the most common flags are the ZERO flag, Z, and the CARRY flag, C. The value of Z will always indicate whether or not the previous data-manipulation instruction produced a result of zero. Normally, a μP control signal sets Z to the HIGH state when the result of an instruction is zero and clears Z to a LOW when the result of an instruction is not zero. The value of C always indicates whether the previous instruction produced a result that exceeded the μP word size. For example, whenever the addition of two 8-bit data words produces a sum that exceeds 8 bits (i.e., a carry-out of the eighth position), the C flag will be set to 1. If the addition produces no carry, C will be 0. The C flag can be thought of as the ninth bit of any arithmetic result.

Most μP instruction sets contain several *conditional branch* instructions which determine the sequence of instructions to be executed, depending on the flag values. The jump-on-zero (JPZ) instruction is a prime example. When the CPU executes the JPZ instruction, it examines the value of the ZERO flag. If the ZERO flag is LOW, indicating that the previous instruction produced a nonzero result, the next instruction to be executed will be taken in normal sequence. If the ZERO flag is HIGH, indicating a zero result, the program will jump to the operand address for its next instruction and will continue in sequence from there. Other examples are jump-on-carry-set (JCS) and jump-on-carry-cleared (JCC), both of which examine the CARRY flag.

The Stack Pointer Register (SP)   Before defining the function of this register, we must first define the *stack*. The stack is a portion of RAM reserved for the temporary storage and retrieval of information, typically the contents of the μP's internal registers. The *stack pointer* register acts as a special memory address register used only for the stack portion of RAM. Whenever a word is to be stored on the stack, it is stored at the address contained in the stack pointer. Similarly, whenever a word is to be read from the stack, it is read from the address specified by the stack pointer. The contents of the stack pointer is initialized by the programmer at the beginning of the program. Thereafter, the SP is automatically decremented *after* a word is stored on the stack and incremented *before* a word is read from the stack. (The incrementing and decrementing is done automatically by the μP control section.)

1. For each statement below, indicate which μP register is being described.
   (a) Holds data address.
   (b) Results of arithmetic operations are placed here.
   (c) Is used in a special type of memory addressing.
   (d) Contains flags that indicate various conditions in the μP.
   (e) Holds instruction addresses.

12.16 THE ARITHMETIC/LOGIC UNIT (ALU)

Most modern μPs have ALUs capable of performing a wide variety of arithmetic and logical operations. These operations can involve two operands, such as the accumulator and a data word from memory, or the accumulator and another μP internal register. Some of the operations involve only a single operand, such as the accumulator, a register, or a word from memory.

A simplified diagram of a typical ALU is shown in Figure 12.15. The ALU block represents all the logic circuitry used to perform arithmetic, logic, and manipulation operations on the operand inputs. Two 8-bit operands are shown as inputs to the ALU, although frequently only one operand is used. Also shown as an input to the ALU is the carry flag, C, from the μP status register. The reason for this will be explained later. The functions that the ALU will perform are determined by the control-signal inputs from the μP control section. The number of these control inputs varies from μP to μP.

The ALU produces two sets of outputs. One set is an 8-bit output, representing the results of the operation performed on the operands. The other is a set of status

FIGURE 12.15 Simplified diagram of a typical μP ALU.

signals which are sent to the μP status register to set or clear various flag bits. For example, if the result of an ALU operation is exactly zero, a signal is sent to the Z flag in the status register to set it to the HIGH state. If the result of the ALU operation produces a carry into the ninth bit position, a signal is sent to set the C flag in the status register. Other signals from the ALU will set or clear other flags in the status register.

The operand inputs A and B can come from several sources. When two operands are to be operated on by the ALU, one of the operands comes from the accumulator and the other operand comes from a data word fetched from memory which is stored in a data buffer register within the μP. In some μPs, the second operand can also be the contents of one of the general-purpose registers. The result of the ALU operation performed on the two operands is normally sent to the accumulator.

When only one operand is to be operated on by the ALU, the operand can be the contents of the accumulator, a general-purpose register, an index register, or a memory data word. The result of the ALU operation is then sent back to the source of the operand. When the single operand is a memory word, the result is sent to a data buffer register, from where the CPU writes it back into the memory location of the operand.

Single-Operand Operations   We will now describe some of the common ALU operations performed on a single operand.

1. Clear: All bits of the operand are cleared to 0. If the operand comes from the accumulator, for example, we can symbolically represent this operation as $0 \rightarrow [\mathbf{A}]$.

2. Complement (or invert):   All bits are changed to their opposite logic level. If the operand comes from register X, for example, this operation is represented as $[\overline{\mathbf{X}}] \rightarrow [\mathbf{X}]$.

3. Increment: The operand is increased by 1. For example, if the operand is 11010011, it will have 00000001 added to it in the ALU, to produce a result of 11010100. This instruction is very useful when the program is using one of the μP registers as a counter. Symbolically, this is represented as $[\mathbf{X} + 1] \rightarrow [\mathbf{X}]$.

4. Decrement: The operand is decreased by 1. In other words, the number 00000001 is subtracted from the operand. This instruction is useful in programs where a register is used to count down from an initial value. Symbolically, this operation is represented as $[\mathbf{X} - 1] \rightarrow [\mathbf{X}]$.

5. Shift: The bits of the operand are shifted to the left or to the right one place and the empty bit is made a 0. This process is illustrated below for a *shift-left* operation.

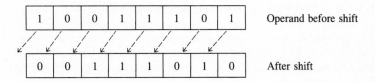

| 1 | 0 | 0 | 1 | 1 | 1 | 0 | 1 | Operand before shift |

| 0 | 0 | 1 | 1 | 1 | 0 | 1 | 0 | After shift |

Note that a 0 is shifted into the rightmost bit. Also note that the original leftmost bit is shifted out and is lost. In most μPs the bit that is shifted out of the operand is not lost; instead, it is shifted into the carry flag bit, C, of the status register. An illustration of a *shift-right* operation follows on p. 538.

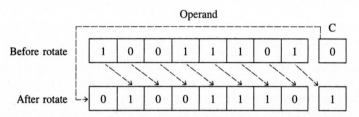

Note that a 0 is shifted into the leftmost bit and also note that the original rightmost bit is shifted into C (the original value of C is lost). For the corresponding shift-left operation, the leftmost bit would be shifted into C. This type of operation is used by the programmer to test the value of a specific bit in an operand. He does this by shifting the operand the required number of times until the bit value is in the C flag. A conditional jump instruction is then used to test the C flag to determine what instruction to execute next.

6. Rotate: This is a modified shift operation in which the C flag becomes part of a circulating shift register along with the operand; that is, the value shifted out of the operand is shifted into C and the previous value of C is shifted into the empty bit of the operand. This is illustrated below for a "rotate-right" operation.

Note that the original C bit is shifted into the leftmost bit of the operand and that the rightmost bit of the operand is shifted into C. A rotate-left operates in the same manner except in the opposite direction.

Two-Operand ALU Operations We will now describe some of the ALU operations performed on two operands.

1. Add: The ALU produces the binary sum of the two operands. Generally, one of the operands comes from the accumulator, the other from memory, and the result is sent to the accumulator. Symbolically, this operation is written $[A] + [M] \rightarrow [A]$. If the addition of the two operands produces a carry-out of the MSB position, the carry flag, C, in the status register is set to 1. Otherwise, C is cleared to 0. In other words, C serves as the ninth bit of the result.

2. Subtract: The ALU subtracts one operand (obtained from memory) from the second operand (the accumulator) and places the result in the accumulator. Symbolically, this is written $[A] - [M] \rightarrow [A]$. Most μPs use the 2's-complement method of subtraction (Chapter 6), whereby the operand from memory is 2's-complemented and then added to the operand from the accumulator. Once again, the carry bit generated by this operation is stored in the C flag.

3. Compare: This operation is the same as subtraction except that the result is not placed in the accumulator. Symbolically, this is written $[A] - [M]$. The subtraction is performed solely as a means for determining which operand is larger without affecting the contents of the accumulator. Depending on whether the result is positive, negative, or zero, various condition flags will be affected. The programmer can then use conditional jump instructions to test these flags.

4. Logical AND: The corresponding bits of the two operands are ANDed and the result is placed in the accumulator. Symbolically, this is written $[A] \cdot [M] \rightarrow [A]$. One of the operands is always the accumulator and the other comes from memory. As an example, let's assume that $[A] = 10110101$ and $[M] = 01100001$. The AND operation is performed as follows:

Note that each bit of the result is obtained by ANDing the corresponding bits of the operands. For example, bit 7 of the accumulator is a 1 and bit 7 of the memory word is a 0. Thus, bit 7 of the result is $1 \cdot 0 = 0$. Similarly, bit 5 of each operand is a 1, so that bit 5 of the result is $1 \cdot 1 = 1$.

5. Logical OR: The corresponding bits of the two operands are ORed and the result placed in the accumulator. Symbolically, this is shown as $[A]$ OR $[M] \rightarrow [A]$. A plus sign $(+)$ can be used in place of OR but it might cause confusion with the binary addition operation. Using the same two operands used in the illustration of the AND operation above, the result of the OR operation will be 11110101.

6. Exclusive-OR: The corresponding bits of the two operands are exclusive-ORed and the result is placed in the accumulator. Symbolically, this is written $[A] \oplus [M] \rightarrow [A]$. Using the same two operands used in the prior illustrations, the result of the EX-OR operation will be 11010100. The logic AND, OR, and EX-OR operations are very useful to a programmer.

7. Decimal (BCD) arithmetic: Many μPs have some provision for performing addition and subtraction in the BCD system whereby an 8-bit data word is treated as two BCD-coded decimal digits. As we pointed out in Chapter 6, arithmetic operations on BCD numbers require extra steps to obtain the correct results. While in some μPs these extra steps are performed automatically whenever BCD arithmetic instructions are executed, other μPs require the programmer to insert special instructions to correct the results.

This list of arithmetic/logic operations does not include the more complex operations of multiplication, division, square roots, and so on. These operations are not explicitly performed by any currently available μP because of the extra circuitry they would require. To perform such operations, the programmer can instruct the μP to execute an appropriate series of simple arithmetic operations. For example, the multiplication operation can be obtained through a series of shifting and adding operations, and the division operation by a series of shifting and subtracting operations.

The sequences of instructions that the programmer develops for performing such complex arithmetic operations are called *subroutines*. A multiplication subroutine for obtaining the product of two 8-bit numbers might consist of as many as 20 or 30 instruction steps requiring 200 μs to execute. If this long execution time is undesirable, it is possible to use an external LSI chip containing a high-speed multiplier.

Such LSI multipliers are available with execution times of only 100 ns and can be connected to the μP as I/O devices. Another alternative to slow subroutines is to use a ROM as a storage table which can store such data as multiplication tables, trigonometric tables, and logarithmic tables.

Signed Numbers   The circuitry in the ALU performs the add and subtract operations in the binary number system and always treats the two operands as 8-bit binary numbers. This is true even when the program calls for BCD arithmetic, the only difference being the extra steps needed to correct the BCD result. With an 8-bit data word, decimal numbers from 0 to 255 can be represented in binary code, assuming that all 8 bits are used for the numerical value. However, if the programmer wishes to use both positive and negative numbers, the MSB of each data word is used as the *sign bit* with 0 for + and 1 for −. The other 7 bits of each data word represent the magnitude; for positive numbers the magnitude is in true binary form, while for negative numbers the magnitude is in 2's-complement form.

The beauty of this method for representing signed numbers is that it requires no special operations by the ALU. The circuitry in the ALU will perform addition and subtraction on the two operands in the same manner, regardless of whether the operands are unsigned 8-bit data words or 7-bit data words plus a sign bit. As was shown in Chapter 6, the sign bit participates in the add and subtract operations just like the rest of the bits. What this means is that the μP really does not know or care whether the data it is processing are signed or unsigned. Only the programmer is concerned about the distinction, and he must at all times know the format of the data he is using in his program.

As an aid to the programmer, almost all μPs transfer the value of bit 7 of the ALU result to a flag bit in the status register. This flag bit, sometimes called the SIGN flag, S, or the NEGATIVE flag, N, will be set to 1 if the ALU result has a 1 in bit 7 and will be cleared to 0 if bit 7 is a 0. The program can then test to see if the result was positive or negative by performing a conditional jump instruction based on the SIGN flag. For example, a jump-on-negative instruction (JPN) will cause the CPU to examine the SIGN flag in the status register to determine what sequence of instructions to follow next.

REVIEW QUESTIONS

1. Indicate whether one or two operands are required for each of the ALU operations listed below:
   - (a) Shift.
   - (b) Decrement.
   - (c) Exclusive-OR.
   - (d) Rotate.
   - (e) Complement.
   - (f) Compare.
2. Assume the data word in the accumulator is 5D (hex), and a data word in a certain memory location is 16 (hex). Determine the result in the accumulator after these two words are ANDed. (*Ans.* $00010010_2 = 12_{16}$)
3. Repeat for the EX-OR operation. (*Ans.* $01001101_2 = 4D_{16}$)
4. Assume [**A**] = 36 (hex). What will be the accumulator data after two increments and a shift-right operation? (*Ans.* 1C)

12.17 FINAL COMMENTS

Our discussion has been, by necessity, only a brief introduction to the basic principles, terminology, and operations common to most microprocessers and microcomputers. We have not even begun to gain an insight into the capabilities and applications of these devices. Almost all areas of technology have started taking advantage of the inexpensive computer control which microprocessors can provide. Some typical applications include microwave ovens, traffic controllers, home computers, electronic measuring instruments, industrial process control, electronic games, automobile emission control, and a rapidly growing number of new products.

With the revolutionary impact that microprocessors have had on the electronics industry, it is not unreasonable to expect that everyone working in electronics and related areas will have to become knowledgeable in the operation of microprocessors. Hopefully, our introduction will serve as a firm foundation for further study in this important area.

APPENDIX I: POWERS of 2

2^n	n	2^{-n}
1	0	1.0
2	1	0.5
4	2	0.25
8	3	0.125
16	4	0.062 5
32	5	0.031 25
64	6	0.015 625
128	7	0.070 812 5
256	8	0.003 906 25
512	9	0.001 953 125
1 024	10	0.000 976 562 5
2 048	11	0.000 488 281 25
4 096	12	0.000 244 140 625
8 192	13	0.000 122 070 312 5
16 384	14	0.000 061 035 156 25
32 768	15	0.000 030 517 578 125
65 536	16	0.000 015 258 789 062 5
131 072	17	0.000 007 629 394 531 25
262 144	18	0.000 003 814 697 265 625
524 288	19	0.000 001 907 348 632 812 5
1 048 576	20	0.000 000 953 674 316 406 25
2 097 152	21	0.000 000 476 837 158 203 125
4 194 304	22	0.000 000 238 418 579 101 562 5
8 388 608	23	0.000 000 119 209 289 550 781 25
16 777 216	24	0.000 000 059 604 644 775 390 625
33 554 432	25	0.000 000 029 802 322 387 695 312 5
67 108 864	26	0.000 000 014 901 161 193 847 656 25
134 217 728	27	0.000 000 007 450 580 596 923 828 125

APPENDIX II: MANUFACTURERS' IC DATA SHEETS

These data sheets are presented through the courtesy of Fairchild Semi-Conductor.

54/7400
54H/74H00
54S/74S00
54LS/74LS00

QUAD 2-INPUT NAND GATE

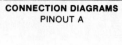

ORDERING CODE: See Section 9

PKGS	PIN OUT	COMMERCIAL GRADE V_{CC} = +5.0 V ±5%, T_A = 0°C to +70°C	MILITARY GRADE V_{CC} = +5.0 V ±10%, T_A = -55°C to +125°C	PKG TYPE
Plastic DIP (P)	A	7400PC, 74H00PC 74LS00PC, 74S00PC		9A
Ceramic DIP (D)	A	7400DC, 74H00DC 74LS00DC, 74S00DC	5400DM, 54H00DM 54LS00DM, 54S00DM	6A
Flatpak (F)	A	74LS00FC, 74S00FC	54LS00FM, 54S00FM	3I
	B	7400FC, 74H00FC	5400FM, 54H00FM	

INPUT LOADING/FAN-OUT: See Section 3 for U.L. definitions

PINS	54/74 (U.L.) HIGH/LOW	54/74H (U.L.) HIGH/LOW	54/74S (U.L.) HIGH/LOW	54/74LS (U.L.) HIGH/LOW
Inputs	1.0/1.0	1.25/1.25	1.25/1.25	0.5/0.25
Outputs	20/10	12.5/12.5	25/12.5	10/5.0 (2.5)

DC AND AC CHARACTERISTICS: See Section 3*

SYMBOL	PARAMETER	54/74 Min	54/74 Max	54/74H Min	54/74H Max	54/74S Min	54/74S Max	54/74LS Min	54/74LS Max	UNITS	CONDITIONS
I_{CCH}	Power Supply		8.0		16.8		16		1.6	mA	V_{IN} = Gnd, V_{CC} = Max
I_{CCL}	Current		22		40		36		4.4		V_{IN} = Open
t_{PLH}	Propagation Delay		22		10	2.0	4.5		10	ns	Figs. 3-1, 3-4
t_{PHL}			15		10	2.0	5.0		10		

*DC limits apply over operating temperature range; AC limits apply at T_A = +25°C and V_{CC} = +5.0 V.

CONNECTION DIAGRAMS
PINOUT A

54/7401
54H/74H01
QUAD 2-INPUT NAND GATE
(With Open-Collector Output)

PINOUT B

ORDERING CODE: See Section 9

PKGS	PIN OUT	COMMERCIAL GRADE $V_{CC} = +5.0$ V $\pm5\%$, $T_A = 0°$C to $+70°$C	MILITARY GRADE $V_{CC} = +5.0$ V $\pm10\%$, $T_A = -55°$C to $+125°$C	PKG TYPE
Plastic	A	7401PC		9A
DIP (P)	B	74H01PC		
Ceramic	A	7401DC	5401DM	6A
DIP (D)	B	74H01DC	54H01DM	
Flatpak (F)	C	7401FC, 74H01FC	5401FM, 54H01FM	3I

PINOUT C

INPUT LOADING/FAN-OUT: See Section 3 for U.L. definitions

PINS	54/74 (U.L.) HIGH/LOW	54/74H (U.L.) HIGH/LOW
Inputs	1.0/1.0	1.25/1.25
Outputs	OC**/10	OC**/12.5

DC AND AC CHARACTERISTICS: See Section 3*

SYMBOL	PARAMETER	54/74 Min	54/74 Max	54/74H Min	54/74H Max	UNITS	CONDITIONS	
I_{CCH}	Power Supply		8.0		10	mA	V_{IN} = Gnd	V_{CC} = Max
I_{CCL}	Current		22		40		V_{IN} = Open	
t_{PLH}	Propagation Delay		45		15	ns	Figs. 3-2, 3-4	
t_{PHL}			15		12			

*DC limits apply over operating temperature range; AC limits apply at T_A = +25°C and V_{CC} = +5.0 V.

**OC — Open Collector

CONNECTION DIAGRAMS
PINOUT A

54/7404
54H/74H04
54S/74S04
54S/74S04A
54LS/74LS04
HEX INVERTER

ORDERING CODE: See Section 9

PKGS	PIN OUT	COMMERCIAL GRADE $V_{CC} = +5.0\ V\ \pm5\%$, $T_A = 0°C$ to $+70°C$		MILITARY GRADE $V_{CC} = +5.0\ V\ \pm10\%$, $T_A = -55°C$ to $+125°C$	PKG TYPE
Plastic DIP (P)	A	7404PC, 74H04PC 74S04PC, 74S04APC 74LS04PC			9A
Ceramic DIP (D)	A	7404DC, 74H04DC 74S04DC, 74S04ADC 74LS04DC		5404DM, 54H04DM 54S04DM, 54S04ADM 54LS04DM	6A
Flatpak (F)	A	74S04FC, 74S04AFC 74LS04FC		54S04FM, 54S04AFM 54LS04FM	3I
	B	7404FC, 74H04FC		5404FM, 54H04FM	

PINOUT B

INPUT LOADING/FAN-OUT: See Section 3 for U.L. definitions

PINS	54/74 (U.L.) HIGH/LOW	54/74H (U.L.) HIGH/LOW	54/74S (U.L.) HIGH/LOW	54/74LS (U.L.) HIGH/LOW
Inputs	1.0/1.0	1.25/1.25	1.25/1.25	0.5/0.25
Outputs	20/10	12.5/12.5	25/12.5	10/5.0 (2.5)

DC AND AC CHARACTERISTICS: See Section 3*

SYMBOL	PARAMETER	54/74 Min	54/74 Max	54/74H Min	54/74H Max	54/74S Min	54/74S Max	54/74LS Min	54/74LS Max	UNITS	CONDITIONS
I_{CCH}	Power Supply		12		26		24		2.4	mA	$V_{IN} = Gnd$ $V_{CC} = Max$
I_{CCL}	Current		33		58		54		6.6		$V_{IN} = Open$
t_{PLH}	Propagation Delay		22		10	2.0	4.5		10	ns	Fig. 3-1, 3-4
t_{PHL}			15		10	2.0	5.0		10		
t_{PLH}	Propagation Delay					1.0	3.5			ns	Fig. 3-1, 3-4
t_{PHL}	(54/74S04A only)					1.0	4.0				

*DC limits apply over operating temperature range; AC limits apply at $T_A = +25°C$ and $V_{CC} = +5.0\ V$.

54/7405
54H/74H05
54S/74S05
54S/74S05A
54LS/74LS05

HEX INVERTER
(With Open-Collector Output)

ORDERING CODE: See Section 9

PKGS	PIN OUT	COMMERCIAL GRADE Vcc = +5.0 V ±5%, TA = 0° C to +70° C	MILITARY GRADE Vcc = +5.0 V, ±10%, TA = -55° C to +125° C	PKG TYPE
Plastic DIP (P)	A	7405PC, 74H05PC 74S05PC, 74S05APC 74LS05PC		9A
Ceramic DIP (D)	A	7405DC, 74H05DC 74S05DC, 74S05ADC 74LS05DC	5405DM, 54H05DM 54S05DM, 54S05ADM 54LS05DM	6A
Flatpak (F)	A	74S05FC, 74S05AFC 74LS05FC	54S05FM, 54S05AFM 54LS05FM	3I
	B	7405FC, 74H05FC	5405FM, 54H05FM	

PINOUT B

INPUT LOADING/FAN-OUT: See Section 3 for U.L. definitions

PINS	54/74 (U.L.) HIGH/LOW	54/74H (U.L.) HIGH/LOW	54/74S (U.L.) HIGH/LOW	54/74LS (U.L.) HIGH/LOW
Inputs	1.0/1.0	1.25/1.25	1.25/1.25	0.5/0.25
Outputs	OC**/10	OC**/12.5	OC**/12.5	OC**/5.0 (2.5)

DC AND AC CHARACTERISTICS: See Section 3*

SYMBOL	PARAMETER	54/74 Min	54/74 Max	54/74H Min	54/74H Max	54/74S Min	54/74S Max	54/74LS Min	54/74LS Max	UNITS	CONDITIONS
ICCH	Power Supply		12		26		19.8		2.4	mA	VIN = Gnd, Vcc = Max
ICCL	Current		33		58		54		6.6		VIN = Open
tPLH	Propagation Delay		55		18	2.0	7.5		22	ns	Fig. 3-2, 3-4
tPHL			15		15	2.0	7.0		18		
tPLH	Propagation Delay					2.0	5.5			ns	Fig. 3-2, 3-4
tPHL	(54S/74S05A only)					1.5	5.0				

*DC limits apply over operating temperature range; AC limits apply at TA = +25°C and Vcc = +5.0 V.
**OC — Collector

CONNECTION DIAGRAM
PINOUT A

54/7406

HEX INVERTER BUFFER/DRIVER
(With Open-Collector High-Voltage Output)

ORDERING CODE: See Section 9

PKGS	PIN OUT	COMMERCIAL GRADE V_{CC} = +5.0 V ±5%, T_A = 0°C to +70°C	MILITARY GRADE V_{CC} = +5.0 V ±10%, T_A = -55°C to +125°C	PKG TYPE
Plastic DIP (P)	A	7406PC		9A
Ceramic DIP (D)	A	7406DC	5406DM	6A
Flatpak (F)	A	7406FC	5406FM	3I

INPUT LOADING/FAN-OUT: See Section 3 for U.L. definitions

PINS	54/74 (U.L.) HIGH/LOW
Inputs	1.0/1.0
Outputs	OC**/10

DC AND AC CHARATERISTICS: See Section 3*

SYMBOL	PARAMETER		54/74 Min	54/74 Max	UNITS	CONDITIONS	
V_{OL}	Output LOW Voltage	XC		0.7	V	I_{OL} = 40 mA	V_{CC} = Min
		XM		0.7		I_{OL} = 30 mA	V_{IN} = V_{IH}
		XC, XM		0.4		I_{OL} = 16 mA	
I_{OH}	Output HIGH Current			0.25	mA	V_{OH} = 30 V, V_{CC} = Min V_{IN} = V_{IL}	
I_{CCH}	Power Supply Current			48	mA	V_{IN} = Gnd	V_{CC} = Max
I_{CCL}				51		V_{IN} = Open	
t_{PLH}	Propagation Delay			15	ns	Fig. 3-2, 3-4	
t_{PHL}				23			

*DC limits apply over operating temperature range; AC limits apply at T_A = +25°C and V_{CC} = +5.0 V.
**OC—Open Collector

54/7414
54LS/74LS14

HEX SCHMITT TRIGGER INVERTER

ORDERING CODE: See Section 9

PKGS	PIN OUT	COMMERCIAL GRADE Vcc = +5.0 V ±5%, TA = 0°C to +70°C	MILITARY GRADE Vcc = +5.0 V ±10%, TA = -55°C to +125°C	PKG TYPE
Plastic DIP (P)	A	7414PC, 74LS14PC		9A
Ceramic DIP (D)	A	7414DC, 74LS14DC	5414DM, 54LS14DM	6A
Flatpak (F)	A	7414FC, 74LS14FC	5414FM, 54LS14FM	3I

INPUT LOADING/FAN-OUT: See Section 3 for U.L. definitions

PINS	54/74 (U.L.) HIGH/LOW	54/74LS (U.L.) HIGH/LOW
Inputs	1.0/1.0	0.5/0.25
Outputs	20/10	10/5.0 (2.5)

DC AND AC CHARACTERISTICS: See Section 3*

SYMBOL	PARAMETER	54/74 Min Max	54/74LS Min Max	UNITS	CONDITIONS	
V_{T+}	Positive-going Threshold Voltage	1.5 2.0	1.5 2.0	V	V_{CC} = +5.0 V	
V_{T-}	Negative-going Threshold Voltage	0.6 1.1	0.6 1.1	V	V_{CC} = +5.0 V	
$V_{T+} - V_{T-}$	Hysteresis Voltage	0.4	0.4	V	V_{CC} = +5.0 V	
I_{T+}	Input Current at Positive-going Threshold	-0.43**	-0.14**	mA	V_{CC} = +5.0 V, V_{IN} = V_{T+}	
I_{T-}	Input Current at Negative-going Threshold	-0.56**	-0.18**	mA	V_{CC} = +5.0 V, V_{IN} = V_{T-}	
I_{IL}	Input LOW Current	-1.2	-0.4	mA	V_{CC} = Max, V_{IN} = 0.4 V	
I_{OS}	Output Short Circuit Current	-18 -55	-20 -100	mA	V_{CC} = Max, V_{OUT} = 0 V	
I_{CCH}	Power Supply Current	36	16	mA	V_{IN} = Gnd	V_{CC} = Max
I_{CCL}		60	21		V_{IN} = Open	
t_{PLH}	Propagation Delay	22	22	ns	Figs. 3-1, 3-15	
t_{PHL}		22	22			

*DC limits apply over operating temperature range; AC limits apply at TA = +25°C and Vcc = +5.0 V. **Typical Value

54/7420
54H/74H20
54S/74S20
54LS/74LS20
DUAL 4-INPUT NAND GATE

ORDERING CODE: See Section 9

PKGS	PIN OUT	COMMERCIAL GRADE V_{CC} = +5.0 V ±5%, T_A = 0°C to +70°C	MILITARY GRADE V_{CC} = +5.0 V ±10%, T_A = -55°C to +125°C	PKG TYPE
Plastic DIP (P)	A	7420PC, 74H20PC 74S20PC, 74LS20PC		9A
Ceramic DIP (D)	A	7420DC, 74H20DC 74S20DC, 74LS20DC	5420DM, 54H20DM 54S20DM, 54LS20DM	6A
Flatpak (F)	A	74S20FC, 74LS20FC	54S20FM, 54LS20FM	3I
	B	7420FC, 74H20FC	5420FM, 54H20FM	

INPUT LOADING/FAN-OUT: See Section 3 for U.L. definitions

PINS	54/74 (U.L.) HIGH/LOW	54/74H (U.L.) HIGH/LOW	54/74S (U.L.) HIGH/LOW	54/74LS (U.L.) HIGH/LOW
Inputs	1.0/1.0	1.25/1.25	1.25/1.25	0.5/0.25
Outputs	20/10	12.5/12.5	25/12.5	10/5.0 (2.5)

DC AND AC CHARACTERISTICS: See Section 3*

SYMBOL	PARAMETER	'54/74 Min Max	54/74H Min Max	54/74S Min Max	54/74LS Min Max	UNITS	CONDITIONS
I_{CCH}	Power Supply	4.0	8.4	8.0	0.8	mA	V_{IN} = Gnd V_{CC} = Max
I_{CCL}	Current	11	20	18	2.2		V_{IN} = Open
t_{PLH}	Propagation Delay	22	10	2.0 4.5	15	ns	Figs. 3-1, 3-4
t_{PHL}		15	10	2.0 5.0	15		

*DC limits apply over operating temperature range; AC limits apply at T_A = +25°C and V_{CC} = +5.0 V.

54/7432
54S/74S32
54LS/74LS32

QUAD 2-INPUT OR GATE

ORDERING CODE: See Section 9

PKGS	PIN OUT	COMMERCIAL GRADE Vcc = +5.0 V ±5%, TA = 0°C to +70°C	MILITARY GRADE Vcc = +5.0 V ±10%, TA = -55°C to +125°C	PKG TYPE
Plastic DIP (P)	A	7432PC, 74S32PC 74LS32PC		9A
Ceramic DIP (D)	A	7432DC, 74S32DC 74LS32DC	5432DM, 54S32DM 54LS32DM	6A
Flatpak (F)	A	7432FC, 74S32FC 74LS32FC	5432FM, 54S32FM 54LS32FM	3I

INPUT LOADING/FAN-OUT: See Section 3 for U.L. definitions

PINS	54/74 (U.L.) HIGH/LOW	54/74S (U.L.) HIGH/LOW	54/74LS (U.L.) HIGH/LOW
Inputs	1.0/1.0	1.25/1.25	0.5/0.25
Outputs	20/10	25/12.5	10/5.0 (2.5)

DC AND AC CHARACTERISTICS: See Section 3 for U.L. definitions

SYMBOL	PARAMETER	54/74 Min	54/74 Max	54/74S Min	54/74S Max	54/74LS Min	54/74LS Max	UNITS	CONDITIONS
I_{CCH}	Power Supply Current		22		32		6.2	mA	V_{IN} = Open, V_{CC} = Max
I_{CCL}			38		68		9.8		V_{IN} = Gnd
t_{PLH}	Propagation Delay		15	2.0	7.0		15	ns	Figs. 3-1, 3-5
t_{PHL}			22	2.0	7.0		15		

*DC limits apply over operating temperature range; AC limits apply at TA = +25°C and Vcc = +5.0 V.

54/7437
54LS/74LS37

QUAD 2-INPUT NAND BUFFER

ORDERING CODE: See Section 9

PKGS	PIN OUT	COMMERCIAL GRADE $V_{CC} = +5.0\ V\ \pm5\%$, $T_A = 0°C$ to $+70°C$	MILITARY GRADE $V_{CC} = +5.0\ V\ \pm10\%$, $T_A = -55°c$ to $+125°C$	PKG TYPE
Plastic DIP (P)	A	7437PC, 74LS37PC		9A
Ceramic DIP (D)	A	7437DC, 74LS37DC	5437DM, 54LS37DM	6A
Flatpak (F)	A	7437FC, 74LS37FC	5437FM, 54LS37FM	3I

INPUT LOADING/FAN-OUT: See Section 3 for U.L. definitions

PINS	54/74 (U.L.) HIGH/LOW	54/74LS (U.L.) HIGH/LOW
Inputs	1.0/1.0	0.5/0.25
Outputs	30/30	30/15 (7.5)

DC AND AC CHARACTERISTICS: See Section 3*

SYMBOL	PARAMETER		54/74 Min Max	54/74LS Min Max	UNITS	CONDITIONS	
V_{OH}	Output HIGH Voltage	XM	2.4	2.5	V	V_{CC} = Max, I_{OH} = -1.2 mA	
		XC	2.4	2.7		$V_{IN} = V_{IL}$	
V_{OL}	Output LOW Voltage	XM, XC	0.4		V	I_{OL} = 48 mA	V_{CC} = Min
		XM		0.4		I_{OL} = 12 mA	V_{IN} = 2.0 V
		XC		0.5		I_{OL} = 24 mA	
I_{OS}	Output Short Circuit Current	XM	-20 -70	-30 -130	mA	V_{CC} = Min, V_{OUT} = 0 V	
		XC	-18 -70	-30 -130			
I_{CCH}	Power Supply Current		15.5	2.0	mA	V_{IN} = Gnd	V_{CC} = Max
I_{CCL}			54	12		V_{IN} = Open	
t_{PLH}	Propagation Delay		22	20	ns	Figs. 3-1, 3-4	
t_{PHL}			15	20			

*DC limits apply over operating temperature range; AC limits apply at T_A = +25°C and V_{CC} = +5.0 V.

54/7486
54S/74S86
54LS/74LS86

QUAD 2-INPUT EXCLUSIVE-OR GATE

ORDERING CODE: See Section 9

PKGS	PIN OUT	COMMERCIAL GRADE V_{CC} = +5.0 V ±5%, T_A = 0°C to +70°C	MILITARY GRADE V_{CC} = +5.0 V ±10%, T_A = -55°C to +125°C	PKG TYPE
Plastic DIP (P)	A	7486PC, 74S86PC 74LS86PC		9A
Ceramic DIP (D)	A	7486DC, 74S86DC 74LS86DC	5486DM, 54S86DM 54LS86DM	6A
Flatpak (F)	A	7486FC, 74S86FC 74LS86FC	5486FM, 54S86FM 54LS86FM	3I

INPUT LOADING/FAN-OUT: See Section 3 for U.L. definitions

PINS	54/74 (U.L.) HIGH/LOW	54/74S (U.L.) HIGH/LOW	54/74LS (U.L.) HIGH/LOW
Inputs Outputs	1.0/1.0 20/10	1.25/1.25 25/12.5	1.0/0.375 10/5.0 (2.5)

DC AND AC CHARACTERISTICS: See Section 3*

SYMBOL	PARAMETER		54/74 Min Max	54/74S Min Max	54/74LS Min Max	UNITS	CONDITIONS
I_{CC}	Power Supply Current	XM	43	75	10	mA	V_{CC} = Max, V_{IN} = Gnd
		XC	50	75	10		
t_{PLH} t_{PHL}	Propagation Delay		23 17	3.5 10.5 3.0 10	12 17	ns	Other Input LOW Figs. 3-1, 3-5
t_{PLH} t_{PHL}	Propagation Delay		30 22	3.5 10.5 3.0 10	13 12	ns	Other Input HIGH Figs. 3-1, 3-4

*DC limits apply over operating temperature range; AC limits apply at T_A = +25°C and V_{CC} = +5.0 V.

54/7490A
54LS/74LS90
DECADE COUNTER

DESCRIPTION — The '90 is a 4-stage ripple counter containing a high speed flip-flop acting as a divide-by-two and three flip-flops connected as a divide-by-five counter. It can be connected to operate with a conventional BCD output pattern or it can be connected to provide a 50% duty cycle output. In the BCD mode, HIGH signals on the Master Set (MS) inputs set the outputs to BCD nine. HIGH signals on the Master Reset (MR) inputs force all outputs LOW. For a similar counter with corner power pins, see the 'LS290; for dual versions, see the 'LS390 and 'LS490.

LOGIC SYMBOL

V_{CC} = Pin 5
GND = Pin 10
NC = Pins 4, 13

ORDERING CODE: See Section 9

PKGS	PIN OUT	COMMERCIAL GRADE V_{CC} = +5.0 V ±5%, T_A = 0°C to +70°C	MILITARY GRADE V_{CC} = +5.0 V ±10%, T_A = -55°C to +125°C	PKG TYPE
Plastic DIP (P)	A	7490APC, 74LS90PC		9A
Ceramic DIP (D)	A	7490ADC, 74LS90DC	5490ADM, 54LS90DM	6A
Flatpak (F)	A	7490AFC, 74LS90FC	5490AFM, 54LS90FM	3I

INPUT LOADING/FAN-OUT: See Section 3 for U.L. definitions

PIN NAMES	DESCRIPTION	54/74 (U.L.) HIGH/LOW	54/74LS (U.L.) HIGH/LOW
$\overline{CP}_0$	÷2 Section Clock Input (Active Falling Edge)	2.0/2.0	0.125/1.5
$\overline{CP}_1$	÷5 Section Clock Input (Active Falling Edge)	3.0/3.0	0.250/2.0
MR_1, MR_2	Asynchronous Master Reset Inputs (Active HIGH)	1.0/1.0	0.5/0.25
MS_1, MS_2	Asynchronous Master Set (Preset 9) Inputs (Active HIGH)	1.0/1.0	0.5/0.25
Q_0	÷2 Section Output*	20/10	10/5.0 (2.5)
$Q_1 - Q_3$	÷5 Section Outputs	20/10	10/5.0 (2.5)

*The Q_0 output is guaranteed to drive the full rated fan-out plus the $\overline{CP}_1$ input.

90

FUNCTIONAL DESCRIPTION — The '90 is a 4-bit ripple type decade counter. It consists of four master/slave flip-flops which are internally connected to provide a divide-by-two section and a divide-by-five section. Each section has a separate clock input which initiates state changes of the counter on the HIGH-to-LOW clock transition. State changes of the Q outputs do not occur simultaneously because of internal ripple delays. Therefore, decoded output signals are subject to decoding spikes and should not be used for clocks or strobes. The Q_0 output of each device is designed and specified to drive the rated fan-out plus the $\overline{CP}_1$ input. A gated AND asynchronous Master Reset (MR_1, MR_2) is provided which overrides the clocks and resets (clears) all the flip-flops. A gated AND asynchronous Master Set (MS_1, MS_2) is provided which overrides the clocks and the MR inputs and sets the outputs to nine (HLLH). Since the output from the divide-by-two section is not internally connected to the succeeding stages, the devices may be operated in various counting modes.:

A. BCD Decade (8421) Counter — The $\overline{CP}_1$ input must be externally connected to the Q_0 output. The $\overline{CP}_0$ input receives the incoming count and a BCD count sequence is produced.

B. Symmetrical Bi-quinary Divide-By-Ten Counter — The Q_3 output must be externally connected to the $\overline{CP}_0$ input. The input count is then applied to the $\overline{CP}_1$ input and a divide-by-ten square wave is obtained at output Q_0.

C. Divide-By-Two and Divide-By-Five Counter — No external interconnections are required. The first flip-flop is used as a binary element for the divide-by-two function ($\overline{CP}_0$ as the input and Q_0 as the output). The $\overline{CP}_1$ input is used to obtain binary divide-by-five operation at the Q_3 output.

MODE SELECTION

RESET/SET INPUTS				OUTPUTS			
MR_1	MR_2	MS_1	MS_2	Q_0	Q_1	Q_3	Q_3
H	H	L	X	L	L	L	L
H	H	X	L	L	L	L	L
X	X	H	H	H	L	L	H
L	X	L	X	Count			
X	L	X	L	Count			
L	X	X	L	Count			
X	L	L	X	Count			

H = HIGH Voltage Level
L = LOW Voltage Level
X = Immaterial

BCD COUNT SEQUENCE

COUNT	OUTPUTS			
	Q_0	Q_1	Q_2	Q_3
0	L	L	L	L
1	H	L	L	L
2	L	H	L	L
3	H	H	L	L
4	L	L	H	L
5	H	L	H	L
6	L	H	H	L
7	H	H	H	L
8	L	L	L	H
9	H	L	L	H

NOTE: Output Q_0 is connected to Input $\overline{CP}_1$ for BCD count.

LOGIC DIAGRAM

54/7493A
54LS/74LS93
DIVIDE-BY-SIXTEEN COUNTER

CONNECTION DIAGRAM
PINOUT A

$\overline{CP}_1$	1	14 $\overline{CP}_0$
MR$_1$	2	13 NC
MR$_2$	3	12 Q$_0$
NC	4	11 Q$_3$
V$_{CC}$	5	10 GND
NC	6	9 Q$_1$
NC	7	8 Q$_2$

DESCRIPTION — The '93 is a 4-stage ripple counter containing a high speed flip-flop acting as a divide-by-two and three flip-flops connected as a divide-by-eight. HIGH signals on the Master Reset (MR) inputs override the clocks and force all outputs to the LOW state.

LOGIC SYMBOL

14 —○ CP$_0$
1 —○ CP$_1$
MR   Q$_0$ Q$_1$ Q$_2$ Q$_3$
1 2
2 3   12 9 8 11

V$_{CC}$ = Pin 5
GND = Pin 10
NC = Pins 4, 6, 7, 13

ORDERING CODE: See Section 9

PKGS	PIN OUT	COMMERCIAL GRADE V$_{CC}$ = +5.0 V ±5%, T$_A$ = 0° C to +70° C	MILITARY GRADE V$_{CC}$ = +5.0 V ±10%, T$_A$ = -55° C to +125° C	PKG TYPE
Plastic DIP (P)	A	7493APC, 74LS93PC		9A
Ceramic DIP (D)	A	7493ADC, 74LS93DC	5493ADM, 54LS93DM	6A
Flatpak (F)	A	7493AFC, 74LS93FC	5493AFM, 54LS93FM	3I

INPUT LOADING/FAN-OUT: See Section 3 for U.L. definitions

PIN NAMES	DESCRIPTION	54/74 (U.L.) HIGH/LOW	54/74LS (U.L.) HIGH/LOW
$\overline{CP}_0$	÷2 Section Clock Input (Active Falling Edge)	2.0/2.0	1.0/1.5
$\overline{CP}_1$	÷5 Section Clock Input (Active Falling Edge)	2.0/2.0	1.0/1.0
MR$_1$, MR$_2$	Asynchronous Master Reset Inputs (Active HIGH)	1.0/1.0	0.5/0.25
Q$_0$	÷2 Section Output*	20/10	10/5.0 (2.5)
Q$_1$ — Q$_3$	÷8 Section Outputs	20/10	10/5.0 (2.5)

*The Q$_0$ output is guaranteed to drive the full rated fan-out plus the $\overline{CP}_1$ input.

FUNCTIONAL DESCRIPTION — The '93 is a 4-bit ripple type binary counter. It consists of four master/slave flip-flops which are internally connected to provide a divide-by-two section and a divide-by-eight section. Each section has a separate clock input which initiates state changes of the counter on the HIGH-to-LOW clock transition. State changes of the Q outputs do not occur simultaneously because of internal ripple delays. Therefore, decoded output signals are subject to decoding spikes and should not be used for clocks or strobes. The Q_0 output of each device is designed and specified to drive the rated fan-out plus the $\overline{CP}_1$ input of the device. A gated AND asynchronous Master Reset (MR_1, MR_2) is provided which overrides the clocks and resets (clears) all the flip-flops. Since the output from the divide-by-two section is not internally connected to the succeeding stages, the devices may be operated in various counting modes.

A. 4-Bit Ripple Counter — The output Q_0 must be externally connected to input $\overline{CP}_1$. The input count pulses are applied to input $\overline{CP}_0$. Simultaneous divisions of 2, 4, 8, and 16 are performed at the Q_0, Q_1, Q_2, and Q_3 outputs as shown in the Truth Table.

B. 3-Bit Ripple Counter — The input count pulses are applied to input $\overline{CP}_1$. Simultaneous frequency divisions of 2, 4, and 8 are available at the Q_1, Q_2, and Q_3 outputs. Independent use of the first flip-flop is available if the reset function coincides with reset of the 3-bit ripple-through counter.

MODE SELECTION

RESET INPUTS		OUTPUTS			
MR_1	MR_2	Q_0	Q_1	Q_2	Q_3
H	H	L	L	L	L
L	H	Count			
H	L	Count			
L	L	Count			

H = HIGH Voltage Level
L = LOW Voltage Level

TRUTH TABLE

COUNT	OUTPUTS			
	Q_0	Q_1	Q_2	Q_3
0	L	L	L	L
1	H	L	L	L
2	L	H	L	L
3	H	H	L	L
4	L	L	H	L
5	H	L	H	L
6	L	H	H	L
7	H	H	H	L
8	L	L	L	H
9	H	L	L	H
10	L	H	L	H
11	H	H	L	H
12	L	L	H	H
13	H	L	H	H
14	L	H	H	H
15	H	H	H	H

NOTE: Output Q_0 connected to $\overline{CP}_1$.

LOGIC DIAGRAM

54/74107
54LS/74LS107

DUAL JK FLIP-FLOP
(With Separate Clears and Clocks)

DESCRIPTION — The '107 dual JK master/slave flip-flops have a separate clock for each flip-flop. Inputs to the master section are controlled by the clock pulse. The clock pulse also regulates the state of the coupling transistors which connect the master and slave sections. The sequence of operation is as follows: 1) isolate slave from master; 2) enter information from J and K inputs to master; 3) disable J and K inputs; 4) transfer information from master to slave.

TRUTH TABLE

INPUTS		OUTPUT
@ t_n		@ t_{n+1}
J	K	Q
L	L	Q_n
L	H	L
H	L	H
H	H	$\overline{Q}_n$

H = HIGH Voltage Level
L = LOW Voltage Level
t_n = Bit time before clock pulse.
t_{n+1} = Bit time after clock pulse.

CLOCK WAVEFORM

HIGH
LOW

Asynchronous Input:
LOW input to $\overline{C}_D$ sets Q to LOW level
Clear is independent of clock

The 'LS107 offers individual J, K, clear, and clock inputs. These dual flip-flops are designed so that when the clock goes HIGH, the inputs are enabled and data will be accepted. The logic level of the J and K inputs may be allowed to change when the clock is HIGH and the bistable will perform according to the Truth Table as long as minimum setup times are observed. Input data is transferred to the outputs on the negative-going edge of the clock pulse.

LOGIC SYMBOL

V_{CC} = Pin 14
GND = Pin 7

ORDERING CODE: See Section 9

PKGS	PIN OUT	COMMERCIAL GRADE V_{CC} = +5.0 V ±5%, T_A = 0°C to +125°C	MILITARY GRADE V_{CC} = +5.0 V ±10%, T_A = -55°C to +125°C	PKG TYPE
Plastic DIP (P)	A	74107PC, 74LS107PC		9A
Ceramic DIP (D)	A	74107DC, 74LS107DC	54107DM, 54LS107DM	6A
Flatpak (F)	A	74107FC, 74LS107FC	54107FM, 54LS107FM	3I

INPUT LOADING/FAN-OUT: See Section 3 for U.L. definitions

PIN NAMES	DESCRIPTION	54/74 (U.L.) HIGH/LOW	54/74LS (U.L.) HIGH/LOW
J_1, J_2, K_1, K_2	Data Inputs	1.0/1.0	0.5/0.25
$\overline{CP}_1$, $\overline{CP}_2$	Clock Pulse Inputs (Active Falling Edge)	2.0/2.0	2.0/0.5
$\overline{C}_{D1}$, $\overline{C}_{D2}$	Direct Clear Inputs (Active LOW)	2.0/2.0	1.5/0.5
Q_1, Q_2, $\overline{Q}_1$, $\overline{Q}_2$	Outputs	20/10	10/5.0 (2.5)

LOGIC DIAGRAM (one half shown)

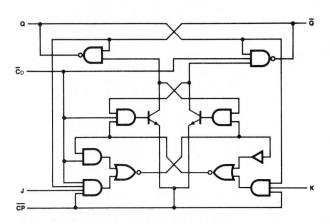

DC CHARACTERISTICS OVER OPERATING TEMPERATURE RANGE (unless otherwise specified)

SYMBOL	PARAMETER	54/74		54/74LS		UNITS	CONDITIONS
		Min	Max	Min	Max		
I_{CC}	Power Supply Current		40		8.0	mA	V_{CC} = Max, V_{CP} = 0 V

AC CHARACTERISTICS: V_{CC} = +5.0 V, T_A = +25°C (See Section 3 for waveforms and load configurations)

SYMBOL	PARAMETER	54/74 C_L = 15 pF R_L = 400 Ω		54/74LS C_L = 15 pF		UNITS	CONDITIONS
		Min	Max	Min	Max		
f_{max}	Maximum Clock Frequency	15		30		MHz	Figs. 3-1, 3-9
t_{PLH} t_{PHL}	Propagation Delay $\overline{CP}_n$ to Q_n or $\overline{Q}_n$		25 40		20 30	ns	Figs. 3-1, 3-9
t_{PLH} t_{PHL}	Propagation Delay $\overline{C}_{Dn}$ to Q_n or $\overline{Q}_n$		25 40		20 30	ns	Figs. 3-1, 3-10

107

AC OPERATING REQUIREMENTS: V_{CC} = +5.0 V, T_A = +25° C

SYMBOL	PARAMETER	54/74		54/74LS		UNITS	CONDITIONS
		Min	Max	Min	Max		
t_s (H)	Setup Time HIGH J_n or K_n to $\overline{CP}_n$	0		20		ns	Fig. 3-18 ('107) Fig. 3-7 ('LS107)
t_h (H)	Hold Time HIGH J_n or K_n to $\overline{CP}_n$	0		0		ns	
t_s (L)	Setup Time LOW J_n or K_n to $\overline{CP}_n$	0		20		ns	
t_h (L)	Hold Time LOW J_n or K_n to $\overline{CP}_n$	0		0		ns	
t_w (H) t_w (L)	$\overline{CP}_n$ Pulse Width	20 47		13.5 20		ns	Fig. 3-9
t_w (L)	$\overline{C}_{Dn}$ Pulse Width LOW	25		25		ns	Fig. 3-10

54S/74S112
54LS/74LS112

DUAL JK NEGATIVE
EDGE-TRIGGERED FLIP-FLOP

DESCRIPTION — The '112 features individual J, K, Clock and asynchronous Set and Clear inputs to each flip-flop. When the clock goes HIGH, the inputs are enabled and data will be accepted. The logic level of the J and K inputs may change when the clock is HIGH and the bistable will perform according to the Truth Table as long as minimum setup and hold times are observed. Input data is transferred to the outputs on the falling edge of the clock pulse.

TRUTH TABLE

INPUTS		OUTPUT
@ t_n		@ t_{n+1}
J	K	Q
L	L	Q_n
L	H	L
H	L	H
H	H	$\overline{Q}_n$

Asynchronous Inputs:
LOW input to $\overline{S}_D$ sets Q to HIGH level
LOW input to $\overline{C}_D$ sets Q to LOW level
Clear and Set are independent of clock
Simultaneous LOW on $\overline{C}_D$ and $\overline{S}_D$
makes both Q and $\overline{Q}$ HIGH

t_n = Bit time before clock pulse.
t_{n+1} = Bit time after clock pulse.
H = HIGH Voltage Level
L = LOW Voltage Level

LOGIC SYMBOL

V_{CC} = Pin 16
GND = Pin 8

ORDERING CODE: See Section 9

PKGS	PIN OUT	COMMERCIAL GRADE V_{CC} = +5.0 V ±5%, T_A = 0°C to +70°C	MILITARY GRADE V_{CC} = +5.0 V ±10%, T_A = -55°C to +125°C	PKG TYPE
Plastic DIP (P)	A	74S112PC, 74LS112PC		9B
Ceramic DIP (D)	A	74S112DC, 74LS112DC	54S112DM, 54LS112DM	6B
Flatpak (F)	A	74S112FC, 74LS112FC	54S112FM, 54LS112FM	4L

INPUT LOADING/FAN-OUT: See Section 3 for U.L. definitions

PIN NAMES	DESCRIPTION	54/74S (U.L.) HIGH/LOW	54/74LS (U.L.) HIGH/LOW
J_1, J_2, K_1, K_2	Data Inputs	1.25/1.0	0.5/0.25
$\overline{CP}_1$, $\overline{CP}_2$	Clock Pulse Inputs (Active Falling Edge)	2.5/2.5	2.0/0.5
$\overline{C}_{D1}$, $\overline{C}_{D2}$	Direct Clear Inputs (Active LOW)	2.5/4.375	1.5/0.5
$\overline{S}_{D1}$, $\overline{S}_{D2}$	Direct Set Inputs (Active LOW)	2.5/4.375	1.5/0.5
Q_1, Q_2, $\overline{Q}_1$, $\overline{Q}_2$	Outputs	25/12.5	10/5.0 (2.5)

LOGIC DIAGRAM (one half shown)

DC CHARACTERISTICS OVER OPERATING TEMPERATURE RANGE (unless otherwise specified)

SYMBOL	PARAMETER	54/74S		54/74LS		UNITS	CONDITIONS
		Min	Max	Min	Max		
Icc	Power Supply Current		50		8.0	mA	V_{CC} = Max, V_{CP} = 0 V

AC CHARACTERISTICS: V_{CC} = +5.0 V, T_A = +25° C (See Section 3 for waveforms and load configurations)

SYMBOL	PARAMETER	54/74S C_L = 15 pF R_L = 280 Ω		54/74LS C_L = 15 pF		UNITS	CONDITIONS
		Min	Max	Min	Max		
f_{max}	Maximum Clock Frequency	80		30		MHz	Figs. 3-1, 3-9
t_{PLH} t_{PHL}	Propagation Delay $\overline{CP}_n$ to Q_n or $\overline{Q}_n$		7.0 7.0		16 24	ns	Figs. 3-1, 3-9
t_{PLH} t_{PHL}	Propagation Delay $\overline{C}_{Dn}$ or $\overline{S}_{Dn}$ to Q_n or $\overline{Q}_n$		7.0 7.0		16 24	ns	Figs. 3-1, 3-10

AC OPERATING REQUIREMENTS: V_{CC} = +5.0 V, T_A =+25° C

SYMBOL	PARAMETER	54/74S		54/74LS		UNITS	CONDITIONS
		Min	Max	Min	Max		
t_s (H) t_s (L)	Setup Time J_n or K_n to $\overline{CP}_n$	7.0 7.0		20 15		ns	Fig. 3-7
t_h (H) t_h (L)	Hold Time J_n or K_n to $\overline{CP}_n$	0 0		0 0		ns	
t_w (H) t_w (L)	$\overline{CP}_n$ Pulse Width	6.0 6.5		20 15		ns	Fig. 3-9
t_w (L)	$\overline{C}_{Dn}$ or $\overline{S}_{Dn}$ Pulse Width LOW	8.0		15		ns	Fig. 3-10

54/74121
MONOSTABLE MULTIVIBRATOR

DESCRIPTION — The '121 features positive and negative dc level triggering inputs and complementary outputs. Input pin 5 directly activates a Schmitt circuit which provides temperature compensated level detection, increases immunity to positive-going noise and assures jitter-free response to slowly rising triggers.

When triggering occurs, internal feedback latches the circuit, prevents re-triggering while the output pulse is in progress and increases immunity to negative-going noise. Noise immunity is typically 1.2 V at the inputs and 1.5 V on V_{CC}.

Output pulse width stability is primarily a function of the external R_x and C_x chosen for the application. A 2 kΩ internal resistor is provided for optional use where output pulse width stability requirements are less stringent. Maximum duty cycle capability ranges from 67% with a 2 kΩ resistor to 90% with a 40 kΩ resistor. Duty cycles beyond this range tend to reduce the output pulse width. Otherwise, output pulse width follows the relationship:

$$t_w = 0.69\ R_x C_x$$

LOGIC SYMBOL

V_{CC} = Pin 14
GND = Pin 7
NC = Pins 2,8,12,13

ORDERING CODE: See Section 9

PKGS	PIN OUT	COMMERCIAL GRADE V_{CC} = +5.0 V ±5%, T_A = 0° C to +70° C	MILITARY GRADE V_{CC} = +5.0 V ±10%, T_A = -55° C to +125° C	PKG TYPE
Plastic DIP (P)	A	74121PC		9A
Ceramic DIP (D)	A	74121DC	54121DM	6A
Flatpak (F)	A	74121FC	54121FM	3I

INPUT LOADING/FAN-OUT: See Section 3 for U.L.definitions

PIN NAMES	DESCRIPTION	54/74 (U.L.) HIGH/LOW
$\overline{A}_1$, $\overline{A}_2$	Trigger Inputs (Active Falling Edge)	1.0/1.0
B	Schmitt Trigger Input (Active Rising Edge)	2.0/2.0
Q, $\overline{Q}$	Outputs	20/10

CONNECTION DIAGRAM
PINOUT A

54/74193
54LS/74LS193

UP/DOWN BINARY COUNTER
(With Separate Up/down Clocks)

DESCRIPTION — The '193 is an up/down modulo-16 binary counter. Separate Count Up and Count Down Clocks are used and in either counting mode the circuits operate synchronously. The outputs change state synchronous with the LOW-to-HIGH transitions on the clock inputs. Separate Terminal Count Up and Terminal Count Down outputs are provided which are used as the clocks for subsequent stages without extra logic, thus simplifying multistage counter designs. Individual preset inputs allow the circuits to be used as programmable counters. Both the Parallel Load ($\overline{PL}$) and the Master Reset (MR) inputs asynchronously override the clocks. For functional description and detail specifications please refer to the '192 data sheet.

LOGIC SYMBOL

V_{CC} = Pin 16
GND = Pin 8

ORDERING CODE: See Section 9

PKGS	PIN OUT	COMMERCIAL GRADE V_{CC} = +5.0 V ±5%, T_A = 0°C to +70°C	MILITARY GRADE V_{CC} = +5.0 V ±10%, T_A = -55°C to +125°C	PKG TYPE
Plastic DIP (P)	A	74193PC, 74LS193PC		9B
Ceramic DIP (D)	A	74193DC, 74LS193DC	54193DM, 54LS193DM	6B
Flatpak (F)	A	74193FC, 74LS193FC	54193FM, 54LS193FM	4L

INPUT LOADING/FAN-OUT: See Section 3 for U.L. definitions

PIN NAMES	DESCRIPTION	54/74 (U.L.) HIGH/LOW	54/74LS (U.L.) HIGH/LOW
CP_U	Count Up Clock Input (Active Rising Edge)	1.0/1.0	0.5/0.25
CP_D	Count Down Clock Input (Active Rising Edge)	1.0/1.0	0.5/0.25
MR	Asynchronous Master Reset Input (Active HIGH)	1.0/1.0	0.5/0.25
$\overline{PL}$	Asynchronous Parallel Load Input (Active LOW)	1.0/1.0	0.5/0.25
$P_0 — P_3$	Parallel Data Inputs	1.0/1.0	0.5/0.25
$Q_0 — Q_3$	Flip-flop Outputs	20/10	10/5.0 (2.5)
$\overline{TC_D}$	Terminal Count Down (Borrow) Output (Active LOW)	20/10	10/5.0 (2.5)
$\overline{TC_U}$	Terminal Count Up (Carry) Output (Active LOW)	20/10	10/5.0 (2.5)

193

MODE SELECT TABLE

MR	$\overline{PL}$	CP_U	CP_D	MODE
H	X	X	X	Reset (Asyn.)
L	L	X	X	Preset (Asyn.)
L	H	H	H	No Change
L	H	⌐	H	Count Up
L	H	H	⌐	Count Down

H = HIGH Voltage Level
L = LOW Voltage Level
X = Immaterial
Z = High Impedance

STATE DIAGRAM

——► COUNT UP
----► COUNT DOWN

LOGIC EQUATIONS
FOR TERMINAL COUNT

$$\overline{TC}_U = Q_0 \bullet Q_1 \bullet Q_2 \bullet Q_3 \bullet \overline{CP}_U$$
$$\overline{TC}_D = \overline{Q}_0 \bullet \overline{Q}_1 \bullet \overline{Q}_2 \bullet \overline{Q}_3 \bullet \overline{CP}_D$$

LOGIC DIAGRAM

4001B 4002B

QUAD 2-INPUT NOR GATE ● DUAL 4-INPUT NOR GATE

DESCRIPTION — These CMOS logic elements provide the positive input NOR function. The outputs are fully buffered for highest noise immunity and pattern insensitivity of output impedance.

4001B
LOGIC AND CONNECTION DIAGRAM
DIP (TOP VIEW)

NOTE:
The Flatpak versions have the same pinouts (Connection Diagram) as the Dual In-line Package.

4002B
LOGIC AND CONNECTION DIAGRAM
DIP (TOP VIEW)

DC CHARACTERISTICS: V_{DD} as shown, V_{SS} = 0 V

SYMBOL	PARAMETER		LIMITS									UNITS	TEMP	TEST CONDITIONS
			V_{DD} = 5 V			V_{DD} = 10 V			V_{DD} = 15 V					See Note 1
			MIN	TYP	MAX	MIN	TYP	MAX	MIN	TYP	MAX			
I_{DD}	Quiescent Power Supply Current	XC			1			2			4	μA	MIN, 25°C	All inputs at 0 V or V_{DD}
					7.5			15			30		MAX	
		XM			0.25			0.5			1	μA	MIN, 25°C	
					7.5			15			30		MAX	

AC CHARACTERISTICS: V_{DD} as shown, V_{SS} = 0 V, T_A = 25°C, 4001B only (See Note 2)

SYMBOL	PARAMETER		LIMITS								UNITS	TEST CONDITIONS
			V_{DD} = 5 V			V_{DD} = 10 V			V_{DD} = 15V			See Note 2
		MIN	TYP	MAX	MIN	TYP	MAX	MIN	TYP	MAX		
t_{PLH}	Propagation Delay		60	110		25	60		20	48	ns	C_L = 50 pF,
t_{PHL}			60	110		25	60		20	48	ns	R_L = 200 kΩ
t_{TLH}	Output Transition Time		60	135		30	70		20	45	ns	Input Transition
t_{THL}			60	135		30	70		20	45	ns	Times ⩽ 20 ns

AC CHARACTERISTICS: V_{DD} as shown, V_{SS} = 0 V, T_A = 25°C, 4002B only

SYMBOL	PARAMETER		LIMITS								UNITS	TEST CONDITIONS
			V_{DD} = 5 V			V_{DD} = 10 V			V_{DD} = 15V			See Note 2
		MIN	TYP	MAX	MIN	TYP	MAX	MIN	TYP	MAX		
t_{PLH}	Propagation Delay		65	110		30	60		20	48	ns	C_L = 50 pF,
t_{PHL}			70	110		30	60		23	48	ns	R_L = 200 kΩ
t_{TLH}	Output Transition Time		75	135		40	70		30	45	ns	Input Transition
t_{THL}			60	135		23	70		15	45	ns	Times ⩽ 20 ns

NOTES:
1. Additional DC Characteristics are listed in this section under 4000B Series CMOS Family Characteristics.
2. Propagation Delays and Output Transition Times are graphically described in this section under 4000B Series CMOS Family Characteristics.

4001B   FAIRCHILD 4000B SERIES CMOS FAMILY CHARACTERISTICS

DC CHARACTERISTICS FOR THE 4000B SERIES CMOS FAMILY — Parametric Limits listed below are guaranteed for the entire Fairchild CMOS Family unless otherwise specified on the individual data sheets.

DC CHARACTERISTICS: V_{DD} = 5 V, V_{SS} = 0 V

SYMBOL	PARAMETER			LIMITS MIN	LIMITS TYP	LIMITS MAX	UNITS	TEMP	TEST CONDITIONS	
V_{IH}	Input HIGH Voltage			3.5			V	All	Guaranteed Input HIGH Voltage	
V_{IL}	Input LOW Voltage					1.5	V	All	Guaranteed Input Low Voltage	
V_{OH}	Output HIGH Voltage			4.95			V	Min, 25°C	I_{OH} < 1 µA, Inputs at 0 or 5 V per	
				4.95				MAX	the Logic Function or Truth Table	
				4.5			V	All	I_{OH} < 1 µA, Inputs at 1.5 or 3.5 V	
V_{OL}	Output LOW Voltage					0.05	V	MIN, 25°C	I_{OL} < 1 µA, Inputs at 0 or 5 V per	
						0.05		MAX	the Logic Function or Truth Table	
						0.5	V	All	I_{OL} < 1 µA, Inputs at 1.5 or 3.5 V	
I_{OH}	Output HIGH Current			−0.63			mA	MIN, 25°C	V_{OUT} =	Inputs at 0 or 5 V per the Logic Function or Truth Table
				−0.36				MAX	4.6 V	
I_{OL}	Output LOW Current			1			mA	MIN, 25°C	V_{OUT} = 0.4 V	
				0.8						
				0.4				MAX		
C_{IN}	Input Capacitance Per Unit Load					7.5	pF	25°C	Any Input	
I_{DD}	Quiescent Power Supply Current	Gates	XC			1	µA	MIN, 25°C	All Inputs at 0 V or V_{DD} for all Valid Input Combinations	
						7.5		MAX		
			XM			0.25	µA	MIN, 25°C		
						7.5		MAX		
		Buffers and Flip-Flops	XC			4	µA	MIN, 25°C		
						30		MAX		
			XM			1	µA	MIN, 25°C		
						30		MAX		
		MSI	XC			20	µA	MIN, 25°C		
						150		MAX		
			XM			5	µA	MIN, 25°C		
						150		MAX		

Appendix II   **569**

4049B • 4050B
4049B HEX INVERTING BUFFER • 4050B HEX NON-INVERTING BUFFER

DESCRIPTION — These CMOS buffers provide high current output capability suitable for driving TTL or high capacitance loads. Since input voltages in excess of the buffers' supply voltage are permitted, these buffers may also be used to convert logic levels of up to 15 V to standard TTL levels. The 4049B provides six inverting buffers, the 4050B six non-inverting buffers. Their guaranteed fan out into common bipolar logic elements is shown in Table 1.

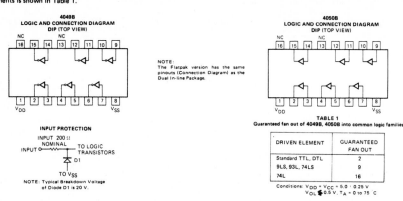

4049B LOGIC AND CONNECTION DIAGRAM DIP (TOP VIEW)

4050B LOGIC AND CONNECTION DIAGRAM DIP (TOP VIEW)

NOTE:
The Flatpak version has the same pinouts (Connection Diagram) as the Dual In-line Package.

INPUT PROTECTION

INPUT 200 Ω NOMINAL
INPUT ○——▷——○ TO LOGIC TRANSISTORS
D1
TO V_{SS}
NOTE: Typical Breakdown Voltage of Diode D1 is 20 V.

TABLE 1
Guaranteed fan out of 4049B, 4050B into common logic families

DRIVEN ELEMENT	GUARANTEED FAN OUT
Standard TTL, DTL	2
9LS, 93L, 74LS	9
74L	16

Conditions: $V_{DD} = V_{CC} = 5.0 \cdot 0.25$ V
$V_{OL} \cdot 0.5$ V, $T_A = 0$ to 75 C

DC CHARACTERISTICS: V_{DD} as shown, $V_{SS} = 0$ V, 4049BXM and 4050BXM (See Note 1)

SYMBOL	PARAMETER	$V_{DD} = 5$ V MIN	TYP	MAX	$V_{DD} = 10$ V MIN	TYP	MAX	$V_{DD} = 15$ V MIN	TYP	MAX	UNITS	TEMP	TEST CONDITIONS
I_{OH}	Output HIGH Current	-1.85 -1.25 -0.9	-2.5								mA	MIN, 25°C MAX	$V_{OUT} = 2.5$ V for $V_{DD} = 5$ V Inputs at 0 or V_{DD} per Function
		-0.62 -0.5 -0.35	-1		-1.85 -1.25 -0.9	-2.5		-5.5 -3.75 -2.7	-7.5		mA	MIN, 25°C MAX	$V_{OUT} = 4.5$ V for $V_{DD} = 5$ V $V_{OUT} = 9.5$ V for $V_{DD} = 10$ V $V_{OUT} = 13.5$ V for $V_{DD} = 15$ V Inputs at 0 or V_{DD} per Function
I_{OL}	Output LOW Current	3.75 3 2.1	6		10 8 5.6	16		30 24 16.8	48		mA	MIN, 25°C MAX	$V_{OUT} = 0.4$ V for $V_{DD} = 5$ V $V_{OUT} = 0.5$ V for $V_{DD} = 10$ V $V_{OUT} = 1.5$ V for $V_{DD} = 15$ V Inputs at 0 or V_{DD} per Function
		3.3 2.6 1.8	5.2								mA	MIN, 25°C MAX	$V_{OUT} = 0.4$ V for $V_{DD} = 4.5$ V Inputs at 0 V or V_{DD} per Function
I_{DD}	Quiescent Power Supply Current		1 30			2 60			4 120		μA	MIN, 25°C MAX	All Inputs at 0 V or V_{DD}

Notes on the following page.

DC CHARACTERISTICS: V_{DD} as shown, V_{SS} = 0 V, 4049BXC and 4050BXC (Cont'd) (See Note 1)

SYMBOL	PARAMETER	LIMITS									UNITS	TEMP	TEST CONDITIONS
		V_{DD} = 5 V			V_{DD} = 10 V			V_{DD} = 15 V					
		MIN	TYP	MAX	MIN	TYP	MAX	MIN	TYP	MAX			
I_{OH}	Output HIGH Current	-1.5	-2.5								mA	MIN	V_{OUT} = 2.5 V for V_{DD} = 5 V
		-1.25									mA	25°C	Inputs at 0 or V_{DD}
		-1.0									mA	MAX	per Function
		-0.6	-1		-1.5	-2.5		-4.5	-7.5		mA	MIN	V_{OUT} = 4.5 V for V_{DD} = 5 V
		-0.5			-1.25			-3.75			mA	25°C	V_{OUT} = 9.5 V for V_{DD} = 10 V
		-0.4			-1.0			-3			mA	MAX	V_{OUT} = 13.5 V for V_{DD} = 15 V
													Inputs at 0 or V_{DD}
													per Function
I_{OL}	Output LOW Current	3.6	6		9.6	16		28	48		mA	MIN	V_{OUT} = 0.4 V for V_{DD} = 5 V
		[3.0]			8			24			mA	25°C	V_{OUT} = 0.5 V for V_{DD} = 10 V
		2.5			6.6			19			mA	MAX	V_{OUT} = 1.5 V for V_{DD} = 15 V
													Inputs at 0 or V_{DD}
													per Function
		3.1	5.2								mA	MIN	V_{OUT} = 0.4 V for V_{DD} = 4.5 V
		2.6									mA	25°C	Inputs at 0 V or V_{DD}
		2.1									mA	MAX	per Function
I_{DD}	Quiescent Power Supply Current			4			8			16	µA	MIN,25°C	All inputs at 0 V or V_{DD}
				30			60			120		MAX	

AC CHARACTERISTICS AND SET-UP REQUIREMENTS: V_{DD} as shown, V_{SS} = 0 V, T_A = 25°C, 4049B only (See Note 2)

SYMBOL	PARAMETER	LIMITS									UNITS	TEST CONDITIONS
		V_{DD} = 5 V			V_{DD} = 10 V			V_{DD} = 15 V				
		MIN	TYP	MAX	MIN	TYP	MAX	MIN	TYP	MAX		
t_{PLH}	Propagation Delay		65	130		30	65		29	52	ns	C_L = 50 pF, R_L = 200 kΩ
t_{PHL}			50	105		25	50		17	40		
t_{TLH}	Output Transition Time		73	145		40	80		30	60	ns	Input Transition Times ⩽ 20 ns
t_{THL}			33	65		13	25		9	20		

NOTES:
1. Additional DC Characteristics are listed in this section under 4000B Series CMOS Family Characteristics.
2. Propagation Delays and Output Transition Times are graphically described in this section under 4000B Series CMOS Family Characteristics.

Appendix II

ANSWERS to SELECTED PROBLEMS

Chapter One

1.2 (a) 25
    (b) 9.5625
    (c) 1241.6875
1.4 1023
1.5 Nine bits
1.6 (a) four circuits for parallel
    (b) one circuit for serial

Chapter Two

2.1 (a) 22
    (b) 141
    (c) 2313
2.2 (a) 100101
    (b) 1110
    (c) 10111101
    (d) 11001101
    (e) 100100001001
2.3 65,535
2.4 (a) 483
    (b) 30
    (c) 2047
    (d) 175
2.5 (a) 73_8
    (b) 564_8
    (c) 1627_8
    (d) 200000_8

2.6 (a) 111100011
    (b) 011110
    (c) 011111111111
    (c) 010101111
2.7 (a) 26_8
    (b) 215_8
    (c) 4411_8
    (d) 1727_8
2.8 165, 166, 167, 170, 171, 171, 172, 173, 174, 175, 176, 177, 200
2.9 100100001001
2.10 (a) 146
    (b) 422
    (c) 14333
    (d) 448
2.11 (a) 4B
    (b) 13A
    (c) 800
    (d) 6413
2.12 (a) 16
    (b) 8D
    (c) 909
    (d) 3D7
2.13 (a) 10010010
    (b) 000110100110
    (c) 0011011111111101
    (d) 001011000000
2.14 (a) 65,536 addresses
    (b) 000 to FFF (hex)

2.15 280, 281, 282, 283, 284, 285, 286, 287, 288, 289, 28A, 28B, 28C, 28D, 28E, 28F, 290, 291, 292, 293, 294, 295, 296, 297, 298, 299, 29A, 29B, 29C, 29D, 29E, 29F, 2A0

2.16 (a) 01000111
(b) 100101100010
(c) 000110000111
(d) 0100001001101000100101100100111

2.17 10 bits for binary, 12 bits for BCD

2.18 (a) 9752
(b) 184

2.19 X = 25/Y

2.20 11011000 (X); 10111101 (=); 10110010 (2); 00110101 (5); 10101111 (/); 01011001 (Y)

2.21 (a) 110110110 (parity bit is leftmost bit)
(b) 000101000
(c) 111110111

2.22 (a) 101110100 (parity bit on the left)
(b) 000111000
(c) 0000101100101
(d) 11001001000000001

2.23 (a) no single-bit error
(b) single-bit error
(c) double error
(d) no single-bit error

2.24 (a) 10110001001
(b) 11111111
(c) 209
(d) 59,943
(e) 4701
(f) 777
(g) 157
(h) 2254
(i) 1961
(j) 15900
(k) 640
(l) 952B
(m) 100001100101
(n) 947
(o) 135_{16}
(p) 5464_{10}
(q) 1001010
(r) 01011000 (BCD)

Chapter Three

3.2 with A = 0, x will be the same as B

3.3 with A = 1, x will always be 1

3.5 thirty one

3.6 (b) x = 0
(c) x = B

3.7 change the OR gate to an AND gate

3.8 OUT is always LOW

3.10 $x = \overline{\overline{A} + \overline{B}} + B + C$

3.11 $x = \overline{A}\,\overline{B}\,\overline{C} + A\overline{B}\,\overline{C} + \overline{A}\,\overline{B}D$

3.12 $x = D \cdot (AB + C) + E$

3.13 x = 1 only when B = C = 1 and A = D = 0

3.19 $x = \overline{(\overline{A + B}) \cdot (\overline{B}C)}$

3.20 x = 1 only when A = B = 0, C = 1

3.23 $MP\overline{N} + \overline{M}\,\overline{P}N$

3.24 $B\overline{C}$

3.26 (a) $A + \overline{B} + C$
(b) $A \cdot (B + \overline{C})$
(c) $\overline{A} + \overline{B} + CD$
(d) $\overline{A} + B + \overline{C} + \overline{D}$

3.27 $A + B + C$

3.32

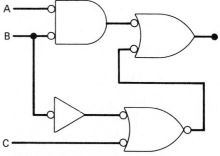

3.33 (a) NOR
(b) AND
(c) NAND

3.34

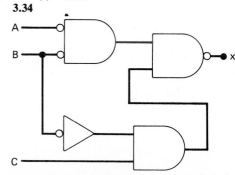

3.35 Z will be LOW when A or B is HIGH, or when C or D is LOW.

3.37 X will go HIGH when E = 1, or D = 1, or B = C = 0, or when B = 1 and A = 0.

Chapter Four

4.1 (a) $BC + \overline{B}(\overline{C} + A)$
(b) $\overline{BC + \overline{B}(\overline{C} + A)}$
(c) $\overline{D} + AB\,\overline{C} + \overline{A}\,BC$

4.2 $Q(M + N)$

4.3 MN + Q

4.4 one solution: $x = \overline{B}C + AB\overline{C}$

another: $x = \overline{A}B + \overline{B}C + BC$

4.5 $x = \overline{A}\,\overline{B} + \overline{A}C + \overline{B}\,\overline{C}$

4.6 $x = \overline{A}\,\overline{B}\,\overline{C} + ABC$

4.7 $x = \overline{A}_3(A_2 + A_1A_0)$

4.11 $x = BC + \overline{B}\,\overline{C} + AC$; or $BC + \overline{B}\,\overline{C} + A\overline{B}$

4.12 $y = \overline{D} + \overline{A}\,\overline{B}C + \overline{A}B\overline{C}$

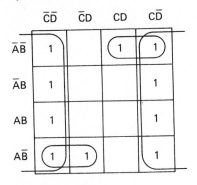

4.13 $x = \overline{A}_3A_2 + \overline{A}_3A_1A_0$

4.14 (a)$x = \overline{A}\,\overline{B} + \overline{A}C + \overline{B}D + ABC\overline{D}$

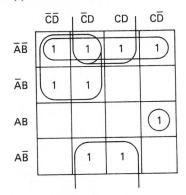

(b) $x = \overline{A}\,\overline{D} + \overline{B}C + \overline{B}\,\overline{D}$

(c) $y = \overline{B} + AC$

4.16 $x = \overline{A}$ as long as B = 0

4.17 $x = \overline{A}$ as long as B = 1

4.18 A = 0, B = C = 1

4.20 four outputs where Z_3 is the MSB

$Z_3 = Y_1Y_0X_1X_0$

$Z_2 = Y_1X_1(\overline{Y}_0 + \overline{X}_0)$

$Z_1 = Y_0X_1(\overline{Y}_1 + \overline{X}_0) + Y_1X_0(\overline{Y}_0 + \overline{X}_1)$

$Z_0 = Y_0X_0$

4.21 $x = A_3A_2 + A_3A_1$

4.26 $N/S = \overline{C}\,\overline{D}(A + B) + AB(\overline{C} + \overline{D})$

$E/W = \overline{N/S}$

4.27 use an AND gate

4.28 use an OR gate

4.30 $Z = A_0\overline{S} + A_1S$

4.31 $Z = X_1X_0Y_1Y_0 + X_1\overline{X}_0Y_1\overline{Y}_0$
$+ \overline{X}_1X_0\overline{Y}_1Y_0 + \overline{X}_1\overline{X}_0\overline{Y}_1\overline{Y}_0$

Chapter Five

5.1

5.2 same as 5.1

5.3

5.6 control inputs have to remain stable for $t_s = 20$ ns prior to the active clock transition.

5.7

5.8 Q is a 500-Hz squarewave

5.9

5.11 frequency will be 2.5 kHz

5.13 Q is a 500-Hz squarewave

5.17

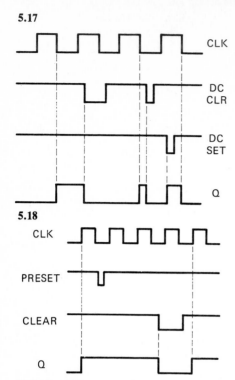

5.18

5.19 (a) 200 ns
(b) 7474
(c) 30 ns

5.20 48 ns

5.21 minimum squarewave period is 40 ns

5.24 correct sequence is A, B, C

5.27 [x] = 1011 after fourth clock pulse

5.31 (a) ten
(b) 1953 Hz

5.32 (a) 128
(b) 0 to 127

5.35 (a) $\overline{Q}$ will stay LOW
(b) PGT at $\overline{Q}$ will set X = 1
(c) make t_p = 200μs

5.38 all circuit outputs stop changing after 14th clock pulse

Chapter Six

6.1 (a) 10100
(b) 10010
(c) 1111.0101
(d) 1.1010
(e) 100111000

6.2 (a) 00010000 (incl. sign bit)
(b) 11110010
(c) 00111111
(d) 10011000

(e) 11111111
(f) 10000000

6.3 (a) +13
(b) −3
(c) +123
(d) −103
(e) +127
(f) −32
(g) −1
(h) −127

6.4 (a) −2048 to +2047
(b) sixteen bits including sign bit

6.5 (a) 00001111
(b) 11111101
(c) 11111011
(d) 10000000
(e) 00000001
(f) 11011110
(g) 00000000
(h) 00010101

6.7 (a) 100011
(b) 1111001
(c) 100011.00101
(d) .10001111

6.8 (a) 11
(b) 111
(c) 101.11
(d) 1111.0011

6.10 (a) 6E24
(b) 100D

6.11 (a) OEFE
(b) 229
(c) 02A6

6.12 16,849 locations

6.18 [A] = 0000

6.19 80 ns

6.20 (a) SUM = 0111
(b) SUM = 1010

6.24 100001000101

Chapter Seven

7.1 250 kHz, 50%

7.2 same as 7.1

7.3 10,000

7.5 100 Hz

7.7 100(four) to 000(zero)

7.8 (b) 000, 001, 010, 100, 101, 110, then repeats

7.9 connect Q3 to MR1; connect Q2, Q1 and Q0 to a 3-input AND; connect AND output to MR2

7.11 60 pps

7.12 (a) 12.5 MHz
(b) 8.33 MHz

7.14 (b) 33 MHz

(c) 16.7 MHz

7.15 counts as a BCD counter

7.19 mod-9

7.24 (a) glitch can cause a problem when counter goes from 111 to 000

7.26 four, sixteen, thirteen

7.30 frequency at Z is 5Hz

7.32 (a) 22

(b) 450

(c) 0 or 1

7.39 $64\mu s$

Chapter Eight

8.1 (a) A, B

(b) A

(c) A

8.3 (a) 157.5 mW, 18.5 ns

(b) 262.5 mW, 7 ns

(c) 42 mW, 15 ns

8.4 (a) 0.5 U.L. (HIGH); 0.25 U.L. (LOW)

(c) five

8.7 fanout is not exceeded

8.8 67 ns

8.9 250 ohms

8.14 AB + CD + FG

8.15 R_c(min) = 359 ohms

8.16 (a) 5V

8.20 AND

8.25 3.6V

8.27 −1; −2

Chapter Nine

9.1 six inputs, 64 outputs

9.2 [A] = 110, E_3 = 1, $\overline{E_1}$ = $\overline{E_2}$ = 0

9.4

$\overline{O_3}$

t_{28} t_{30}

9.5 connect $\overline{E_2}$ to GROUND, and take output from $\overline{O_2}$

9.7 0010, 0011, 0100, 0101, 0110, 1000, 1001

9.11 0111

9.14 *b* and *c*

9.15 *a* and *d*

9.21

1 cycle

9.22

At top:

1 cycle

9.28 *c* and *d*

9.30 At t_1 and t_2: [A] = 1011, [B] = [C] = 1000

At t_3: each register = 1011

9.32 (a) At t_3, all registers will be 1001

Chapter Ten

10.1 3.58 V

10.2 LSB = 20 mV

10.3 20 mV; approx. 0.4%

10.4 approx. 5 mV

10.5 eight

10.6 14.3%, 0.286 V

10.9 10 mV, 0.1%, 6.95 V

10.11 800 ohms; no

10.12 (a) −0.3125 V, −4.6875 V

(b) 4.27 kilohms

(c) −.0267

10.13 V_{out} = 2Vp − p for B = 1111

10.14 20 μA; yes

10.15 (a) seven

10.16 (a) 10010111

(b) 10010111

(c) 102 μs, 51 μs

10.19 (a) 1.2 mV

(b) 2.7 mV

10.20 (a) 0111110110

(b) 0111110111

10.22 80 μs

10.23 100101

Chapter Eleven

11.1 16,384; 32; 524,288

11.2 16,384

11.3 64K × 4

11.4 (a) Hi-Z

(b) 11101101

11.5 (a) register 11

(b) 0100

11.6 (a) 16,384

(b) four

(c) two 1-of-128 decoders

11.7 120 ns

11.8 180 ns

11.9 0111

11.12

1 1 1 1   1   1   1 1   1

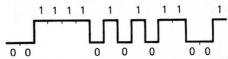

0 0   0   0   0   0 0

Answers to Selected Problems

11.15 (a) 100 ns
    (b) 30 ns
    (c) ten million
    (d) 20 ns
    (e) 30 ns

    (f) 40 ns
    (g) ten million
11.23 32×8
11.25 64 2114 ICs; 16 2118 ICs

INDEX